# IFIP Advances in Information and Communication Technology

319

T0181259

# IFIP – The International Federation for Information Processing

IFIP was founded in 1960 under the auspices of UNESCO, following the First World Computer Congress held in Paris the previous year. An umbrella organization for societies working in information processing, IFIP's aim is two-fold: to support information processing within its member countries and to encourage technology transfer to developing nations. As its mission statement clearly states,

> *IFIP's mission is to be the leading, truly international, apolitical organization which encourages and assists in the development, exploitation and application of information technology for the benefit of all people.*

IFIP is a non-profitmaking organization, run almost solely by 2500 volunteers. It operates through a number of technical committees, which organize events and publications. IFIP's events range from an international congress to local seminars, but the most important are:

- The IFIP World Computer Congress, held every second year;
- Open conferences;
- Working conferences.

The flagship event is the IFIP World Computer Congress, at which both invited and contributed papers are presented. Contributed papers are rigorously refereed and the rejection rate is high.

As with the Congress, participation in the open conferences is open to all and papers may be invited or submitted. Again, submitted papers are stringently refereed.

The working conferences are structured differently. They are usually run by a working group and attendance is small and by invitation only. Their purpose is to create an atmosphere conducive to innovation and development. Refereeing is less rigorous and papers are subjected to extensive group discussion.

Publications arising from IFIP events vary. The papers presented at the IFIP World Computer Congress and at open conferences are published as conference proceedings, while the results of the working conferences are often published as collections of selected and edited papers.

Any national society whose primary activity is in information may apply to become a full member of IFIP, although full membership is restricted to one society per country. Full members are entitled to vote at the annual General Assembly, National societies preferring a less committed involvement may apply for associate or corresponding membership. Associate members enjoy the same benefits as full members, but without voting rights. Corresponding members are not represented in IFIP bodies. Affiliated membership is open to non-national societies, and individual and honorary membership schemes are also offered.

Pär Ågerfalk   Cornelia Boldyreff
Jesús M. González-Barahona
Gregory R. Madey   John Noll (Eds.)

# Open Source Software: New Horizons

6th International IFIP WG 2.13 Conference
on Open Source Systems, OSS 2010
Notre Dame, IN, USA, May 30 – June 2, 2010
Proceedings

 Springer

Volume Editors

Pär Ågerfalk
Uppsala University
75120 Uppsala, Sweden
E-mail: par.agerfalk@im.uu.se

Cornelia Boldyreff
University of East London
London E16 2RD, UK
E-mail: c.boldyreff@uel.ac.uk

Jesús M. González-Barahona
Universidad Rey Juan Carlos de Madrid
28933 Móstoles, Spain
E-mail: jgb@gsyc.escet.urjc.es

Gregory R. Madey
University of Notre Dame
Notre Dame, IN 46556, USA
E-mail: gmadey@nd.edu

John Noll
Lero - The Irish Software Engineering Research Centre
Limerick, Ireland
E-mail: john.noll@lero.ie

CR Subject Classification (1998): D.2, D.3, C.2.4, D.1, K.6.3, D.2.4

| | |
|---|---|
| ISSN | 1868-4238 |
| ISBN-10 | 3-642-42244-6 Springer Berlin Heidelberg New York |
| ISBN-13 | 978-3-642-42244-7 Springer Berlin Heidelberg New York |

springer.com

© IFIP International Federation for Information Processing 2010

Softcover re-print of the Hardcover 1st edition 2010
Typesetting: Camera-ready by author, data conversion by Scientific Publishing Services, Chennai, India
Printed on acid-free paper      06/3180

# General Chairs' Foreword

Welcome to the 6th International Conference on Open Source Systems of the IFIP Working Group 2.13. This year was the first time this international conference was held in North America. We had a large number of high-quality papers, highly relevant panels and workshops, a continuation of the popular doctoral consortium, and multiple distinguished invited speakers. The success of OSS 2010 was only possible because an Organizing Committee, a Program Committee, Workshop and Doctoral Committees, and authors of research manuscripts from over 25 countries contributed their time and interest to OSS 2010. In the spirit of the communities we study, you self-organized, volunteered, and contributed to this important research forum studying free, libre, open source software and systems. We thank you!

Despite our modest success, we have room to improve and grow our conference and community. At OSS 2010 we saw little or no participation from large portions of the world, including Latin America, Africa, China, and India. But opportunities to expand are possible. In Japan, we see a hot spot of participation led by Tetsuo Noda and his colleagues, both with full-paper submissions and a workshop on "Open Source Policy and Promotion of IT Industries in East Asia." The location of OSS 2011 in Salvador, Brazil, will hopefully result in significant participation from researchers in Brazil – already a strong user of OSS – and other South American countries. Under the leadership of Megan Squire, Publicity Chair, we recruited Regional Publicity Co-chairs covering Japan (Tetsuo Noda), Africa (Sulayman Sowe), the Middle East and South Asia (Faheen Ahmed), Russia and Eastern Europe (Alexey Khoroshilov), Western Europe (Yeliz Eseryel), UK and Ireland (Andrea Capiluppi), and the Nordic countries (Björn Lundell). Finally, the future of our community is in the newly emerging researchers, the doctoral students learning the art and craft of scholarly research on OSS topics. They had an opportunity to present their research plans and work-in-progress papers at the doctoral consortium, organized by Kris Ven, Walt Scacchi, and Jen Verelst. Special thanks go to Kevin Crowston for obtaining National Science Foundation (USA) support for the doctoral consortium.

Our planning and organization of OSS 2010 benefited greatly from the experience and advice of organizers of previous conferences. We thank Scott Hissan, Björn Lundell, Anthony Wasserman, Walt Scacchi, Joseph Feller, and Kevin Crowston for their generous time, encouragement, and advice. We also thank Petrinja Etiel and Joseph Feller who served as Tutorial and Panels Chairs, respectively. We also thank the officers of IFIP Working Group 2.13 (Giancarlo Succi, Walt Scacchi, Ernesto Damiani, Scott Hissam, and Pär J. Ågerfalk) for permitting us to move OSS 2010 to North America, given some of the challenges of doing so.

Our greatest attribution and appreciation for the success of OSS 2010 goes to John Noll and his Co-program Chairs (Cornelia Boldyreff and Pär J. Ågerfalk) for the work they did managing the review process, the paper submission server at Lero, and the preparation of the proceedings and program.

Finally, we still have to mention the main contributor to OSS 2010: the FLOSS (free, open source software) development community. Without it, our conference would not be possible. Our thanks go to all those developers for offering all of us such an interesting field of study.

Gregory R. Madey
Jesús M. González-Barahona

# Program Chairs' Foreword

We are very pleased to present the proceedings of the 6th International Conference on Open Source Systems. This year's proceedings include 23 full papers selected from 51 submissions. As in previous years, this year's papers represent a broad range of perspectives on open source systems, ranging from software engineering through organizational issues to law. In a nod to the highly successful conference in Limerick (OSS 2007), this year we included 17 short papers in the program as well (Part II). Five workshop abstracts (Part III) and four panel descriptions (Part IV) round out the proceedings contents, for what we hope will be a highly engaging and useful volume.

We would like to thank the members of the Program Committee for their hard work reviewing papers under this year's shortened review schedule, and especially those who agreed to do additional reviews after the first reviewing cycle: Matthew van Antwerp, Andres Baravalle, Sarah Beecham, Karl Beecher, Padmanabhan Krishnan, Gregory R. Madey, Felipe Ortega, Gregorio Robles, Maha Shaik, Carlos Solis, Klaas-Jan Stol, and Andrea Wiggins. We would also like to thank the General Chairs - Gregory R. Madey and Jesús M. González-Barahona - for organizing the first IFIP WG 2.13 meeting in North America.

John Noll
Cornelia Boldyreff
Pär Ågerfalk

# Organization

## Conference Officials

| | |
|---|---|
| General Chairs: | Gregory R. Madey, University of Notre Dame, USA |
| | Jesús M. González-Barahona, Universidad Rey Juan Carlos de Madrid, Spain |
| Program Chairs: | John Noll, Lero - The Irish Software Engineering Research Centre, Ireland |
| | Cornelia Boldyreff, University of East London, UK |
| | Pär Ågerfalk, Uppsala University, Sweden |
| Doctoral Consortium Chairs: | Kris Ven, University of Antwerp, Belgium |
| | Walt Scacchi, University of California, Irvine, USA |
| | Jen Verelst, University of Antwerp, Belgium |
| Publicity Chair: | Megan Squire, Elon University, USA |
| Publicity Co-chairs: | |
| W. Europe: | Yeliz Eseryel, University of Groningen, The Netherlands |
| UK and Ireland: | Andrea Capiluppi, University of East London, UK |
| Nordic Countries: | Björn Lundell, University of Skovde, Sweden |
| Japan: | Tetsuo Noda, Shimane University, Japan |
| Africa: | Sulayman Sowe, UNU-Merit, The Netherlands |
| Middle East and S. Asia: | Faheen Ahmed, United Arab Emirates University, UAE |
| Russia and E. Europe: | Alexey Khoroshilov, Institute for System Programming of the Russian Academy of Sciences, Russia |
| Panel Chair: | Joseph Feller, University College Cork, Ireland |
| Tutorial Chair: | Petrinja Etiel, Free University of Bolzno-Bozen, Italy |
| Corporate Sponsorships Chair: | Gregory R. Madey, University of Notre Dame, USA |

# IFIP Working Group 2.13 Officers

General Chair:              Giancarlo Succi, Free University of
                            Bolzano-Bozen, Italy
Vice Chair:                 Walt Scacchi, University of California, Irvine,
                            USA
Vice Chair:                 Ernesto Damiani, University of Milan, Italy
Secretary:                  Scott Hissam, Software Engineering Institute,
                            USA
Secretary:                  Pär J. Ågerfalk, Uppsala University, Sweden

# Program Committee

| | |
|---|---|
| Andres Baravalle | University of East London, UK |
| Sarah Beecham | Lero - The Irish Software Engineering Research Centre, Ireland |
| Karl Beecher | Free University of Berlin, Germany |
| Andrea Bonaccorsi | Universitá di Pisa, Italy |
| Andrea Capiluppi | University of East London, UK |
| Antonio Cerone | United Nations University, Macau SAR, China |
| Gabriella Coleman | New York University, USA |
| Jean-Michel Dalle | University Pierre-et-Marie Curie, Paris, France |
| Ernesto Damiani | University of Milan, Italy |
| Paul David | Stanford/Oxford University, USA/UK |
| Francesco Di Cerbo | University of Genoa, Italy |
| Chris DiBona | Google, USA |
| Gabriella Dodero | University of Bolzano-Bozen, Italy |
| Justin Erenkrantz | Apache Software Foundation, USA |
| Joseph Feller | University College Cork, Ireland |
| Fulvio Frati | University of Milan, Italy |
| Daniel German | University of Victoria, Canada |
| Rishab Aiyer Ghosh | UNI-Merit, The Netherlands |
| Jesús M. González-Barahona | University Rey Juan Carlos, Spain |
| Stefan Haefliger | ETH Zurich, Switzerland |
| Jeremy Hayes | University College Cork, Ireland |
| Joachim Henkel | Technische Universität München, Germany |
| James Herbsleb | Carnegie-Mellon University, USA |
| Scott Hissam | Software Engineering Institute, Carnegie Mellon University, USA |
| James Howison | Carnegie Mellon University, USA |
| Chris Jensen | University of California, Irvine, USA |
| Alexey Khoroshilov | Institute for System Programming, Russian Academy of Sciences, Russia |
| Joseph Kiniry | IT University of Copenhagen, Denmark |
| Stefan Koch | Bogazici University, Turkey |

## Acknowledgments

OSS 2010 was organized under the auspices of the International Federation of Information Processsing (IFIP) Working Group 2.13.

The Doctoral Consortium was supported, in part, by the US National Science Foundation, under grant IIS 1005183.

# Table of Contents

# Part II: Short Papers

## Part III: Workshops

## Part IV: Panels

# Spago4Q and the QEST $n$D Model:
# An Open Source Solution
# for Software Performance Measurement

Claudio A. Ardagna[1], Ernesto Damiani[1], Fulvio Frati[1], Sergio Oltolina[2],
Mauro Regoli[1], and Gabriele Ruffatti[2]

[1] Dipartimento di Tecnologie dell'Informazione
Università degli Studi di Milano
via Bramante, 65 – 26013 Crema (CR)
{claudio.ardagna,ernesto.damiani,fulvio.frati}@unimi.it,
mregoli@crema.unimi.it
[2] Engineering Ingegneria Informatica
Via San Martino della Battaglia, 56 – 00185 Roma, Italy
(sergio.oltolina,gabriele.ruffatti}@eng.it

**Abstract.** Improving the software development process requires tools and
model of increasing complexity, capable of satisfying project managers' and
analyzers' needs. In that paper we present a solution integrating a formalized
and established model for performance evaluation like QEST $n$D, and an open
source Business Intelligence platform like Spago4Q. We obtain a new environment that can produce immediate snapshots of projects' status without any constraint on the number of projects and the type of development process.

**Keywords:** QEST $n$D, Spago4Q, Performance Indicator, process monitoring.

## 1 Introduction

The availability of detailed and updated information on the software development
process is of paramount importance for organizations to maintain their competitiveness level and operate in new and more challenging markets. Such a scenario of integrated information is known as *Business Intelligence* and encloses all the business
processes and tools used by organizations for data acquisition.

Within this context, a number of structured process models are adopted by enterprises, depending on their application domain and size, to collect specific knowledge
about their development processes, strengthening, at the same time, their *know-how* in
terms of more efficiency and quality.

In this paper, we describe our experience in the deployment of an integrated and
complete environment for software performance evaluation. To this aim, we exploit a
formal model for process performance evaluation (QEST $n$D) and we connect it with
an open source Business Intelligence application (Spago4Q). In particular, the QEST
$n$D model (**Q**uality factor + **E**conomic, **S**ocial, and **T**echnological **D**imension) is a
multi-dimensional model, proposed in [1] and [3], for the performance evaluation of
software processes. Its multi-dimensional nature is based on three important concepts:

P. Ågerfalk et al. (Eds.): OSS 2010, IFIP AICT 319, pp. 1–14, 2010.

1. A number of measurable concepts derived from different business areas (called *dimensions*) including economic, social, and technological ones.
2. The number of business areas interested by the analysis. This number may change for each single project, without any limit (from here comes the acronym $n$D – $n$ Dimensions).
3. Organizations are allowed to choose the dimensions of each analysis with respect to their needs.

Such a philosophy makes QEST $n$D an open model, decoupled from any specific development process, allowing multi-process, multi-project performance analysis.

The final objective of the model is to express the overall process performance ($P$) as a combination of the performance of any considered dimension, calculated as the weighted sum of the applied metrics. The global performance value approach gives an immediate and accurate snapshot of the current state of the project, and allows a top-down analysis starting from the global value, which includes all the single measurements, to the analysis of the performance of the single dimension. The performance indicator is calculated by the integration of instrumental measurements (called *RP – Rough Productivity*) and the subjective perception of the overall quality (express as *QF – Quality Factor*).

A problem characterizing the QEST $n$D model and preventing its diffusion in the Business Intelligence context was the lack of reliable and flexible environments where the model could be implemented and distributed. As described in the following sections, we deliver QEST $n$D on an open source business intelligence platform, Spago4Q (SpagoBI for Quality) [2][5]. Spago4Q is a platform for maturity assessment, effectiveness of development software processes and application services, and quality inspection of the released software, achieved by the evaluation of data and measures collected from the process management and development tools with non-invasive techniques.

The tool is easily adaptable to different organizational contexts, independently from the development process adopted by the single projects (i.e. waterfall, XP, Scrum, etc.), meeting exactly the multi-process multi-project approach of QEST $n$D. Although the initial vision of Spago4Q was focused on the software development process, the implementation of the QEST $n$D model, and consequently of a global multi-dimensional performance value, could extend the performance and quality evaluation to services and business areas that are typical of software organizations.

The paper is organized as follows. Section 2 describes more in detail the two frameworks (QEST $n$D and Spago4Q). Section 3 shows the steps to build the implemented integration. Section 4 presents two real case studies. Finally, Section 5 contains our conclusions.

## 2   The Context: QEST nD and Spago4Q

In the following sections, we give an overview of the two frameworks describing the mathematical formalization of QEST $n$D, and the main characteristics of Spago4Q.

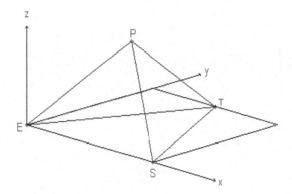

**Fig. 1.** Geometrical representation of QEST model using E, S, T dimensions and P value as edges of the figure [1]

## 2.1 The QEST nD Model

In the software engineering context, several one-dimensional performance models are available which integrate individual measurements into a single performance index. By comparison, in more traditional domains such as Business Modeling, there exist a number of multi-dimensional models that take into account data derived directly from their accounting systems, which means that multiple viewpoints are, in fact, considered [1].

Furthermore, models currently available in the software engineering domains are too oversimplified to properly reflect the different aspects of performance when various perspectives, or viewpoints, must be taken into account at the same time. Therefore, to manage simultaneous multi-dimensional constraints in development projects, managers need to estimate the status of current projects based on their own interpretation of rough data, due to the lack of reliable measurement models.

In multi-dimensional analysis, complex viewpoints are taken into account simultaneously, each one analyzing a distinct aspect of the overall process performance. Therefore, an extension to the traditional single-dimension approach is needed, to consider both quality and performance of the development process.

The QEST nD model [3] is aimed at measuring software project performances addressing the aspects of multidimensionality and qualitative-quantitative assessment. With respect to the original QEST model that was initially designed for measurements to be done at the end of a project, QEST nD provides a dynamic extension to analyze software process data throughout all the development phases. In particular, in the QEST model the quality can be viewed as the integration of at least three different viewpoints.

- **Economical (E):** expresses the viewpoint of *management*, interested in measurements focused on the overall quality level, rather than the quality of specific features or process areas.
- **Social (S):** measures the perspective of the *user*, where the quality is intended as the characteristic of a product to satisfy present and future needs.

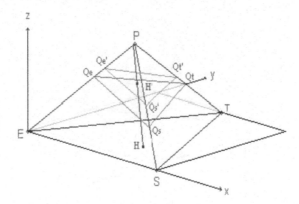

**Fig. 2.** ($Q_e$, $Q_s$, $Q_t$) and ($Q'_e$, $Q'_s$, $Q'_t$) plane sections [1]

- **Technical (T):** relative to *developers*, for whom software quality is achieved by conforming to specific, explicitly stated standards and requirements explicitly stated.

In the QEST model the measurement of performance (P) is given by combining a quantitative measurement, indicated by the component *RP-Rough Productivity*, and a qualitative measurement, calculated as a perception-based measurement of the overall product quality (*QF-Quality Factor*). A detailed explanation of the model, that has been formalized in [3] [4], is out of the scope of this paper. For the initial QEST model, the three-dimension geometrical representation of a regular tetrahedron (see Fig. 1) was selected and studied to help the model formalization. In particular:

- the three dimensions (E, S, T) in the space d to the corners of the pyramid's base, and the convergence of the edges to the P vertex describes the top performance level;
- the three sides are of equal length: the solid shape that represents this 3D concept is therefore a pyramid with a triangular base and sides of equal length (tetrahedron). The figure represents the ranges to which the dimensions performances will belong to.

The geometrical approach permits the representation of the measurement of perform-ance in a simple and visual way, assisting the global performance computation. Thanks to this representation, it is possible to express the performance value in term of geometrical concepts like distance, surface, and volume. The value of each dimen-sion is seen as the weighted sum of a list of *n* distinct measures, each one representing single measurable concepts for each perspective. Then, the values of the three dimen-sions E, S, T are placed on the respective tetrahedron side, describing a sloped plane section and representing the three dimensional performance measurements [3]. Fig. 2 better explains such a geometrical aspect, indicating the three dimensions' values as $Q_e$, $Q_s$, and $Q_t$. Finally, the QF, with respect to each dimension, is added to the previ-ous values describing an upward or downward translation of the plan ($Q_e$, $Q_s$, $Q_t$) finding the new plan ($Q'_e$, $Q'_s$, $Q'_t$).

On the other side, the value RP can simply be expressed as the distance between each single corner (E, S, T) with the specific point $Q_e$, $Q_s$, and $Q_t$. Please note that the maximum value of each edge is 1, consequently all the values that are placed on the tetrahedron have to be normalized. Then, the performance value is calculated as the distance between the center of gravity of the original tetrahedron and the center of the described plane section along the tetrahedron height (see Fig. 2).

The explanation above is valid for the QEST *n*D case, where more than 3 dimensions are taken into account. Through computational geometry, it is possible to develop a generic representation of it with a generalization of a tetrahedral region of space to *n* dimensions describing it with a *simplex* [4]. To conclude, the geometrical formalization of the model allows to describe it with a simple formula for the computation of the global performance value P:

$$P = 1 - \prod_{i=1}^{n} (1 - p_i) \tag{1}$$

Where $p_i$ represents the single dimension performance value added with the respective QF value. Section 3 and 4 deal with the definition of the environment and, in particular, of the metrics model that will be used to compute the value of the single dimensions.

## 2.2 Spago4Q

Spago4Q (SpagoBI for Quality) [2] is an open source platform for the continuous monitoring of software quality. Its most important characteristic is the total independence from the adopted development process and from the number of monitored projects. Spago4Q can then be described as a *multi-process multi-project* monitoring platform.

In Spago4Q, the evaluation of metrics and the collection of data are executed in a fully-transparent way, without any action due by programmers and designers and any change in their typical working tasks.

Spago4Q includes in its package a number of extractors for the main environments that are exploited during the software lifecycle (IDE, text editing tools, requirements management frameworks, and the like) that collect data directly from process workproducts (e.g. java classes or logs).

Since Spago4Q relies exclusively on open frameworks and it is released under an open source license, its structure could be enriched with the implementation of additional extractors for particular work-product types. In any case, the extractors will be executed at specified time intervals and store data directly in the application data warehouse.

Spago4Q is a vertical adaptation of SpagoBI [10], a more complex framework for Business Intelligence analysis, whose structure was enriched with the use of a complex meta-model (see Fig. 3) for the representation and description of the generic development process, the measurement framework, the extractors, and the assessment framework [5], that defines the entities that play a role in the monitoring process the relations between them. The *Development Process* meta-model has been designed to be as generic as possible, allowing the modeling of virtually all process models, from waterfall to XP. It is connected with the *Measurement Framework* meta-model, which defines a skeletal generic framework and is used to obtain measures from most development processes. Then, the *Assessment* meta-model allows to model a generic evaluation structure with a simple classification in terms of *Category, Target,* and *Practice*. Finally, the *Extractors* meta-model is used to formalize and define the extractors used to retrieve data from process module and supply it to the measurement module.

In particular, the inclusion of a specific meta-model for the assessment framework allows the tool to implement metrics that are specific to a particular assessment model. Originally Spago4Q was studied to fully support the CMMi framework [6] for the maturity assessment of the development process. However, the intrinsic generality of the meta-model approach allows adapting it to any assessment model, for instance ISO 9000 or Balanced Scorecards. In fact, Spago4Q provides support for the definition and implementation of measurement frameworks based on the GQM (Goal-Question-Metric) approach [7], which categorizes the metrics in terms of *(i)* generic goals to be measured, *(ii)* the particular aspects that the metric has to measure, and *(iii)* the metrics that implement the actual measurement.

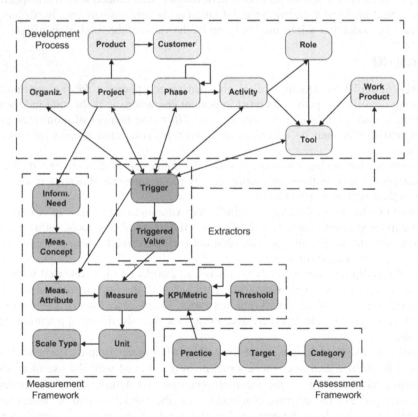

**Fig. 3.** The Spago4Q meta-model [5]

# 3  An Integrated Environment

The below-described work has been the subject of a Master thesis carried out within the Department of Information Technology, Università degli Studi di Milano. The definition of a QEST $n$D model is a five-step procedure that will be described in the following paragraphs, without entering in the implementation details for the sake of conciseness.

The steps, all executable through the graphical interface of Spago4Q, are coherent with the PMAI (*Plan-Measure-Assess-Improve*) cycle [8], which is composed of four logical phases:

1. **PLAN,** which consists in defining a set of metrics basing on the GQM approach, defining the dimensions that characterize the analysis, the mathematic formalization of the metrics and the weight to assign to each metric.
2. **MEASURE,** which includes the collection of data, the computation of metrics values, the normalization of them (values must be $\leq 1$), and the computation of global Performance value using the Eq. 1.
3. **ASSESS:** results are presented in dashboards and reports, and analyzed by the management and analyzers.
4. **IMPROVE:** every negative or low value is deeply analyzed to find problems in the processes and consequently find solutions to improve the overall quality.

## 3.1  Step 1: Metrics and Model Definition

The first step deals with the declaration of a complete GQM, with the definition of the analysis dimensions, the concepts to measure, and the metrics to apply to project work-products.

The GQM will be defined using the *Model Definition* interface, while metrics (or KPI – Key Performance Indicator) are defined through the *KPI Definition* section.

In particular, the definition of KPIs involves the specification of an algorithm for the computation of the metric. This algorithm will exploit the SQL mathematical library for simple computations, or call an *ad-hoc* Java class for more complex ones. The KPI will collect data directly from the Spago4Q data warehouse, which contains all the data that the extractors get from the project work-products.

## 3.2  Step 2: Weights and Threshold Definition

The QEST $n$D model requires each metric to be coupled with the respective *weight*, which indicates the importance that such a concept plays in the dimension it belongs to. A complete analysis of the GQM should be performed prior to define the weights for each KPI. Thus, for each metric it is important to define the specific thresholds, which allow to evaluate the value with respect to the organization policies. The thresholds have to take into account the normalization of metrics and are also important for the creation of complete and understandable reports. The *KPI Definition* interface helps to define such aspects.

Finally, although the use of the QF is optional and its absence does not preclude the entire Global Performance value, in this step it is possible to assign the QF to each specific dimension. The definition of the QF is subjected to the analysis of pool of experts that define the value that will be added to the respective RP. A complete guide for the definition of the QF of a software product can be found in [3].

### 3.3 Step 3: Measures Collection

Measures are taken directly from Spago4Q data warehouse, which in turn is filled by data collected by extractors accessing process work-products. The collection process is defined in the configuration phase, where a specific *dataset*, that contains the description of the metric itself, is defined for each KPI or group of KPIs.

Metrics are described in terms of *(i)* the name of the model to which the metric is assigned to, *(ii)* default value, *(iii)* minimum and maximum values (for normalization), and *(iv)* the algorithm for the metric computation.

In particular, the algorithm can be implemented using the common mathematic library of SQL, as a separate Web Services or, for computations that involve complex operations, as a Java package. Furthermore, the application supplies to users KPI-specific drivers to be used in the metric formula to help the definition of it in the selected programming language, supplying methods for the direct access to data warehouse fields. Finally, specific fields for the KPI results will be added to the data warehouse and supplied to other KPIs and components of Spago4Q.

### 3.4 Step 4: Global Performance Computation

In our approach, both global and dimension-wise performance indexes are computed as simple KPIs that take in input configuration data and results of the metrics at the bottom of the GQM tree. First of all, the performance value of each dimension is calculated as the sums of the product of each metric, belonging to the dimension, with its specific weight:

$$D = \sum_{i=1}^{n} (V_i * w_i) \tag{2}$$

where $D$ is the dimension performance value, $V$ is the result of the $i$-th metric, and $w$ is the assigned weight. If the QF is provided by the model, its value is added to the results:

$$D' = D + QF = \sum_{i=1}^{n} (V_i * w_i) + QF \tag{3}$$

Note that Eq. 2 and 3 are computed for each of the $n$ dimensions that compose the QEST $n$D model that has been specified. Finally, the KPI that computes the global performance indicator could be defined using the Eq. 1, defined in Section 3.1 and it is valid in the case the QF is defined or not.

$$P = 1 - \prod_{j=1}^{n} (1 - D_j) \tag{4}$$

The global performance $P$ value will be assigned to the root node of the created model.

### 3.5  Step 5: Reports

As last step of the process, a set of reports and dashboards could be defined and configured to satisfy any reporting and managerial need. One of the open source reporting tools we integrated with Spago4Q is Eclipse BIRT [9]. The generation of reports requires the creation of a specific dataset that includes all the data that will be described by the report.

Such data is collected from the application data warehouse and, in particular, from the fields that contain the metrics values. Several pre-defined templates and layouts are available.

Spago4Q provides methods and interfaces to directly configure and create a new report using all the facilities provided by BIRT.

## 4  Case Studies

The application of the QEST *n*D model in Spago4Q has been tested using two real case studies on data taken from real development projects realized by Engineering Ingegneria Informatica, a major player within the community of Spago4Q.

The two projects are of increasing complexity: the first one deals with a little project consisting of a single measurement of project data, while the second one measures the complex performance of three big projects with several measurements in a three-month time slot.

For the sake of conciseness, in this section we focus on the second case study only, providing information about the steps that were described in Section 3. The realized QEST *n*D model was called *Business-Service Model* and takes into consideration four specific analysis dimensions:

1. QEST-EC: Economic performance indicator;
2. QEST-RS: Resource performance indicator;
3. QEST-TE: Technical performance indicator;
4. QEST-CS: Customer Satisfaction performance indicator.

Fig. 4 shows the complete model, highlighting the GQM structure of the metrics. The root node is the global performance indicator (QEST-BS), which includes the four **goals** describing the analysis dimensions. For each dimension, a set of **questions** (i.e., the concepts to measure) has been defined, which in turn includes the **metric**, or the metrics, which evaluates the concept. Table 1 summarizes the metrics, along with the respective weights, that compose the model.

Each metric is associated to specific SQL queries or Java classes that directly access to the data warehouse to collect the input for the computation. For each metric, a field in the data warehouse has been created to store its output, to be used by other KPIs or reports.

It is important to remark that the definition of weights and thresholds has to be very careful and must involve skilled experts that have a solid background in the enterprise scenario. In fact, an overestimation, or underestimation, of metrics weights

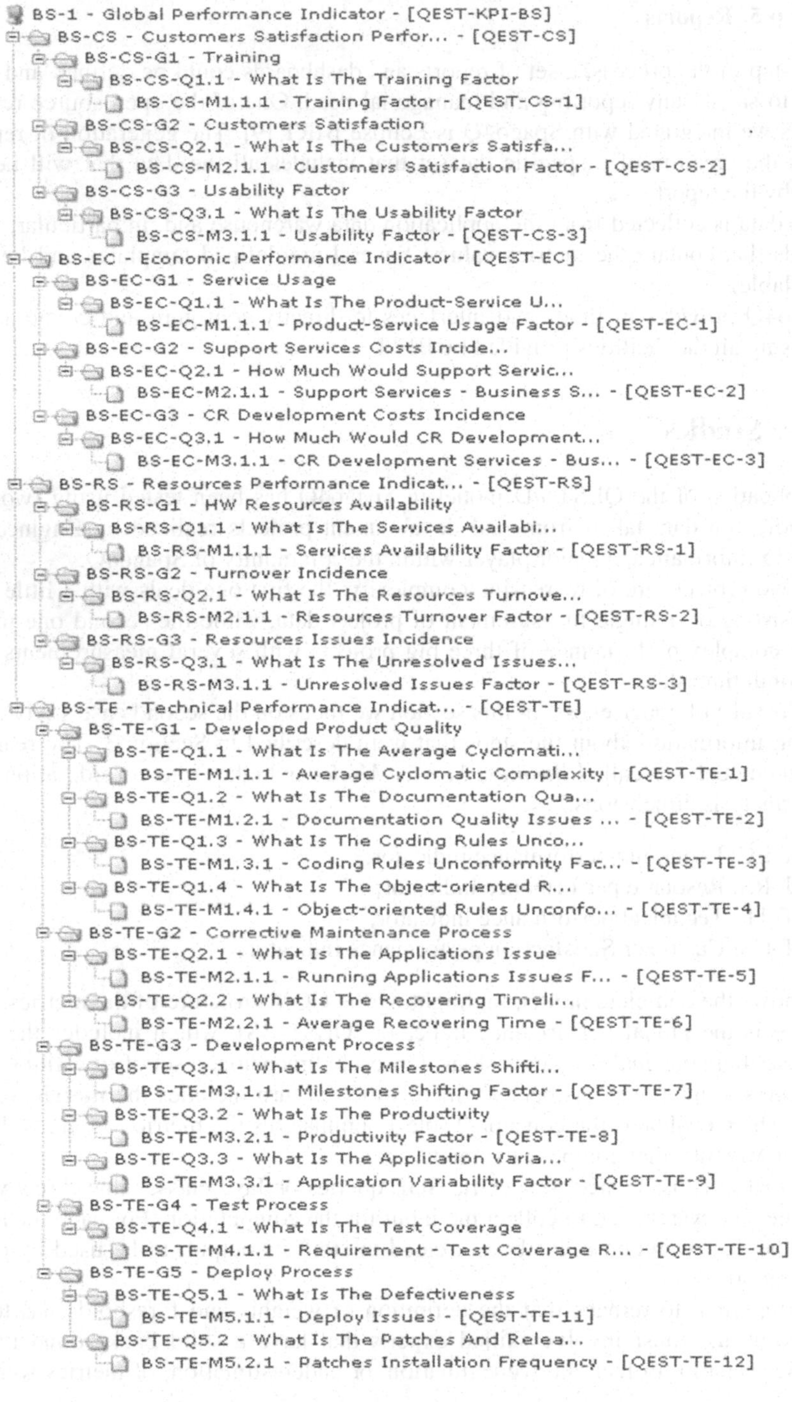

```
▓ BS-1 - Global Performance Indicator - [QEST-KPI-BS]
⊟⊕ BS-CS - Customers Satisfaction Perfor... - [QEST-CS]
   ⊟⊕ BS-CS-G1 - Training
      ⊟⊕ BS-CS-Q1.1 - What Is The Training Factor
         □ BS-CS-M1.1.1 - Training Factor - [QEST-CS-1]
   ⊟⊕ BS-CS-G2 - Customers Satisfaction
      ⊟⊕ BS-CS-Q2.1 - What Is The Customers Satisfa...
         □ BS-CS-M2.1.1 - Customers Satisfaction Factor - [QEST-CS-2]
   ⊟⊕ BS-CS-G3 - Usability Factor
      ⊟⊕ BS-CS-Q3.1 - What Is The Usability Factor
         □ BS-CS-M3.1.1 - Usability Factor - [QEST-CS-3]
⊟⊕ BS-EC - Economic Performance Indicator - [QEST-EC]
   ⊟⊕ BS-EC-G1 - Service Usage
      ⊟⊕ BS-EC-Q1.1 - What Is The Product-Service U...
         □ BS-EC-M1.1.1 - Product-Service Usage Factor - [QEST-EC-1]
   ⊟⊕ BS-EC-G2 - Support Services Costs Incide...
      ⊟⊕ BS-EC-Q2.1 - How Much Would Support Servic...
         □ BS-EC-M2.1.1 - Support Services - Business S... - [QEST-EC-2]
   ⊟⊕ BS-EC-G3 - CR Development Costs Incidence
      ⊟⊕ BS-EC-Q3.1 - How Much Would CR Development...
         □ BS-EC-M3.1.1 - CR Development Services - Bus... - [QEST-EC-3]
⊟⊕ BS-RS - Resources Performance Indicat... - [QEST-RS]
   ⊟⊕ BS-RS-G1 - HW Resources Availability
      ⊟⊕ BS-RS-Q1.1 - What Is The Services Availabi...
         □ BS-RS-M1.1.1 - Services Availability Factor - [QEST-RS-1]
   ⊟⊕ BS-RS-G2 - Turnover Incidence
      ⊟⊕ BS-RS-Q2.1 - What Is The Resources Turnove...
         □ BS-RS-M2.1.1 - Resources Turnover Factor - [QEST-RS-2]
   ⊟⊕ BS-RS-G3 - Resources Issues Incidence
      ⊟⊕ BS-RS-Q3.1 - What Is The Unresolved Issues...
         □ BS-RS-M3.1.1 - Unresolved Issues Factor - [QEST-RS-3]
⊟⊕ BS-TE - Technical Performance Indicat... - [QEST-TE]
   ⊟⊕ BS-TE-G1 - Developed Product Quality
      ⊟⊕ BS-TE-Q1.1 - What Is The Average Cyclomati...
         □ BS-TE-M1.1.1 - Average Cyclomatic Complexity - [QEST-TE-1]
      ⊟⊕ BS-TE-Q1.2 - What Is The Documentation Qua...
         □ BS-TE-M1.2.1 - Documentation Quality Issues ... - [QEST-TE-2]
      ⊟⊕ BS-TE-Q1.3 - What Is The Coding Rules Unco...
         □ BS-TE-M1.3.1 - Coding Rules Unconformity Fac... - [QEST-TE-3]
      ⊟⊕ BS-TE-Q1.4 - What Is The Object-oriented R...
         □ BS-TE-M1.4.1 - Object-oriented Rules Unconfo... - [QEST-TE-4]
   ⊟⊕ BS-TE-G2 - Corrective Maintenance Process
      ⊟⊕ BS-TE-Q2.1 - What Is The Applications Issue
         □ BS-TE-M2.1.1 - Running Applications Issues F... - [QEST-TE-5]
      ⊟⊕ BS-TE-Q2.2 - What Is The Recovering Timeli...
         □ BS-TE-M2.2.1 - Average Recovering Time - [QEST-TE-6]
   ⊟⊕ BS-TE-G3 - Development Process
      ⊟⊕ BS-TE-Q3.1 - What Is The Milestones Shifti...
         □ BS-TE-M3.1.1 - Milestones Shifting Factor - [QEST-TE-7]
      ⊟⊕ BS-TE-Q3.2 - What Is The Productivity
         □ BS-TE-M3.2.1 - Productivity Factor - [QEST-TE-8]
      ⊟⊕ BS-TE-Q3.3 - What Is The Application Varia...
         □ BS-TE-M3.3.1 - Application Variability Factor - [QEST-TE-9]
   ⊟⊕ BS-TE-G4 - Test Process
      ⊟⊕ BS-TE-Q4.1 - What Is The Test Coverage
         □ BS-TE-M4.1.1 - Requirement - Test Coverage R... - [QEST-TE-10]
   ⊟⊕ BS-TE-G5 - Deploy Process
      ⊟⊕ BS-TE-Q5.1 - What Is The Defectiveness
         □ BS-TE-M5.1.1 - Deploy Issues - [QEST-TE-11]
      ⊟⊕ BS-TE-Q5.2 - What Is The Patches And Relea...
         □ BS-TE-M5.2.1 - Patches Installation Frequency - [QEST-TE-12]
```

**Fig. 4.** The complete GQM model defined in the case studies

**Table 1.** List of KPIs defined for the case study

| KPI | Description | Weight |
|---|---|---|
| QEST-KPI-BS | Global Performance Indicator | |
| QEST-EC | Economic Performance Indicator | 1.0 |
| QEST-RS | Resources Performance Indicator | 1.0 |
| QEST-TE | Technical Performance Indicator | 1.0 |
| QEST-CS | Customers Satisfaction Performance Indicator | 1.0 |
| QEST-EC-1 | Product/Service Usage Factor | 0.2 |
| QEST-EC-2 | Support Services – Business Services Costs Ratio | 0.4 |
| QEST-EC-3 | CR Development Services – Business Services Costs Ratio | 0.4 |
| QEST-RS-1 | Services Availability Factor | 0.7 |
| QEST-RS-2 | Resources Turnover Factor | 0.15 |
| QEST-RS-3 | Unresolved Issues Factor | 0.15 |
| QEST-TE-1 | Average Cyclomatic Complexity | 0.1 |
| QEST-TE-2 | Documentation Quality Issues Factor | 0.05 |
| QEST-TE-3 | Coding Rules Unconformity Factor | 0.05 |
| QEST-TE-4 | Object-oriented Rules Unconformity Factor | 0.1 |
| QEST-TE-5 | Running Applications Issues Factor | 0.15 |
| QEST-TE-6 | Average Recovering Time | 0.1 |
| QEST-TE-7 | Milestones Shifting Factor | 0.1 |
| QEST-TE-8 | Productivity Factor | 0.1 |
| QEST-TE-9 | Application Variability Factor | 0.05 |
| QEST-TE-10 | Requirement - Test Coverage Ratio | 0.08 |
| QEST-TE-11 | Deploy Issues | 0.07 |
| QEST-TE-12 | Patches Installation Frequency | 0.05 |
| QEST-CS-1 | Training Factor | 0.1 |
| QEST-CS-2 | Customers Satisfaction Factor | 0.6 |
| QEST-CS-3 | Usability Factor | 0.3 |

will result in a global value that does not reflect the process state, as well as, the definition of incorrect thresholds will imply an incorrect analysis of the organization status. In these case studies, for the sake of conciseness, the QF has not been taken into consideration; hence the performance computation takes into account only the weighted sums of the metric results. The experimentation covered three months of development. Raw data have been collected by the application from process work products and stored in the data warehouse. In particular, Table 2 shows a snapshot of values collected at the end of the three months for each project.

Finally, metric values were normalized and used as inputs to Eq. 2 and Eq. 4, for the computation of single dimension and global performance indicators. The values of indicators can be represented using the internal Spago4Q dashboards (Fig. 5) or the user can create *ad-hoc* reports using BIRT functionalities. In particular, dashboard gives an immediate snapshot of the situation, highlighting problems and suggesting to project managers the areas where the effort should be concentrated or where a quality improvement of the process is needed. By contrast, reports can give a more detailed analysis of the data, describing in details the results of indicators and better targeting the improvement actions.

**Table 2.** Metrics values collected for the three projects at the end of each month. Note that data are cumulative and have to be normalized before performance computation.

| KPI | Project 1 | | | Project 2 | | | Project 3 | | |
|---|---|---|---|---|---|---|---|---|---|
| | M1 | M2 | M3 | M1 | M2 | M3 | M1 | M2 | M3 |
| QEST-EC-1 | 80 | 78 | 82 | 50 | 68 | 72 | 25 | 49 | 70 |
| QEST-EC-2 | 25 | 45 | 30 | 5 | 7 | 6 | 8 | 9 | 7 |
| QEST-EC-3 | 50 | 65 | 55 | 2 | 4 | 5 | 14 | 11 | 10 |
| QEST-RS-1 | 99 | 97 | 98 | 60 | 70 | 87 | 97 | 95 | 96 |
| QEST-RS-2 | 91 | 94 | 95 | 96 | 93 | 94 | 98 | 97 | 97 |
| QEST-RS-3 | 50 | 62 | 82 | 35 | 37 | 68 | 10 | 35 | 88 |
| QEST-TE-1 | 20 | 18 | 15 | 30 | 32 | 25 | 50 | 39 | 35 |
| QEST-TE-2 | 1.1 | 1.9 | 1.8 | 1.8 | 2.3 | 2.1 | 4.5 | 3.2 | 1.9 |
| QEST-TE-3 | 95 | 93 | 94 | 96 | 92 | 93 | 82 | 91 | 92 |
| QEST-TE-4 | 99 | 97 | 95 | 30 | 49 | 67 | 85 | 89 | 93 |
| QEST-TE-5 | 2 | 1 | 2 | 1 | 2 | 1 | 1 | 2 | 2 |
| QEST-TE-6 | 3,5 | 3.2 | 3.1 | 2,4 | 1.3 | 1.1 | 16 | 12 | 8 |
| QEST-TE-7 | 87 | 78 | 83 | 88 | 92 | 93 | 50 | 80 | 85 |
| QEST-TE-8 | 10 | 20 | 17 | 55 | 40 | 35 | 68 | 65 | 55 |
| QEST-TE-9 | 2 | 122 | 9 | 25 | 15 | 12 | 29 | 22 | 18 |
| QEST-TE-10 | 375 | 390 | 410 | 178 | 230 | 245 | 210 | 240 | 255 |
| QEST-TE-11 | 1 | 2 | 2 | 2 | 1 | 1 | 3 | 2 | 2 |
| QEST-TE-12 | 2 | 3 | 4 | 9 | 6 | 5 | 10 | 8 | 7 |
| QEST-CS-1 | 60 | 75 | 89 | 95 | 91 | 92 | 94 | 92 | 91 |
| QEST-CS-2 | 75 | 85 | 90 | 91 | 87 | 90 | 93 | 95 | 94 |
| QEST-CS-3 | 91 | 83 | 92 | 96 | 93 | 92 | 80 | 83 | 93 |

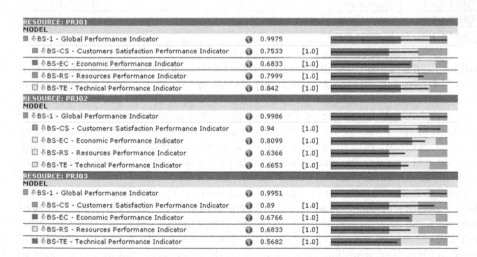

**Fig. 5.** Spago4Q dashboards for projects state at month 3

Looking at the results of our case study, the graphs in Fig. 6 and 7 show that all projects were concluded with an excellent global performance (close to one), showing some issues in the process that is worth analyzing.

For instance, the Economic dimension of Project 1 is characterized by a red square, indicating that the value is within the bad area, hence a deep analysis of that area is needed for next implementations. In fact, project managers discovered that the financial resources assigned to the project were overestimated for the needed effort, suggesting an adjustment to the enterprise criteria for projects financing.

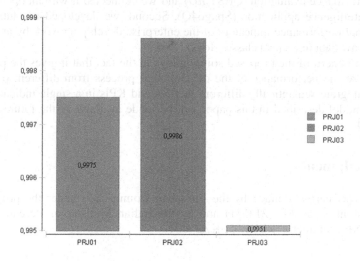

**Fig. 6.** Spago4Q graph for projects global performance comparison

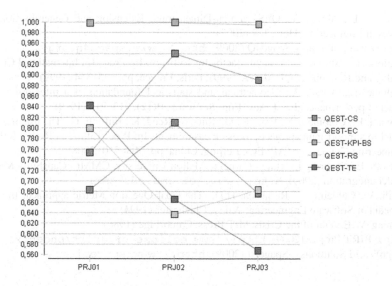

**Fig. 7.** Spago4Q graph for projects single dimensions performance comparison

## 5 Conclusions

The integration of the QEST $n$D model and Spago4Q allows implementing a complete multi-project multi-process performance evaluation environment that combines the mathematical formalization of the QEST $n$D model and the facilities offered by Spago4Q. In this paper, we analyzed such an integration, and described a complete case study that shows the high configurability and reliability of the framework.

The contribution is twofold. First, we implemented and used a formal model for process performance evaluation (QEST $n$D) and we connected it with an open source Business Intelligence application (Spago4Q). Second, we developed a solution that derives global performance indicators of the enterprise developing work by analyzing its process raw data (e.g., java classes, logs).

The main benefit of the proposed solution lays in the fact that it gives the possibility to analyze the performance of the development process from different points of view and integrate semantically different metrics and KPIs in a single indicator. The QEST $n$D model described in this paper will be made available in the future version of Spago4Q.

## Acknowledgments

This work was partly founded by the European Commission under the project SecureSCM (contract n. FP7-213531) and by the Italian Ministry of Research under FIRB TEKNE (contract n. RBNE05FKZ2_004).

## References

1. Buglione, L., Abran, A.: QEST nD: n-Dimensional Extension and Generalisation. Advances in Engineering 33(1), 1–7 (2002)
2. SpagoWorld Solutions - Spago4Q (2009), http://www.spago4q.org
3. Buglione, L., Abran, A.: A quality factor for software. In: Proc. 3rd International Conf. on Quality and Reliability (QUALITA 1999), Paris, France, pp. 335–344 (1999)
4. Buglione, L., Abran, A.: Geometrical and statistical foundation of a three-dimensional model of performance. Int. J. Adv. Eng. Software 30(12), 913–919 (1999)
5. Damiani, E., Colombo, A., Frati, F., Oltolina, S., Reed, K., Ruffatti, G.: The use of a meta-model to support multi-project process measurement. In: Proc. 15th Asia-Pacific Software Engineering Conference (APSEC), Beijing, China, pp. 503–510 (2008)
6. Software Engineering Institute - Carnegie Mellon University, CMMi - Capability Maturity Model integration (2009), http://www.sei.cmu.edu/cmmi/
7. Basili, V., Caldeira, G., Rombach, H.D.: The Goal Question Metric approach. In: Encyclopedia of Software Engineering. Wiley, Chichester (1994)
8. Deming, W.E.: Out of the Crisis. MIT Press, Cambridge (1986)
9. Eclipse BIRT Project (2009), http://www.eclipse.org/birt/phoenix/
10. SpagoWorld Solutions – SpagoBI (2009), http://www.spagobi.org

# An Investigation of the Users' Perception
# of OSS Quality

Vieri del Bianco[1], Luigi Lavazza[2], Sandro Morasca[2], Davide Taibi[2], and Davide Tosi[2]

[1] University College Dublin, Systems Research Group, CASL,
Dublin, Ireland
`vieri.delbianco@ucd.ie`
[2] Università degli Studi dell'Insubria, Dipartimento di Informatica e Comunicazione,
Via Mazzini, 5 – 21100 Varese, Italy
`{luigi.lavazza,sandro.morasca,davide.taibi,`
`davide.tosi}@uninsubria.it`
`http://www.dicom.uninsubria.it`

**Abstract.** The quality of Open Source Software (OSS) is generally much debated. Some state that it is generally higher than closed-source counterparts, while others are more skeptical. The authors have collected the opinions of the users concerning the quality of 44 OSS products in a systematic manner, so that it is now possible to present the actual opinions of real users about the quality of OSS products. Among the results reported in the paper are: the distribution of trustworthiness of OSS based on our survey; a comparison of the trustworthiness of the surveyed products with respect to both open and closed-source competitors; the identification of the qualities that affect the perception of trustworthiness, based on rigorous statistical analysis.

**Keywords:** Open Source Software quality, perceived quality, trustworthiness.

## 1   Introduction

Quality is often an elusive concept in Software Engineering. First, many attributes exist that may be used to describe software quality. For instance, the ISO9126 standard [1] views software quality as a multi-attribute concept, and different people may place different emphasis on the same attribute, depending on their experience, goals, and software at hand. In addition, the actual quantification of even some specific attribute may be problematic, as measures for that attribute may not be mature enough to have reached a sufficient degree of consensus, or may provide inconclusive results. For instance, two measures for the same software attribute may rank two software applications A and B in a conflicting way, i.e., one measure may rank A better than B while the other measure may reverse the ranking.

So, software stakeholders often choose to adopt one application over another based on the quality they perceive, instead of an objective quality evaluation. To some degree, this happens with several different types of products, if not all. For instance, a prospective buyer may choose one car over another based on his or her own perception of the

P. Ågerfalk et al. (Eds.): OSS 2010, IFIP AICT 319, pp. 15–28, 2010.

overall quality of the car, or some of the car's characteristics, or even the characteristics of the car's manufacturer. At any rate, in the case of cars, a number of objective measures exist, like length, width, height, volume of the engine, maximum speed, number of miles per gallon, number of seconds needed to get to some specified speed, so a prospective buyer can make informed decisions. When it comes to software, however, the lack of consensus measures makes decisions even more based on perceptions. Perceptions may be even more important for Open Source Software (OSS) than for other types of software. OSS has often suffered from some kind of biased perception, probably based on the idea that OSS is built by amateur developers in their spare time. It took a few years and a few success cases to dispel at least some of these perceptions about OSS, but some of that stigma is believed to still taint the reputation of OSS vs. Closed Source Software (CSS) at least in some environments. So, it is important to study how various attributes of OSS are perceived, to check if those perceptions about OSS qualities are still valid today and which specific qualities are believed to need improving more than others.

In this paper, we report on an empirical study about the perception of OSS qualities. We carried out the study in the framework of the QualiPSo project [9], which is funded by the European Union in the 6th Framework Program. Trustworthiness is the main focus of the QualiPSo project as for OSS product evaluation. However, OSS trustworthiness itself is a broad concept. On the one hand, trustworthiness is closely related to the idea of overall OSS quality: an OSS product is adopted only if stakeholders have sufficient trust in its quality. On the other hand, as OSS trustworthiness is influenced by a number of diverse factors which may include product- and process-related ones, several concepts and sources of information may need to be taken into account when studying OSS trustworthiness.

In the QualiPSo project we investigated the factors that are *believed* to affect trustworthiness [3] by OSS stakeholders. Then, we defined a conceptual model that represents the dependence of trustworthiness on other qualities and characteristics of the software [4]. To prove the validity of such conceptual model and provide it with quantitative models of trustworthiness, we collected both subjective evaluations and objective measures of OSS. Specifically, the subjective evaluations concerned how users evaluate the trustworthiness and other qualities of OSS. These evaluations are here analyzed in a rigorous way to derive indications concerning the quality of OSS that are both quantitative and reliable, since they are rooted on a reasonably wide sample of users' opinions.

Our investigation has shown that the majority (56%) of OSS products are considered very trustworthy and that the surveyed OSS products are generally considered better than the competitors (both OSS and CSS) by their users. Finally, we discovered statistically significant models that quantitatively describe the dependence of trustworthiness on qualities like reliability, usability, interoperability, efficiency, and documentation.

The paper is organized as follows. Section 2 describes data collection. Section 3 reports the results of the analysis, while the threats to the validity of the results are discussed in Section 4. Section 5 discusses the related work, and Section 6 draws some conclusions and sketches future work.

## 2  The Investigation

We carried out a survey to collect OSS stakeholders' evaluations of several OSS products according to a number of qualities. We actually selected just a few of all the qualities identified in the GQM plan that defines the QualiPSo notion of trustworthiness [4], because we knew that users may not be able to evaluate many OSS products, and, for each OSS product, too many of its qualities. Thus, in addition to a few questions characterizing the users (including how familiar they were with the product), we asked them to evaluate the overall trustworthiness of the products and the following qualities, which are believed to be the ones that most affect trustworthiness, based on a previous survey that we carried out among OSS stakeholders [3]:

- Usability;
- Portability;
- Functional requirements satisfaction;
- Interoperability;
- Reliability;
- Security;
- Developer community utility;
- Efficiency;
- Documentation;
- Trustworthiness vs. OSS competitors;
- Trustworthiness vs. CSS competitors.

We used a questionnaire to ask our respondents how they would rate the qualities of up to 22 Java and 22 C++ OSS products. The list of products appears in Fig. 1. We used a 1 to 6 ordinal scale, where 1 was the worst evaluation and 6 the best evaluation for a specific quality of a product with the following possible answers:

1 = absolutely not;
2 = little;
3 = just enough;
4 = more than enough;
5 = very/a lot;
6 = completely.

For illustration's sake, one of the questions was "How usable is the product?" with reference to some specified product. All other questions about all other qualities were asked in a similar fashion.

Up to the end of August 2009, we collected 100 questionnaires, containing 722 product evaluations, of which about 36% concerned Java products, while the remaining ones concerned C++ products.

The questionnaires were collected at major international events, not necessarily dealing with OSS topics, as summarized in Table 1.

We did not screen our respondents beforehand, so we used what is known as a convenience sample. The possible effects that this may have had on our empirical study are addressed in Section 4.

**Table 1.** Events where data where collected

| Event | Date (in year 2009) and location | Collected questionnaires | Product evaluations |
|---|---|---|---|
| Apache Conference | March 24-27, Amsterdam, The Netherlands | 15 | 31 |
| OW2 Conference | April 1-2, Paris, France | 20 | 31 |
| XP 2009 | April 24-30, Pula, Italy | 12 | 95 |
| OSS 2009 | June 2-5, Skovde, Sweden | 2 | 5 |
| ICSE 2009 | May 15-20, Vancouver, Canada | 9 | 69 |
| CONFSL 2009 | June 12-13, Bologna, Italy | 3 | 27 |
| QualiPSo Meeting | July 1-2, Madrid, Spain | 6 | 38 |
| ESC | August 30-31, Venice, Italy | 31 | 411 |
| Others | | 2 | 15 |

In the rest of the paper, we often call the respondents "users;" however, in addition to end-users, these "users" include also developers, managers, and stakeholders that are interested in OSS for various reasons.

## 3    The Results of the Investigation

### 3.1    The Popularity of the Products

A first result of our investigation concerns the popularity of the OSS products we selected. Since users were asked to answer about the products they knew well enough, the number of evaluations received by a product may be taken as a reasonable indication of its popularity.

Fig. 1 shows how many users evaluated each product, and how many of them answered that they have good familiarity with the product (the shorter bars report how many users rated their familiarity > 3).

The results reported in Fig. 1 were quite expected: MySQL, Eclipse and the Linux Kernel appear to be the most popular products.

We then proceeded to evaluate whether the popularity could be explained in terms of the type of the product (end-user oriented vs. programmer oriented, database management systems vs. configuration management systems vs. libraries, etc.). However, we found no such relationships. We consider this a good result, since it seems to indicate that users evaluated the actual qualities of the products, independent of their types and target users.

### 3.2    The Trustworthiness of the Products

We take the median of the evaluations as representative of the evaluations provided by the respondents (including end users and stakeholders) to the overall trustworthiness of OSS products. However, we noticed that some products had very low median grades because of only two or three respondents who had little familiarity with the product. As such evaluations may be deemed unreliable, we removed the evaluations of respondents with familiarity $\leq 3$ from the dataset. Also, we considered only products that were rated by at least 4 respondents.

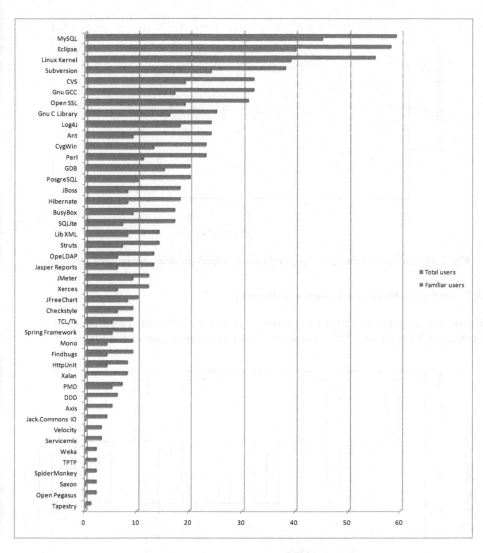

**Fig. 1.** Number and familiarity of respondents per product

Fig. 2 shows the median of users' evaluations for each product. On the x axis are the 32 products for which we collected enough data from users having sufficient familiarity. In other words, "Products" is a nominal variable.

It appears that users are generally very satisfied with the OSS products. Nevertheless, the facts that no product's median reached the maximum, that in several cases the median was only 4, and in one case even 3, shows that the users were not 'fanatic' of OSS. Rather, they seem to have provided well-balanced and reliable evaluations.

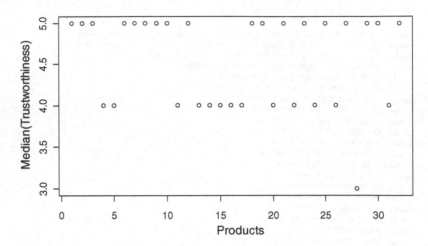

**Fig. 2.** Overall trustworthiness of the evaluated products (medians per product considered)

### 3.3 OSS vs. CSS (Closed-Source Software)

Users were asked to rate the trustworthiness of every product in comparison to similar OSS and CSS products. The medians of these ratings are illustrated in Fig. 3.

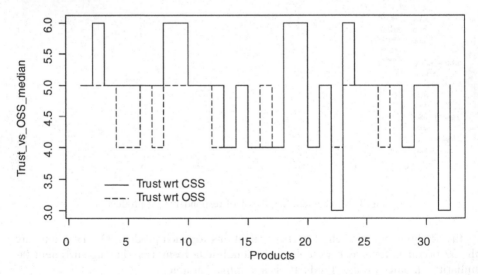

**Fig. 3.** Overall trustworthiness of the evaluated products compared to similar OSS and CSS products (medians per product considered)

A first observation suggested by the figure is that in general the evaluated projects are highly trusted with respect to both OSS and CSS competition. In fact, both scores are positioned mainly in the 4–5 range.

It is then possible to observe that in 21 out of 32 cases (65.6%) the CSS and OSS alternatives are considered of equivalent quality, though generally lower than the considered OSS product. This result seems to confirm that the choice of OSS products to evaluate was actually a good one, in that the considered products are generally considered very well positioned with respect to OSS and CSS alternatives.

In only 9 cases out of 32 (33%), the OSS alternatives are better than the CSS alternatives (this is the case whenever the red solid line is above the blue dashed line), while in only 2 cases out of 32 (6%) the CSS alternatives are considered better than the OSS ones.

The overall impression that is conveyed by Fig. 3 is that the OSS user community does trust OSS, but not in a fanatic manner, since the quality of CSS is also acknowledged, e.g., by considering OSS and CSS alternatives to leading OSS products as substantially equivalent.

### 3.4 The Quality of OSS Products

In the QualiPSo project, we have investigated the qualities that –according to OSS users– most affect the overall notion of OSS trustworthiness [3]. Following such indications, we have built a conceptual model of OSS trustworthiness that proposes an explanation of how OSS product sub-qualities (like as-is utility, exploitability in development, functionality, reliability) contribute to determining the overall trustworthiness as perceived by users. Such model is defined via a GQM plan [4], which involves also objectively measurable characteristics of OSS. The idea is that, when enough data are available, we can build a quantitative model that explains to what extent the subjectively perceived qualities of OSS depend on its internal characteristics.

The data reported here are the result of the data collection (concerning exclusively the subjective evaluations) performed as part of the execution of the GQM plan.

Fig. 4 illustrates the distributions of the median evaluations of different products for each surveyed quality. For this analysis, only the products evaluated by at least 10 users with sufficient familiarity have been considered. For each quality, the box represents the range comprising half the population, the thick segment represents the median, the dashed lines extend to the most extreme data point which is no more than 1.5 times the interquartile range from the box, while the small circles indicate outliers.

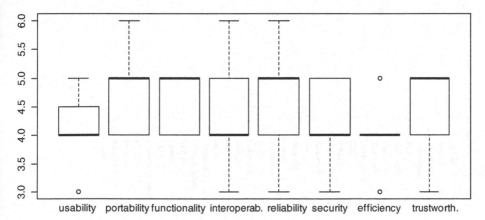

**Fig. 4.** Boxplots and medians for the qualities of the evaluated products

A first observation suggested by Fig. 4 is that all qualities are given quite high grades in general, as most evaluations are between 4.0 and 5.0. Also, while overall trustworthiness is rated very well (see also Fig. 2), the users have been more critical with other qualities –like usability, interoperability, security and efficiency– which are given lower grades than trustworthiness. It is interesting to note that security, which is usually very positively correlated with trustworthiness, is not rated particularly well, even though most products are graded "more than enough secure."

Fig. 5 reports the box-plots that synthesize the distribution of grades across products for every quality concerning the support to the end user, namely the available documentation and the support from the developer community. This is a rather relevant aspect, since users of OSS products often need to rely exclusively on the available documentation or the support from the developer community in order to get information or resolve problems concerning OSS products.

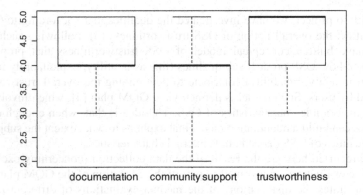

**Fig. 5.** User-support qualities (medians) of the evaluated products

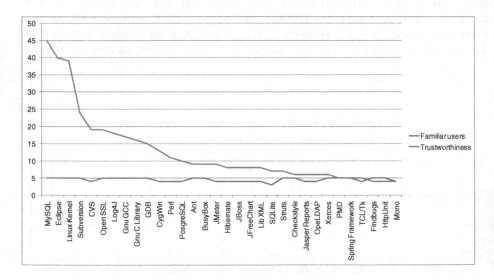

**Fig. 6.** Trustworthiness and popularity

The median value of user support is rated "more than enough" in general. However, user evaluations are not aligned with conventional wisdom: documentation, which is often considered a weak point of OSS, is rated quite well, while the support by the developer community, which is generally believed to be a strong point of OSS is not rated very well for several products.

Fig. 6 shows that the four most popular products are also among those considered most trustworthy. Anyway, there is clearly no correlation between popularity and trustworthiness: for instance, several products having relatively little popularity are considered very trustworthy.

### 3.5 Influence of the Implementation Language on the User-Perceivable Trustworthiness

We collected users' opinions on products written in Java or C++. We investigated if the implementation language affects user-perceived trustworthiness. The box-plots in Fig. 7 summarize the distributions of the trustworthiness evaluations for C++ and Java programs

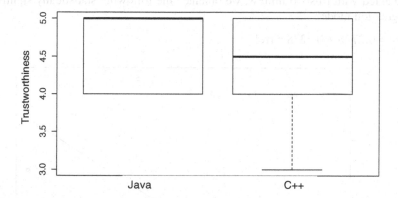

**Fig. 7.** Comparison of the overall (median) trustworthiness of the evaluated products depending on the implementation language

While 64% of the Java products are rated "very trustworthy," only 50% of the C++ ones appear to get such rating. According to Fig. 7, Java programs appear to be slightly better than C++ ones. However, just one additional C++ project rated 5 would move the median trustworthiness of C++ programs to 5, thus making it equal to the one of Java programs. In conclusion, it is not possible to state that there is a dependence of trustworthiness on the implementation language.

### 3.6 Which Factors Affect OSS Trustworthiness?

Here, we investigate whether there is a statistically significant dependence of trustworthiness on other subjective qualities. To this end, we consider the fractions of users that are satisfied with the given qualities. For instance, the fraction of users satisfied with a product's trustworthiness is computed as the number of users that rated the product's trustworthiness above a given threshold, divided by the total number of users who evaluated the product's trustworthiness. The satisfaction threshold was set at 4, i.e., the satisfied users are those who rated the product 5 (very good) or 6 (completely satisfactory).

To have reasonably significant fractions, we limited the analyses to product that were evaluated by at least 10 users having a good familiarity with the product.

The analyses reported below were conducted using Ordinary Least Squares (OLS) regression. The choice of OLS regressions is justified by the fact that both the dependent and independent variables are in the 0..1 range.

We used 0.05 as the statistical significance threshold, as is customarily done in empirical software engineering studies. Therefore, all the reported models have p-value < 0.05. The normality of the distribution of the residuals, which is a statistical requirement for safely applying OLS regression, was tested by means of the Shapiro-Wilk test [5]: consistent with our statistical significance threshold, p-values > 0.05 do not allow the rejection of the normality hypothesis.

A first result involves the dependence of trustworthiness on reliability. If we denote by *rrel* the fraction of users satisfied with reliability and *rtrust* the fraction of users satisfied with trustworthiness, we obtained the following statistically significant OLS regression model:

rtrust = 0.2726 + 0.7278 * rrel

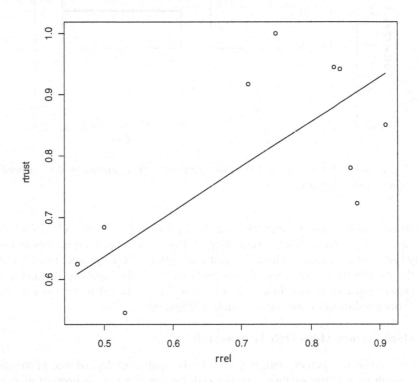

**Fig. 8.** Dependence of trustworthiness on reliability: the linear regression line

The residuals are normally distributed (i.e., the normality hypothesis cannot be rejected), and the determination coefficient is reasonably good ($R^2 = 0.6391$, adjusted $R^2 = 0.594$). So, around 60% of the variability in the degree of trustworthiness satisfaction is explained by the degree of reliability satisfaction.

The precision of the fitting is quite good: MMRE=9.2%, pred(25)=90%, with errors in the -13%..+25% range.

No valid OLS regression model could be found between trustworthiness and the satisfaction of functional requirements, portability, security, and the usefulness of the developers' community. On the contrary, we found statistically significant models of the dependence of trustworthiness on usability, interoperability, efficiency and documentation. These results are summarized in Table 2. In the "Line equation" column, the dependent variable is always *rtrust*, while the independent variable is the fraction of users that were satisfied with the quality reported in the 'Quality' column.

**Table 2.** Correlations found

| Quality | Line equation | $R^2$ | MMRE | Pred(25) | Error range |
|---|---|---|---|---|---|
| Usability | y= 0.3686 + 0.7252 x | 0.4634 | 11.8% | 90% | -16% .. +26% |
| Interoperability | y = 0.1370 * 0.9479 x | 0.6817 | 9.2% | 100% | -16% .. +19% |
| Efficiency | y = 0.3642 * 0.7172 x | 0.6579 | 8.6% | 100% | -21% .. +14% |
| Documentation | y = 0.3712 * 0.7289 x | 0.6256 | 8.9% | 100% | -18% .. +18% |

## 4  Threats to Validity

A number of threats may exist to the validity of a correlational study like ours. We now examine some of the most relevant ones.

### 4.1  Internal Validity

We checked whether variables are normally distributed when carrying out OLS regressions, as required by the theory of OLS regression. Consistent with the literature, we used a 0.05 statistical significance threshold, the same we used for all statistical tests in our paper. The vast majority of statistical tests we carried out to this end provided quite strong evidence that the variables are indeed normally distributed, with the exception of Reliability and Usability, for which the p-values obtained with the Shapiro-Wilk normality test are 0.0612 and 0.09, respectively. These values are close to the 0.05 statistical significance threshold, but, based on these values, we could still not reject the hypotheses that Reliability and Usability are normally distributed, so we could carry out OLS regression. At any rate, the statistical tests used in OLS regression are somewhat robust and they can be practically used even when the variables' distributions are not that close to normal.

### 4.2  External Validity

Like with any other correlational study, the threats to the external validity of our study need to be identified and assessed. The most important issue is about the fact that our sample may not be fully "balanced," and that may have somewhat influenced the results. While this may be true, the following points need to be taken into account.

- It was not possible to interview several additional people that could have made our sample more "balanced," because they were not available or had no or little interest in answering our questionnaire.
- No reliable demographic information about the overall population of OSS "users" is available, so it would be impossible to know if a sample is "balanced" in any way.
- Like in many correlational studies, we used a so-called "convenience sample," composed of respondents who agreed to answer our questions. We collected information about the respondents' experience, application field, etc., but we did not make any screening. Excluding respondents based on some criteria, which must have been perforce subjective, may have resulted in an "unbalanced" sample, which may have biased the results.
- We dealt with motivated interviewees, so this ensured a good level for the quality of responses.
- There is no researcher's bias in our survey, since we simply wanted to collect and analyze data from the field, and not provide evidence supporting or refuting some theory.

### 4.3 Construct Validity

An additional threat concerns the fact that the measures used to quantify the relevant factors may not be adequate. This paper deals with trustworthiness, which is an intrinsically subjective quality, so the only way to measure it is to carry out a survey. As for the the other qualities, we are interested in stakeholders' evaluations and not in objective measures (which do not exist anyway), so, again, a survey is adequate to collect information about them.

## 5  Related Work

Several attempts were made to address the issue of software quality assessment in general, and within OSS in particular (see for instance the seminal papers by Audris Mockus et al. [12][14][15]). The online communication platforms and tools used in the development process (Concurrent Versioning System – such as CVS– Mailing Lists, Bug Tracking Systems – such as Bugzilla – and online discussion forums) contain a considerable amount of evaluations and data about the quality of the software project. Therefore, such repositories have often been used for extracting data concerning the quality of OSS.

In [6] Tawileh et al. define a new approach for quality assessment of F/OSS projects based on social networking. They exploit the use of social networks of users formed around F/OSS projects, in order to collect data about the perception of OSS projects' quality and to make recommendations according to user preferences. Unfortunately, no data about the user' perception of the quality of OSS projects are reported in [6]. In our work, we collect data through a more controlled environment than social networks (i.e., via questionnaires dispensed to OSS users); we statistically analyze the collected data and we report how users perceive the quality of a representative sample of Java and C++ OSS products.

The TOSSAD portal stores an extensive set of surveys about the adoption of OSS products in target countries. For example in [7] OSS users are asked about their perception of quality about OSS in general. In our survey, we investigate the perceived quality of specific OSS products instead of the perceived quality of OSS in general.

In 2009, the Eclipse foundation conducted a deep survey about the overall quality of their IDE as perceived by users [8]. The survey is focalized only on their development environment; moreover, it does not take into account the specific qualities of Eclipse. In our work, we ask OSS users about several specific aspects of quality, thus ranking not only the general perception of trustworthiness, but also the perception of reliability, interoperability, efficiency, usability, and documentation of OSS products.

## 6  Conclusions and Future Work

The evaluation of the trustworthiness of OSS is important because of OSS ever increasing importance in software development and practical applications. However, lacking objective measures, OSS users and stakeholders rely on their own somewhat subjective evaluations when deciding to adopt an OSS product.

We carried out a survey to study the users' perception of trustworthiness and a number of other qualities of OSS products. We selected 22 Java and 22 C++ products, and we studied their popularity, the influence of the implementation language on trustworthiness, and whether OSS products are rated better than CSS products.

In addition, our results seem to provide evidence in favor of the existence of a few relationships between the user evaluations of a number of OSS qualities and trustworthiness. So, it is possible to have an idea of the impact that the evaluations of these qualities have on trustworthiness. Some trustworthiness evaluation methods have been proposed to let potential users assess the quality of OSS products before possibly adopt them. Such methods –like the OpenBQR [2] and the other similar approaches [10][11][12]– face typically two problems: what are the factors that should be taken into consideration, and what is the relative importance of such factors? Generally these decisions are left to the user, who has to choose the qualities in a usually long list and assign weights. So, the work reported here improves our knowledge of the user-perceived qualities and trustworthiness of OSS products and of trustworthiness models.

Our future work will include the following activities.

- Collecting additional data about users' evaluations of OSS.
- Collecting data about additional qualities that may be of interest.
- Carrying out studies to check whether there exist relationships between some structural characteristics of OSS (e.g., size, structural complexity) and the external, user-related qualities we study in this paper.
- Using the profiling information about respondents to build more precise models for specific classes of OSS stakeholders.

The survey of users' opinions is going on through an on-line questionnaire (the Trustworthy Products Questionnaire is accessible via the QualiPSo web page http://qualipso.dscpi.uninsubria.it/limesurvey/index.php?sid=58332&newtest=Y&lang=en) and we invite all interested readers to fill out the questionnaire and contribute to this study).

## Acknowlegments

The research presented in this paper has been partially funded by the IST project
Qualipso (http://www.qualipso.eu/), sponsored by the EU in the 6th FP (IST-034763); the FIRB project ARTDECO, sponsored by the Italian Ministry of Education and University; and the projects "Elementi metodologici per la descrizione e lo sviluppo di sistemi software basati su modelli" and "La qualità nello sviluppo software," funded by the Università degli Studi dell'Insubria.

## References

[1] ISO/IEC 9126-1:2001, Software Engineering—Product Quality—Part 1: Quality model (June 2001)
[2] Taibi, D., Lavazza, L., Morasca, S.: OpenBQR: a framework for the assessment of OSS, Open Source Software 2007, Limerick (June 2007)
[3] del Bianco, V., Chinosi, M., Lavazza, L., Morasca, S., Taibi, D.: How European software industry perceives OSS trustworthiness and what are the specific criteria to establish trust in OSS, QualiPSo report (October 2008)
[4] del Bianco, V., Lavazza, L., Morasca, S., Taibi, D.: Quality of Open Source Software: the QualiPSo Trustworthiness Model. In: The 5th International Conference on Open Source Systems OSS 2009, Skövde, Sweden, June 3-6 (2009)
[5] Shapiro, S.S., Wilk, M.B.: An analysis of variance test for normality (complete samples). Biometrika 52(3-4) (1965)
[6] Tawileh, A., Rana, O., McIntosh, S.: A Social Networking Approach to F/OSS Quality Assessment. In: Proceeding of the International Conference on Computer Mediated Social Networking (ICCMSN 2008), Dunedin, New Zealand (June 2008)
[7] TOSSAD, Annex5 – Survey Report, Web published: http://www.tossad.org/publications/ (Accessed 14/12/2009)
[8] Eclipse, The Open Source Developer Report (May 2009), Web published, http://www.eclipse.org/org/press-release/Eclipse_Survey_2009_final.pdf (Accessed 14/12/2009)
[9] QualiPSo project portal, http://www.qualipso.org
[10] Atos Origin, Method for Qualification and Selection of Open Source software (QSOS), version 1.6, http://www.qsos.org/download/qsos-1.6-en.pdf
[11] Business Readiness Rating for Open Source - A Proposed Open Standard to Facilitate Assessment and Adoption of Open Source Software, BRR 2005 - RFC 1 (2005), http://www.openbrr.org
[12] Making Open Source Ready for the Enterprise: The Open Source Maturity Model, from "Succeeding with Open Source" by Bernard Golden. Addison-Wesley, Reading (2005), http://www.navicasoft.com
[13] Mockus, A., Fielding, R.T., Herbsleb, J.D.: Two case studies of open source software development: Apache and Mozilla. ACM Trans. Softw. Eng. Meth-odol. 11(3) (2002)
[14] Mockus, A., Weiss, D.: Interval quality: relating customer-perceived quality to process quality. In: International Conference on Software Engineering 2008, Leip-zig, Germany (May 2008)
[15] Mockus, A., Zhang, P., Li, P.L.: Predictors of customer perceived soft-ware quality. In: International Conference on Software Engineering 2005, St. Louis (May 2005)

# Engaging without Over-Powering: A Case Study of a FLOSS Project

Andrea Capiluppi[1], Andres Baravalle[1], and Nick W. Heap[2]

[1] Centre of Research on Open Source Software (CROSS)
University of East London, UK
{a.capiluppi,a.baravalle}@uel.ac.uk
[2] The Open University
Walton Hall, Milton Keynes, UK
n.w.heap@open.ac.uk

**Abstract.** The role of Open Source Software (OSS) in the e-learning business has become more and more fundamental in the last 10 years, as long as corporate and government organizations have developed their educational and training programs based on OSS out-of-the-box tools. This paper qualitatively documents the decision of the largest UK e-learning provider, the Open University, to adopt the Moodle e-learning system, and how it has been successfully deployed in its site after a multi-million investment. A further quantitative study also provides evidence of how a commercial stakeholder has been engaged with, and produced outputs for, the Moodle community. Lessons learned from this experience by the stakeholders include the crucial factors of contributing to the OSS community, and adapting to an evolving technology. It also becomes evident how commercial partners helped this OSS system to achieve the transition from an "average" OSS system to a successful multi-site, collaborative and community-based OSS project.

## 1 Introduction

In the first decade of the twenty-first century three factors have been pushing the "e-learning topic" under the spotlight: first, the recognition that it has become, together with the underlying technology, a recognized and sustainable industry. Secondly, the attempts to create Open Data Standards (ODS) for e-learning content, driven by specification organizations such as the IMS Global Learning Consortium, Aviation Industry CBT (Computer-Based Training) Committee (AICC), and Advanced Distributed Learning (ADL) network sponsored by the U.S. Office of the Secretary of Defense, and relevant committees of international standards bodies, such as the IEEE Learning Technology Standards Committee [7]. Finally, the much wider movement that advocates OSS and open data standards. OSS includes highly successful software such as the *Linux* operating system, the *Apache web* server and *OpenOffice.org*.

The increased importance of e-learning, the emergence of ODS for e-learning and the driving push from the Open Source community are creating a fertile environment where innovation can spread more efficiently.

E-learning platforms have been in use for a number of years, but the technological focus has been shifting: early adopters were focusing on e-delivery of teaching material,

P. Ågerfalk et al. (Eds.): OSS 2010, IFIP AICT 319, pp. 29–41, 2010.
© IFIP International Federation for Information Processing 2010

that is just allowing users to download teaching material from the web. At present, academic institutions are also focusing on other areas, and one of the most pressing issues is the packaging and distribution of e-learning resources. Academics, who are used to disseminate the findings of their research to the wider academic community, are less used to use, modify and redistribute teaching material. The complexity of e-learning software, poor interoperability and the elevate cost of commercial e-learning solutions play all a central role in this. OSS and ODS can help to address both interoperability and price, ensuring that teaching material can be exchanged and used more easily and with inferior economical costs.

This paper studies the evolution of the Moodle e-learning platform, and describes the process of its deployment in the the Open University, the largest on-line course provider in the UK. In order to achieve this, this paper uses a mixed qualitative and quantitative approach, and uses a wealth of information sources, ranging from interviews with commercial stakeholders in Moodle, to empirical data contained in the Moodle code repository. It is argued that this system represents a "hybrid" OSS project [6]: since its inception in the early 1980's, OSS projects were purely volunteer-based, heavily relying on personal efforts and non-monetary recognitions, and bearing communication and coordination issues ("Plain OSS", right end of Figure 1, adapted from [6]). Nowadays *Commercial OSS* are also present (more similar to Closed source systems, as in Figure 1), where a commercial company plays a major role in the development and decision making. *Community OSS* instead are more similar to pure OSS systems, since they are driven by the community, but they also often have several commercial stakeholders.

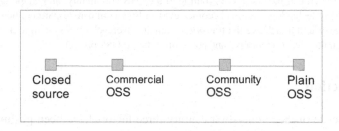

**Fig. 1.** Software licensing continuum

This paper is articulated as follows: section 2 describes the case study from the point of view of its stakeholders. Section 3 focuses on one stakeholder (the largest provider of e-learning resources in UK), and illustrates the process of adopting this OSS solution, and the issues and benefits of doing so. Section 4 focuses on another commercial stakeholder (Catalyst IT Ltd.) and quantifies its contributions to Moodle from the point of view of its developers deployed to Moodle. Section 5ss concludes.

## 2 Moodle

This paper focuses on an extensive analysis (both qualitative and quantitative) of the business and development model of Moodle, a popular Open Source software for e-learning. Given its size, extensive development and user community, a more in-depth appreciation of Moodle, and how it achieved its status, is central to understanding Open Source software and its future among the software competitors.

Moodle's development is centered around various actors:

1. **Moodle core developer:** Martin Dougiamas originally developed Moodle while working at his Ph.D. thesis in Curtin University of Technology, Australia. Now Moodle's development is lead by Moodle Pty Ltd, a company he founded and leads.

2. **Commercial stakeholders and Moodle developers:** the entities that have an interest in the creation and support of Moodle:

(a) Moodle partners: a number of organizations across the world who are directly contributing to the development of Moodle by way of funding or contributing their expertise. As we write (12/2009) there are some 50 partners, distributed across the Americas, Europe, Asia and Oceania. As yet there are no African partners.

(b) *Commercial exponents*, **not** participating in the partnership, but working on the development of modules, plug-ins, themes and language packs.

(c) *Moodle developers*: whilst Moodle's development is lead by Martin Dougiamas through Moodle Pty Ltd, a large number of individuals have been contributing to the development of Moodle. Just over 200 developers have write access at this stage, but not all have been contributing into the source code. Other developers do not have the right to publish their changes in the CVS tree (as quantified in the next sections). A Moodle partner or a commercial exponent may employ a number of developers.

(d) *Commercial exponents* focusing on installation, lightweight customization and support, but **not** providing custom development for Moodle.

3. **Moodle community:** this includes the large number of users of Moodle spread across 204 countries (as of September 2009). The community engages in Moodle's activities though on-line discussions in forums and in other specialised events. While the role of the community is important, Moodle is not led, as other projects, by the community.

It is also important to note that Moodle has been building on existing technologies and research, from a community much wider than the sole Moodle community. For example, Moodle uses PHP and MySQL for server-side development, incorporates previous works on data standards, and is supporting existing technologies as SCORM and LAMS to incorporate teaching objects.

## 2.1 Business Model

Before proceeding, it's important to discuss and analyze the Moodle business model. While Open Source software is free to use, modify and redistribute, it does not mean that successful business modules cannot be created around Open Source software [1]. In the context of Moodle, a number of different strategies are currently pursued by Moodle Pty Ltd, the commercial stakeholders and the developers:

1. **Project lead:** Moodle Pty Ltd, and Moodle developers are in a privileged position to receive funding for additional features to be included in the system.
2. **Partnership synergies:** Moodle partners have a "privileged relationship" with Moodle Pty Ltd, For example, as we will see later, the development on Moodle itself was lead by a UK company rather than by Moodle Pty Ltd, Partners have a privileged access to local markets thanks to customer referrals, and at the same time provide an additional source for funding or resources for Moodle Pty Ltd.
3. **Peripheral development:** commercial exponents who do not have developer or partner roles typically work on the more peripheral areas of Moodle, which do not require changes in the core areas of the code base. As we will see in the next sections, certain stakeholders may find a number of strategies to be ineffective: trying to submit changes in core areas might prove difficult and thus can lead to expensive maintenance costs, as the commercial exponent would have to maintain its own fork of Moodle.

### 2.2 Commercial Stakeholders and Peripheral Development

This subsection summarizes the experience of one of the commercial exponents, but not a Moodle member. Mediamaisteri Group ltd.[1] is a Finish leader in the area of virtual learning environments and as part of its business activities sells Moodle related services, such as maintenance, deployment, content production. In the past years Mediamaisteri has been developing a variety of custom modules (about 15) and components, only some of which are used in the community version of Moodle. Between 2003 and 2006 a number of their modules have been approved and included in Moodle but from 2007 there has been a change in trends.

Although it has been supporting 5 to 10 developers working on Moodle in the previous years, now the investment on Moodle development has been slowing. Their modules are not making it in the official Moodle distribution and they find it hard to support them just with their own workforce. Similarly, their changes to core areas of Moodle are also not making it to the official Moodle release, and they are now put in a situation where they have to support their own version of Moodle.

While active in Moodle development, Mediamaisteri is not an official Moodle partner, nor any of its developers has official developer status in Moodle. The implication is that they are not involved in the planning phases, and they have a competitive disadvantage comparing to other companies.

Mediamaisteri experience shows that, at least in case of Moodle, commercial partners are treated similarly to any other OSS contributor, and their code patches will go through the usual scrutiny from the community. Organizations (and individuals) like Mediamaisteri who have a limited involvement (at least in terms of resources committed to the project) may find it difficult to modify core components.

## 3 Moodle at the Open University

The Open University of the United Kingdom is a centrally funded higher education institution specializing in blended and distance learning, with an established reputation

---

[1] http://www.mediamaisteri.com

for its contributions to educational technologies. Recently, the Open University scored the highest student satisfaction rating in a National Student Satisfaction Survey covering England, Wales and Northern Ireland. Students are not required to satisfy academic entry requirements, which encourages participation from a diverse student body able to enroll and pursue the majority of awards and curricula.

Experiments with e-learning date from the mid-1980s and the spread of home computers. Computer conferencing was introduced to courses of 5000 students as early as 1989 followed by the first web sites in 1993. All these developments were bespoke and hence expensive to develop and maintain.

In November 2005, the Open University's Learning and Teaching Office (LTO) announced it was to commence a £5 million programme to *"build a comprehensive online student learning environment for the 21st century"*. Moodle is just one part of this student learning environment, but is the most visible from a student's perspective. The first courses were hosted in May 2006 at which time it was claimed to be the largest use of Moodle in the world.

There are valuable lessons to be learned from the Open University's experience, such as how the institution arrived at its decision to use Moodle, what were the main issues in its planned development and deployment, and what benefits were gained from the early adoption of this OSS package.

## 3.1 Initial Selection

The selection phase for the core of the VLE platform commenced in 2003 and ran for almost two years. A complete review was undertaken of all existing support and delivery systems along with visits to other institutions to learn about their experiences with various platforms.

Early consideration was given to an in-house development that could tie together the mixed-bag of systems supporting registration, content delivery, and learning support, but it was quickly discounted as prohibitively expensive.

A range of proprietary solutions were also considered, but excluded because they offered limited customization and could not guarantee the scalability required; the Open University has some 150,000 students using its on-line systems. OSS solutions were reviewed and initially rejected because of concerns about the high level of risk and the lack of a viable partner.

Having eliminated all the options the review team went back to investigate a combination of in-house development coupled with an OSS learning platform. By September 2005 the business case was completed and the formal decision to adopt Moodle was announced in November 2005.

The substantive development phase has now drawn to a close and all courses migrated to the new platform. Although some work remains, the time is right to reflect on what has been achieved and what lessons have been learnt. As a consequence a small number of interviews have been undertaken with development staff including the Director of the Learning Innovation Office, various Project Leads, and individual programmers.

The selection phase had established some 23 areas of development work that would be required to add or enhance Moodle features – as they existed in 2005. Of these the following were regarded as potential 'show stoppers':

1. **Existing user model:** the Open University's student support model and administration system requires a hierarchy of user roles (and associated permissions) to support the various combinations of full-time and part-time teaching staff, editorial, production, and technical, and the Help-Desk. A typical course may have as many as 30 user categories whereas Moodle supported just three roles.
2. **Limited database support:** Moodle offered no support for either Microsoft's or Oracle's RDBMS, which were the database servers in use in the university. Furthermore, a database abstraction layer was missing.
3. **Grade-book feature:** a new facility to permit students and teaching staff to review assignment grades.
4. **Data entry forms:** inconsistent coding of data and text entry forms contributing to poor accessibility and difficult maintenance.

One of the greatest challenges for the Open University was to balance the benefits of the Moodle solution against the costs of the enhancements to ensure fitness for purpose. The benefits of adopting an off-the-shelf solution would quickly disappear if too many in-house changes were implemented prior to deployment.

Discussions within the Moodle community concluded that items 2, 3 and 4 would enhance the Moodle core and so the Open University agreed to fund the development costs by contracting out the work to Moodle Pty. As a result of this effort, a database abstraction layer was created (named XMLDB) and the API improved.

On the other hand, the 'user model' changes were regarded as controversial: many in the community considered the changes unnecessary, whilst others were concerned about the potential impact on performance: as a result this effort was undertaken by the Open University.

### 3.2  Role of the Open University as Community Contributor

The challenges faced by the Open University are common to any organisation that contemplates the introduction of an OSS solution into its core business functions. However, the attitude of many is to adopt OSS packages behind the scenes, possibly adapting it to fit some niche requirement, but in general to avoid full participation in the community [20]. They may partake of the community support, which in some cases is extremely fast, or use the OSS brand to distinguish themselves from their competitors. At worst they may be viewed as exploiting the computing skills of the community. What is clear is that their motivations are very different from the individual developers who decide to invest their efforts into an OSS community [3].

Perhaps it is not surprising that organizations are cautious when it comes to community participation, for as the Open University had to learn, there are significant challenges to becoming a full and active member of an OSS community. In this regard the Open University faced two major challenges. The first was to understand the philosophy of an OSS community, how it operates, how consensus is achieved, and the pace at which change occurs. The second was to come to terms with the underlying technology of Moodle.

The concept of "contributing to an OSS community" was new to the Open University. As a well known national organization, it was more familiar with a commercial procurement model of purchasing, based on requirements and specifications, with fixed delivery dates and penalties for non-compliance.

The second challenge was that the Open University's in-house team had very limited experience of developing with Moodle's programming language, PHP. Since PHP is an Object-Oriented (OO) language, the Open University developers erroneously assumed that they could migrate their OOs skills and practices directly to the Moodle community. Instead, the developers found that their solutions either didn't work properly, or impacted Moodle's performance.

The following issues were reported by interviewees when the Open University attempted to contribute code, or proposals for enhancements, to the Moodle community:

- **Coding standards:** early versions of Moodle showed wide ranges of coding skills and practice. In order to ameliorate this, recently the overarching Moodle Pty company has created, and is enforcing, coding and documentation standards. What was also done in this respect was to bring in more stringent review procedures: they are applied both to proposed changes to Moodle and during the implementation phase.
- **Rejection of contributions:** some of the refinements proposed, or completed, by the Open University were rejected after discussion for the core trunk of the Moodle project. For example, the Open University's Wiki development was rejected by the community, even though the change itself was later supported by Moodle Pty, due to cleaner code, and a better usability.
- **Slow uptake of contributions:** although the Blogs and Wiki developments, undertaken by the Open University, have been contributed back to the community (in the "plugins" section), the uptake from the community was low.
- **The contribution process:** in general, it was felt that contributing to the Moodle community is often hard work. Proposals of development must be developed first, and time has to be allowed for their public reviews. In some cases, these reviews may even highlight secondary changes, that eventually increase the costs of the proposed development. Only occasionally it was found that the requests from the Open University and Community coincide, and in those cases the contribution process was facilitated.
- **Support to the contribution:** within the Moodle community it is accepted that the contributions require support during the early stages of testing and deployment. It is also a shared expectation within the community that the original contributor will support his/her changes. Even in this case, this cost may not be that significant, as the Open University would have to test and maintain for its own code-base anyway.

Apart from these aspects, the Open University has been greatly benefiting from Moodle, as summarised in the following points:

- **No license fees:** while the Open University has invested considerable amounts of money, they are now free from license fees.
- **Maintenance:** the code adopted for core is maintained by Moodle Pty so will be retained in future releases.

- **No vendor lock-in:** the Open University has already experienced the problem of vendor lock-ins in the past. With thousands of modules running, vendor lock-ins can be problematic and migration to a different technology prohibitively expensive.

### 3.3 The Open University and the Open Source Community

The Open University avoided to create its own fork of Moodle in order to maximise the interaction with the community. When new versions are released, all that is necessary is to replace those modules that provide connectivity with other systems, but the process has been semi-automated. A recent update of the entire system required only half a day.

Early on in its Moodle development process, the Open University has been releasing too many plug-ins requiring changes to the core, but these were not widely adopted. Familiarity with Moodle's code has reduced this. More recent developments by the Open University have been contributed back and have been better received, for example a new "Session data storage" feature, by using a file server, and fine tuning.

Moodle 2.0 will bring a number of changes, and will reflect the influence of the Open University. It will also reduce the number of Open University changes to the standard core, but is likely to increase problems for the tailored plug-ins that the Open University has been developing.

The Open University has provided a number of benefits for the community:

- Bug reports, code reviews, feed-back.
- Financial help: funding for improvements in a number of critical areas, especially related to scalability and performance.
- Valuable proof of concept: the Open University's Moodle installation is used in a high availability environment and it has to sustain high levels of load. This in turn has proved that Moodle is a commercially viable product for even the most challenging environments.
- An improved image: the Moodle community can count-in a prestigious university. This has clearly benefits for companies providing Moodle consultancy.

On the other hand, the presence of the Open University as a stakeholder within Moodle has also produced some disadvantages:

- Scheduling of development work and release dates: the Open University has a long lead-time for course development and production, typically 2-3 years, so needs to know when features will be available so that they can be incorporated into new courses.
- The Open University might be, at same stage, an intrusive guest. With over 450,000 users of their courses (between OpenLearn and the paid-for courses) and nearly 3,600 active modules, they can have an important influence in the development of the software. Smaller Moodle installations have typically different requirements, priorities, complexities.

## 4  The Catalyst Involvement

In the previous sections, a report of what Moodle achieved in terms of popularity, the tiers of its development, and the involvement of the largest e-learning institute in UK was documented. As a further analysis, it was studied the specific involvement of *Catalyst IT Ltd* ("Catalyst"), a Moodle partner which has so far provided a large number of modifications to the core Moodle, by deploying several of its own developers who became active contributors within the community.

The analysis of Catalyst's involvement was achieved empirically, by analysing the public data pertaining the open development of Moodle. In terms of data sources, it has been established that different development practices have an influence on the best data source([5], [17]), and that both the Configuration Management Systems (CMS) and the ChangeLog files offer more reliable information ([4], [11], [21]).

The steps to extract the information from the Moodle server, and to produce the results regarding Catalyst were i) extraction of raw data, ii) filtering of the raw data, and iii) extraction of metrics. As part of these steps, Perl scripts were written to download, extract the activity logs, and parse the raw data contained in the CMS, and finally to extract pre-defined data fields.

### 4.1  Raw Data Extraction and Filtering

The choice of the information sources was focused on the CMS commits of the system. The Moodle project maintains an own CMS server[2], and the data contained spans some 9 years, between Nov 2001 and Aug 2009. Perl scripts were used to identify and extract every occurrence of the following items:

- *Committer*: contributor responsible for the commit;
- *Commit:* the detailed activity a committer was responsible for;
- *Date:* day, month and year of change.

The field Commit type includes: File affected (the name of the file created or directly modified by a change), and Module (the name of the subsystem a file belongs to). As mentioned above, two types of changes were considered in the present study: the creation of an element (a file or a module), and the modification of existing files or modules. After performing the extraction, we arranged the resulting data on a SQL table. It made up to some 72,000 entries, including new element creations and changes.

Apart from the basic information on the authorized committers to the Moodle CMS, several cases were identified were sporadic contributors (i.e., without a committer ID) submitted their code patches directly to the core Moodle developers. This additional information was also extracted, and some additional cleansing performed: for example, obvious variations of people ID's, in this case their email addresses, were mapped to one unique ID. Finally, the email address ID's relating to a known committer ID were converted into a single ID.

---

[2] The web interface to the Moodle CVS is browsable at http://cvs.moodle.org/

## 4.2 Metrics Choice and Description

The analysis of the Moodle system involved the analysis of *input metrics*: the effort of developers was evaluated by counting the number of unique (or *distinct*, in a SQL-like terminology) developers during a specific interval of time. The chosen granularity of time was based on months: different approaches may be used, as on a weekly or on a daily basis, but it is believed that the month represented a larger grained unit of time to gather the number of active developers (*i.e.*, man-month).

## 4.3 Results

This section presents the main results obtained in the analysis of the Catalyst involvement in the Moodle development. As an high-level objective, it was studied whether it was possible to trace the activity of this commercial stakeholders: in particular, the results of *Commercial OSS* systems (*e.g.*, Eclipse, as reported in [22]) should be compared with Moodle as an example of *Community OSS* system.

Since March 2004, Catalyst had from one developer up to a maximum of 6 developers (March 2005) working on Moodle. The profile of the contributed outputs is visible in Figure 2, and can be defined as a "seasonal" effort pattern, meaning a large contribution on a very specific time interval, and lower levels of effort before and after it. Also the modules developed by Catalyst are specifically targeted to a quite focused part of the core of Moodle: Figure 3 displays the distribution of effort along the modules, and it becomes evident how Catalyst wanted to be involved early on in the development of the SCORM (Sharable Content Object Reference Model) collection of specifications.

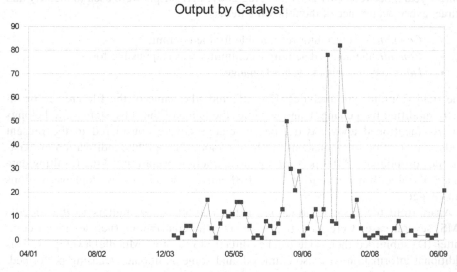

**Fig. 2.** Output produced by one of the partners (Catalyst)

Figure 2 and Figure 3 show that the involvement of commercial entities follows the same principle of attracting individuals into an OSS community: they start to contribute to the periphery, then become more confident with the code, and have a peak of productivity, then leave [19].

Modules worked on by Catalyst

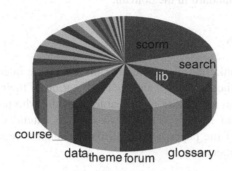

**Fig. 3.** Modules contributed by Catalyst

The second observation shows that the Community OSS projects (from Figure 1) are not overly dependent on specific companies: the reduction of effort and output by Catalyst does not shrink the overall productivity: on the contrary, Commercial OSS projects (e.g., Eclipse), led and managed by specific companies (e.g. IBM) would probably collapse when the company decided to pull away its support.

## 5 Conclusions

This paper has proposed a mixed qualitative and quantitative study in order to study the Moodle e-learning platform. This project started as a small project managed by only one developer on the SourceForge OSS portal, and is now used internationally, sponsored by several commercial partners and supported by even more commercial companies. Its usage and needs have grown to the point to require its own servers, and to gradually being pulled away from the SourceForge hosting.

This paper was essentially two-fold: at first, it proposed the account of the largest e-learning provider in the UK, the Open University, which in 2005 migrated its technology to use the Moodle platform. In turn, this had the effect of becoming an active participant in the development process, and to increase the popularity and visibility of Moodle as a widely-spread solution for e-learning needs. The second strand of research quantitatively studied the quantitative involvement of a Moodle partner (Catalyst IT Ltd) during its evolution, and recognised an established trend for OSS contributions: a first stage of development where a limited output is contributed, then a peak of contributions, finally the abandonment of the commitment. The overall development of Moodle still appears not to be affected, even when Catalyst discontinued its contributions to the Moodle core.

As a corollary, this project achieved a double transition: as mentioned in a previous research work [2], Moodle transited from an Open Forge (*i.e.*, SourceForge) to a more defined, more successful status, as experienced by OSS projects transiting to more renowned and quality-stringent OSS portals. Secondly, starting from a "pure" OSS project, Moodle has become a Community OSS project, where several commercial stakeholders start to act as sponsors of the project, increasing its visibility and establishing it as a de-facto standard in the domain.

## Acknowledgements

The authors wish to thank the Open University staff who were interviewed for this paper. Mediamaisteri for their feedback and the clarification of their involvement in the community development, and the Remote-Learner Canada, who provided feedback on their involvement within Moodle. Finally we wish to extend our gratitude to the anonymous reviewers of the paper, who contributed valuable feedback, apart from a more apt title for this paper.

## References

[1]  Baravalle, A., Chambers, S.: Market Relations. Non-Market Relations and Free Software. PsychNology Journal 5(3), 299–309 (2007)
[2]  Beecher, K., Capiluppi, A., Boldyreff, C.: Identifying exogenous drivers and evolutionary stages in FLOSS projects. Journal of Systems and Software 82(5), 739–750 (2009)
[3]  Bonaccorsi, A., Rossi, C.: Altruistic individuals, selfish firms? The structure of motivation in open source software. First Monday 1(9) (January 2004)
[4]  Capiluppi, A.: Models for the evolution of OS projects. In: Proc. of Intl. Conference on Software Maintenance (ICSM 2003), Amsterdam, Netherlands, pp. 65–74 (2003)
[5]  Capiluppi, A., Michlmayr, M.: From the Cathedral to the Bazaar: An Empirical Study of the Lifecycle of Volunteer Community Projects. In: Feller, J., Fitzgerald, B., Scacchi, W., Silitti, A. (eds.) Open Source Development, Adoption and Innovation, pp. 31–44 (2007)
[6]  Capra, E., Francalanci, C., Merlo, F.: An empirical study on the relationship between software design quality, development effort and governance in open source projects. IEEE Trans. Softw. Eng. 34(6), 765–782 (2008)
[7]  Dalziel, J.: Open standards versus open source in e-learning: The easy answer not be the best answer. Educause Quarterly 4, 4–7 (2003)
[8]  Fang, Y., Neufeld, D.: Understanding Sustained Participation in Open Source Software Projects. Journal of Management Information Systems 25(4), 9–50 (2009)
[9]  Feller, J., Fitzgerald, B., Hecker, F., Hissam, S., Lakhani, K., van der Hoek, A. (eds.): Characterizing the OSS process. ACM, New York (2002)
[10] Fischer, M., Pinzger, M., Gall, H.: Populating a release history database from version control and bug tracking systems. In: Proc. of Intl. Conference on Software Maintenance (ICSM 2003), Amsterdam, Netherlands, pp. 23–32 (2003)
[11] German, D.M.: An Empirical Study of Fine-Grained Software Modifications. In: Proc. of Intl. Conference on Software Maintenance (ICSM 2004), Chicago, US (2004)
[12] German, D.M.: The gnome project: a case study of open source, global software development. Software Process: Improvement and Practice 8(4), 201–215 (2004)
[13] Hemetsberger, A., Reinhardt, C.: Sharing and creating knowledge in open-source communities: The case of kde. In: Procedings of the Fifth European Conference on Organizational Knowledge, Learning and Capabilities (OKLC), Insbruck University (2004)
[14] Koch, S., Schneider, G.: Effort, cooperation and coordination in an open source software project: Gnome. Information Systems Journal 12(1), 27–42 (2002)

[15] Kuniavsky, M., Raghavan, S.: Guidelines are a tool: building a design knowledge management system for programmers. In: DUX '05: Proceedings of the 2005 conference on Designing for User eXperience. AIGA: American Institute of Graphic Arts, New York (2005)

[16] de Laat, P.B.: Governance of open source software: State of the art. Journal of Management and Governance 11(2), 115–117 (2007)

[17] Mens, T., Ramil, J.F., Godfrey, M.W.: Analyzing the evolution of large-scale software: Guest editorial. Journal of Software Maintenance and Evolution 16(6), 363–365 (2004)

[18] Michlmayr, M., Senyard, A.: A statistical analysis of defects in Debian and strategies for improving quality in free software projects. In: Bitzer, J., Schrder, P.J.H. (eds.) The Economics of Open Source Software Development, Elsevier, Amsterdam (2006)

[19] Robles, G., González-Barahona, J.M.: Contributor turnover in libre software projects. In: Damiani, E., Fitzgerald, B., Scacchi, W., Scotto, M., Succi, G. (eds.) OSS. IFIP, vol. 203, pp. 273–286. Springer, Heidelberg (2006)

[20] Robles, G., Duenas, S., González-Barahona, J.M.: Corporate involvement of libre software: Study of presence in Debian code over time. In: Feller, J., Fitzgerald, B., Scacchi, W., Sillitti, A. (eds.) OSS. IFIP, vol. 234. Springer, Heidelberg (2007)

[21] Smith, N., Capiluppi, A., Ramil, J.F.: Agent-based simulation of open source evolution. Software Process: Improvement and Practice 11(4), 423–434 (2006)

[22] Wermelinger, M., Yu, Y., Strohmaier, M.: Using formal concept analysis to construct and visualise hierarchies of socio-technical relations. In: Proc. of the 31st International Conference on Software Engineering, Vancouver, Canada, May 18-24 (2009)

# The Meso-level Structure of F/OSS Collaboration Network: Local Communities and Their Innovativeness

Guido Conaldi[1] and Francesco Rullani[2]

[1] Centre for Organisational Research, Univerisity of Lugano, Via Giuseppe Buffi 13, CH-6900 Lugano, Switzerland
guido.conaldi@usi.ch

[2] Department of Innovation and Organizational Economics,
Copenhagen Business School, Kilevej 14A, 2000 Frederiksberg, Denmark
fr.ino@cbs.dk

**Abstract.** Social networks in Free/Open Source Software (F/OSS) have been usually analyzed at the level of the single project e.g., [6], or at the level of a whole ecology of projects, e.g., [33]. In this paper, we also investigate the social network generated by developers who collaborate to one or multiple F/OSS projects, but we focus on the less-studied meso-level structure emerging when applying to this network a community-detection technique. The network of 'communities' emerging from this analysis links sub-groups of densely connected developers, sub-groups that are smaller than the components of the network but larger than the teams working on single projects. Our results reveal the complexity of this meso-level structure, where several dense sub-groups of developers are connected by sparse collaboration among different sub-groups. We discuss the theoretical implications of our findings with reference to the wider literature on collaboration networks and potential for innovation. We argue that the observed empirical meso-structure in F/OSS collaboration network resembles that associated to the highest levels of innovativeness.

## 1 Introduction

The production of F/OSS is an organizational phenomenon characterized by a strong bottom-up tendency, which hinges upon the creation of social networks of developers freely interacting and collaborating [11,13,26]. Therefore, given the central role as productive infrastructures that social networks play in F/OSS projects, it is not surprising that they have been object of several studies. Indeed, various studies have investigated the social networks generated by developers who take part to F/OSS projects focusing both on the social structure internal to individual projects, e.g.,[6,17], and on the larger network of collaborations linking the wide population of F/OSS projects through common developers, e.g., [33]. Particularly the entire ecology of F/OSS projects hosted on SourceForge has been object of study because of the representativeness of the repository for

P. Ågerfalk et al. (Eds.): OSS 2010, IFIP AICT 319, pp. 42–52, 2010.

the entire population of F/OSS projects and thanks to the availability of rich public data [29].

In this paper we also investigate the social network formed by F/OSS developers who collaborate to one or multiple F/OSS projects. However we concentrate on a different level of analysis. Instead of focusing on 'macro' or 'micro' networks, we investigate the overall collaboration network by looking at the *meso-level* structure of collaboration. We apply a technique able to detect sub-groups of densely connected developers whose connectivity and size is in between that of the whole network and that of single projects. These sub-groups are commonly known as 'communities' in the methodological literature on graph theory and network analysis, and constitute the meso-level structure we will investigate in the following.

We connect our empirical findings to a wider literature on collaboration networks and potential for innovation. More specifically, the theorization on the role of both strongly cohesive teams and brokerage among separated groups [4,9] can be translated into the configurations characterizing the network of communities revealed by our analyses. Therefore we discuss theoretically which configurations of collaboration networks have the potential to foster innovation and argue that the observed empirical F/OSS collaboration network resembles that associated to the highest levels of innovation.

The paper is structured as follows: in the second section we discuss more in detail the existing evidence on the social networks of F/OSS projects. In the third section we firstly describe the data used for reconstructing the F/OSS collaboration network. We then relate our preliminary descriptive findings on the overall collaboration network with the existing empirical evidence. Subsequently we introduce the method adopted to find communities in the collaboration network and present the results. Finally, in the fourth section we discuss our structural findings with reference to the potential for innovation of the overall F/OSS collaboration network which we investigate.

## 2    Background and Related Work

A rapidly growing body of research adopts a network approach for the understanding of the structural characteristics of the F/OSS phenomenon. Several contributions focus on the internal network structure of F/OSS projects, e.g., [6,3,18,17], whereas other contributions reconstruct the networks of collaboration among different projects, e.g., [33].

Several characteristics manifested by the internal communication and collaboration networks of F/OSS projects are already known. Studies investigating the entire spectrum of F/OSS project demonstrate that a significant portion of them are formed by very few developers, or even only one [5], therefore introducing size as a dimension influencing the different structures F/OSS networks can assume. Furthermore, the configuration of F/OSS social networks has been demonstrated to change throughout the life of the projects. F/OSS projects indeed evolve over time. On the one hand they experience a high turnover rate among developers that is negatively correlated with the degree of involvement into the project [25,12,27]. On the other hand their overall structure reflects the different maturity and complexity a F/OSS project can assume over time. To this respect a progression pattern from single hub configurations to core-periphery structures is found in a longitudinal study of several F/OSS projects [17]. The variation over time of F/OSS network structures is also confirmed by a study that tracks the network centralization values of two F/OSS projects and shows how they varied over time [32].

Several studies present evidence of internal hierarchy in F/OSS internal communication and collaboration networks. It has been shown that well-established and large F/OSS communities manifest hierarchical structures [6,17]. Sometimes the project founders assume a great authority on the entire development process [22,28], whereas other equally relevant projects develop a complex meritocratic structure that relies on different status levels and voting procedures [21,8].

Also the overall F/OSS phenomenon has been studied adopting a network perspective. The social network formed by all individuals connected through the F/OSS projects to which they co-collaborate has been shown to represent a prototypical complex evolving network [19]. Furthermore, this global F/OSS collaboration network has been analyzed by sub-dividing it in four subsets of different type of actors (project founders, core-developers, co-developers, and active users) and shown to be a self-organizing system that in all subsets obeys scale-free behaviors [33]. The same study also finds the same network to have a small average distance and a high clustering coefficient, therefore characterizing itself as a small world [31]. Finally, [33] discusses the role of individual actors in the overall F/OSS ecology and stresses the potential impact of co-developers and active users as direct connections among projects that could benefit from the fast sharing of information.

Here we adopt a global perspective on the overall F/OSS phenomenon similar to [33] and we reconstruct a similar F/OSS collaboration network. However, our focus is on the meso-level of the network. Consequently, we concentrate firstly on individuating dense communities of co-collaborating developers. The configuration of the network formed by these communities and their connections will then be at the core of our empirical analysis and theoretical discussion.

## 3   Methods and Analysis

### 3.1   Data

We use data describing the activities of 1,347,698 actors working on 170,706 F/OSS projects hosted on SourceForge [29] in September 2006. The period was chosen in accordance to the availability of information on individuals' emails, necessary for data cleaning. However, this should not be a problem, because we believe that the evolution of a self-organizing social network as that under analysis here follows general rules, such as growth and preferential attachment [2], that are very unlikely to change over a three-year period. SourceForge is likely not representative of the whole universe of F/OSS projects, as it is a company-owned platform and does not host some of the most famous project, such as Linux. However, it represents by far the largest repository of F/OSS projects worldwide, and it hosts extremely heterogenous projects, from very famous, active and large ones to very small or even 'dormant' ones. Thus, data relative to the activities it hosts have been already widely used in previous studies, e.g., [33,16,10]. In this study we will follow the same line of research.

Only projects labeled as 'active' have been retained in the dataset, as well as only 'active' actors registered with at least one project. This assured that we took into account all the projects and individuals relevant for our analysis, excluding only non-active projects or individuals who do not belong to the network. Different virtual identities belonging to the same individual have been aggregated through email address matching. This reduced the number of individuals to 161,983 and the number of projects to 115,112.

In order to reconstruct the F/OSS collaboration network we used individuals as nodes and affiliations to the same projects as ties, weighted by the number of projects in common. In other terms, we projected the weighted two-mode network formed by developers and projects that we originally collected into a one-mode network formed only by developers.

All analyses were performed using the igraph package [7] for the R environment.

## 3.2   The Overall Network Structure of F/OSS Collaboration

As a first step, we investigate the collaboration network similarly to what has been previously done by other studies on F/OSS, e.g., [33], and on other virtual networks, as for example Internet [1]. The main descriptive statistics for the generated network can be found in the first column of Table 1.

**Table 1.** Main characteristics of F/OSS collaboration network

| Indicator | Whole network | Giant Component |
|---|---|---|
| Number of Nodes | 161,983 | 58,481 |
| Number of Ties | 753,421 | 632,046 |
| Global Clustering Coefficient | 0.910 | 0.907 |
| Average Path Length (APL) | 7.105 | 7.106 |
| APL for a comparable random network | 7.804 | 4.612 |
| (Equal size and average number of ties) | | |

We then isolate and analyze a giant component composed by 58,481 individuals, the second component spanning 201 nodes. The statistics relative to the degree distribution (mean: 21.62; standard deviation: 41.23; skewness: 5.12) signal the heterogeneity of the ego-networks of F/OSS project members (see Figure 1).

As the values reported in the second column of Table 1 show, the Global Clustering Coefficient (or, equivalently, Transitivity [30]) is extremely high and

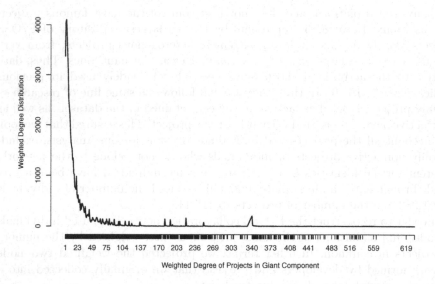

**Fig. 1.** Degree distribution of projects in the giant component of the collaboration network

the Average Path Length is low (50% larger than that of a random graph, a proportion in the range of those reported by [1] for comparable cases). This shows that in the F/OSS world individuals not only gather locally in dense groups of neighboring collaborators, but also establish collaboration ties with members of local groups located elsewhere in the network, thus acting as 'brokers'. Thus, the network clearly resembles a 'small world' [31], a property detected also by [33]) in a similar context.

The mean (0.38), standard deviation (0.28) and skewness (0.79) of the distribution of Burt's measure of constraint [4] confirm this interpretation at the individual level. Indeed, Burt's constraint only focuses on the direct neighborhood of each F/OSS project in the network and captures the proportion of realized ties among its neighbors out of all the possible ones. The low average value of constraint found among the projects in the giant component, 0.38 with the constraint index varying in the range [0,2], confirms that projects tend to connect otherwise disconnected projects, therefore spanning so-called 'structural holes' in their neighborhood and thus acting, in Burt's terms,as brokers.

### 3.3   Finding Communities in the F/OSS Collaboration Network

The evidence at the global level of analysis just presented is consistent with what has been found in the field (e.g., [33]). We now focus on the meso-level of analysis and we test whether sub-groups of densely connected developers (i.e., communities) can be identified and whether they are connected through sparser collaborations.

In order to find communities in a network several algorithms are available, e.g., [23]. We apply the Walktrap algorithm [24]. This algorithm is based on the intuition that short random walks performed on a sparse network will tend to remain trapped in denser local areas of the network corresponding to communities. The Walktrap algorithm makes use of information on the weights of the ties in the network. This is a fundamental property for our purposes because of the wide variation existing in the F/OSS context concerning the number of collaborations in which different developers take part. A characteristic that is coherently reflected by the weighted degree distributions of our collaboration network.

The algorithm induces a sequence of partitions of the original network into communities. It starts with each node representing a community and ends with all nodes in one community. In order to find which partition best represents the community structure of the original network we adopt the most widely used criterion: the modularity $Q$ index [23]. $Q$ relies on the fraction of ties inside a community and the fraction of ties bound to that community: the best partition maximizes $Q$ (that lies in the range [-1,1]) and therefore defines communities which are internally densely connected with only sparse connections among them.

When applying the Walktrap algorithm to the giant component of our F/OSS collaboration network we find that the best partition (with a high $Q$ of 0.865, see Figure 2) individuate 9,931 communities, many of them extremely small, reaffirming the tendency to create dense, i.e., very close, local sub-groups of co-collaborators loosely interconnected. Nonetheless, the community-detection

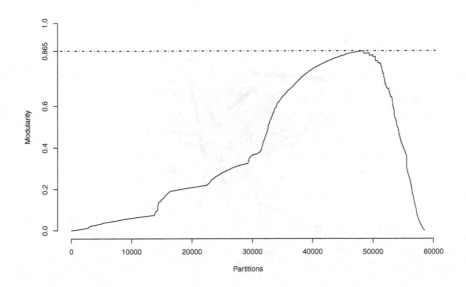

**Fig. 2.** Modularity values for all partitions in the giant component of the F/OSS collaboration network

analysis shows the prominent role also of outward connections: 71% of individuals in communities with at least two members create ties beyond their community borders, and the average ratio between the number of their outward weighted ties and their total weighted degree is 0.28 (standard deviation: 0.19). This means that these many dense communities, whereas clearly distinguishable, are not almost isolated, but connected by a large number of brokers and through important ties. The network of the largest communities (see Figure 3)clearly shows the coexistence of brokerage and closure.

In Figure 3 each node is a sub-group of densely interconnected developers, i.e., a community, identified through the Walktrap algorithm. For simplicity, only communities having 25 or more members are depicted. The size of each node is proportional to the amount of collaborations in the same F/OSS projects among its members. The thickness and color shade of the ties linking the communities are proportional to the total number of F/OSS projects to which the members of the different communities co-collaborate. Figure 3 shows that high levels of collaborations can exist among communities of both comparable and different sizes.

Figure 4 shows the inner structure of the largest community, identified in brighter red with a yellow contour in Figure 3. The community is here magnified to show the connections among its 541 members (the round nodes). The thickness and color shade of the ties are here proportional to the total number of F/OSS projects to which two individuals co-collaborate, while the size of the nodes is proportional to the total number of collaborations each individual has with

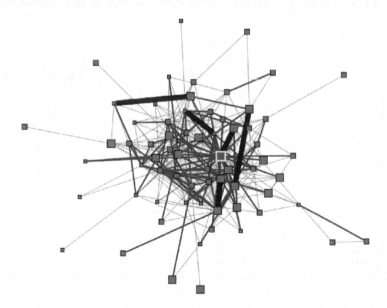

**Fig. 3.** Network of identified communities (with size > 25) in the giant component of the F/OSS collaboration network

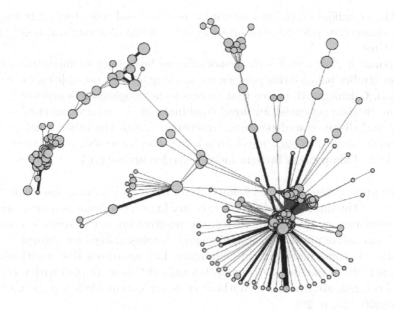

**Fig. 4.** Closer view of the largest identified community in the giant component of the F/OSS collaboration network

members of other communities. Figure 4 shows that inside a community both central and peripheral individuals manifest high levels of external collaboration.

Therefore, we can affirm that both the network of communities and the network of developers belonging to a same community assume a similar configuration that combines densely connected sub-groups with the presence structural holes [4], i.e. lack of ties, isolating the sub-groups and some relevant brokers spanning these borders thanks to inter-community co-collaborations. This combination, in line with the literature on similar phenomena, e.g., the Internet [1], places F/OSS in a sort of 'middle range' between full closure and extensive brokerage.

## 4  Discussion and Conclusions: Community Structure and Innovativeness

Among many others, one question that our analysis raises is whether the community structure that we just described hinders or fosters innovation. In a recent article Mayer-Schönberger [20] warns against the possibility of overestimating the innovation capabilities of open systems such as F/OSS or the Internet itself. If diversity is crucial for producing novelty, a *conditio sine qua non* for ideas recombination, extremely densely connected collaboration networks, such as he assumes the F/OSS collaboration network to be, have difficulties to reach high level of innovation because of their intrinsic tendency to establish many redundant connections, creating thus homogenization and group-thinking. In other words, Mayer-Schönberger puts forward a positive and monotonic relationship between innovative performance and the importance of structural holes [4] in the structure.

This point of view certifies the importance of brokerage for innovation, however other studies balance this perspective warning against possible excessive disconnection. Gilsing et al. notice that 'access to heterogeneous sources of knowledge [that creates] potential for novel combinations ... requires an emphasis on diversity and disintegrated network structures. ... On the other hand [actors] need to make sure that such novel knowledge, once accessed, is evaluated, and ... absorbed. This process favours more homogenous network structures' [9, p. 1718].

Studies undertaken in related fields confirm this view. Laursen and Salter [14] found that as the number of different external knowledge sources a firm can use (e.g., universities, or clients) increases, the positive impact of one more source on firms' innovative performance significantly decreases, because absorption and combination become more difficult. Similarly, Lazear argues that multicultural teams benefit from members' diversity, but only if this is coupled with a certain degree of commonality because 'Without communication, there can be no gains from diversity'[15, p. 20].

Thus, according to the full spectrum of studies just introduced, the relationship between the importance of structural holes and innovative performance should be rather described as an inverted U-shaped curve. This means that structures able to produce the highest innovation rate should be located somewhere in the middle in the continuum between full closure and extensive brokerage [4].

When exploring the social network generated by individuals collaborating in F/OSS projects on SourceForge we discovered a distinguishable meso-level network linking communities of closely co-collaborating individuals. Furthermore, we investigated also the collaboration network internal to the largest of these communities. We found that the structure of these networks both clearly resemble a mixture between densely connected local areas, where information flows pervasively and diversity is reduced, and a great number of structural holes that are spanned by several brokers. As the above-mentioned studies on the structure of innovativeness suggest, in such a structure the idiosyncratic knowledge created in one community can flow throughout the entire network thereby mixing with diverse knowledge, increasing the probability of generating novel recombinations.

This mix of closure and brokerage suggests that the F/OSS collaborative environment is not a fully connected network in which everybody is co-collaborating with everybody else, as [20] assumes. On the contrary, the F/OSS collaborative environment appears to possess the structural characteristics necessary to place itself in that middle-range area that the literature on innovation associates to the highest segments of the innovativeness curve.

Therefore, our results give a preliminary insight on a more complex relationship between the structural dimension of collaboration in the F/OSS world and innovativeness than the one prosed by [20]. Consequently they also call for a more in-depth research on the actual innovative performances achieved by different local areas of the overall F/OSS collaboration network in order to elaborate beyond a first description of the meso-level of community structure that represented the aim of this study.

# References

1. Albert, R., Barabási, A.-L.: Statistical mechanics of complex networks. Reviews of Modern Physics 74(1), 47–97 (2002)
2. Barabási, A.L.: Scale-free networks: a decade and beyond. Science 325(5939), 412–413 (2009)
3. Bird, C., Goey, A., Devanbu, P., Swaminathan, A., Hsu, G.: Open Borders? Immigration in Open Source Projects. In: Proceedings of the Fourth International Workshop on Mining Software Repositories (2007)
4. Burt, R.: Structural Holes: The Social Structure of Competition. Harvard University Press, Cambridge
5. Capiluppi, A., Lago, P., Morisio, M.: Characteristics of Open Source Projects, p. 317 (2003)
6. Crowston, K., Howison, J.: The social structure of free and open source software development. First Monday 10(2)
7. Csardi, G., Nepusz, T.: The igraph software package for complex network research. Inter. Journal Complex Systems, 1695 (2006)
8. Germán, D.M.: The GNOME project: a case study of open source, global software development. Software Process: Improvement and Practice 8(4), 201–215 (2004)
9. Gilsing, V., Nooteboom, B., Vanhaverbeke, W., Duysters, G., Oord, A.V.D.: Network embeddedness and the exploration of novel technologies: Technological distance, betweenness centrality and density. Research Policy 37(10), 1717–1731 (2008)
10. González-Barahona, J.M., Robles, G., Andradas-Izquierdo, R., Ghosh, R.A.: Geographic origin of libre software developers. Information Economics and Policy 20(4), 356–363 (2008)
11. von Hippel, E., von Krogh, G.: Open Source Software and the "Private-Collective" Innovation Model: Issues for Organization Science. Organization Science 14(2), 209–223 (2003)
12. Howison, J., Inoue, K., Crowston, K.: Social dynamics of free and open source team communications. In: Damiani, E., Fitzgerald, B., Scacchi, W., Scotto, M. (eds.) Proceedings of the IFIP Second International Conference on Open Source Software (Lake Como, Italy). IFIP International Federation for Information Processing, vol. 203, pp. 319–330. Springer, Boston (2006)
13. von Krogh, G., von Hippel, E.: The Promise of Research on Open Source Software. Management Science 52(7), 975–983 (2006)
14. Laursen, K., Salter, A.: Open for innovation: the role of openness in explaining innovation performance among U.K. manufacturing firms. Strategic Management Journal 27(2), 131–150 (2006)
15. Lazear, E.P.: Globalisation and the Market for Team-Mates. The Economic Journal 109(454), C15–C40 (1999)
16. Lerner, J., Tirole, J.: The Scope of Open Source Licensing. Journal of Law, Economics, and Organization 21(1), 20–56 (2005)
17. Long, Y., Siau, K.: Social Network Structures in Open Source Software Development Teams. Journal of Database Management 18(2), 25–40 (2007)

18. López-Fernández, L., Robles, G., González-Barahona, J.M., Herraiz, I.: Applying Social Network Analysis Techniques to Community-driven Libre Software Projects. International Journal of Information Technology and Web Engineering 1(3), 27–48 (2006)
19. Madey, G., Freeh, V., Tynan, R.: Modeling the free/open source software community: A quantitative investigation. In: Koch, S. (ed.) Free/Open Source Software Development, pp. 203–220. Idea Group Publishing, USA (2005)
20. Mayer-Schönberger, V.: Can we reinvent the Internet? Science 325(5939), 396–7 (2009)
21. Mockus, A., Fielding, R.T., Herbsleb, J.D.: Two case studies of open source software development: Apache and Mozilla. ACM Transactions on Software Engineering and Methodology (TOSEM) 11(3) (2002)
22. Moon, J.Y., Sproull, L.S.: Essence of Distributed Work: The Case of the Linux Kernel. First Monday 5(11) (2000)
23. Newman, M.E.J., Girvan, M.: Finding and evaluating community structure in networks. Physical Review E 69(2), 026113 (2004)
24. Pons, P., Latapy, M.: Computing Communities in Large Networks Using Random Walks. Journal of Graph Algorithms and Applications 10(2), 191–218 (2006)
25. Robles, G., González-Barahona, J.M.: Developer identification methods for integrated data from various sources. In: Proceedings of the 2005 international workshop on Mining software repositories, vol. 30(4) (2005)
26. Scacchi, W., Feller, J., Fitzgerald, B., Hissam, S., Lakhani, K.: Understanding Free/Open Source Software Development Processes. Software Process: Improvement and Practice 11(2), 95–105 (2006)
27. Shah, S.K.: Motivation, Governance, and the Viability of Hybrid Forms in Open Source Software Development. Management Science 52(7), 1000–1014 (2006)
28. Shaikh, M., Cornford, T.: Version Control Tools: A Collaborative Vehicle for Learning in F/OS. In: Collaboration, Conflict and Control: The 4th Workshop on Open Source Software Engineering 2004, Edimburgh, Scotland (2004)
29. Van Antwerp, M., Madey, G.: Advances in the SourceForge Research Data Archive (SRDA). Paper presented at the Fourth International Conference on Open Source Systems, IFIP 2.13, Milan, Italy (September 2008)
30. Wasserman, S., Faust, K.: Social Network Analysis: Methods and Applications. Cambridge University Press, Cambridge (1994)
31. Watts, D.J., Strogatz, S.H.: Collective dynamics of 'small-world' networks. Nature 393(6684), 440–442 (1998)
32. Wiggins, A., Howison, J., Crowston, K.: Social Dynamics of FLOSS Team Communication Across Channels. In: Proceedings of the Fourth International Conference on Open Source Software, Milan, Italy (2008)
33. Xu, J., Gao, Y., Christley, S., Madey, G.: A Topological Analysis of the Open Souce Software Development Community. In: Proceedings of the 38th Annual Hawaii International Conference on System Sciences (2005)

# To Patent or Not to Patent: A Pilot Experiment on Incentives to Copyright in a Sequential Innovation Setting

Paolo Crosetto[1,2]

[1] DEAS, Università degli Studi di Milano
paolo.crosetto@unimi.it
[2] Dipartimento di Scienze Economiche e Aziendali, Libera Universita Internazionale
degli Studi Sociali, LUISS, Roma
pcrosetto@luiss.it

**Abstract.** This paper presents preliminary results from a pilot experiment dealing with the economic motivations to contribute to Free/Open Source Software (FOSS). Bessen and Maskin [1] argue that in a dynamic sequential innovation framework the standard argument for granting patent protection is no more valid and the innovator has at certain conditions an incentive to fully disclose the results of his works; in these same conditions, a copyright strategy could result in a *tragedy of the anticommons* [5,2].

We study in the lab the choice of copyrighting or copylefting subsequent innovations in a dynamic setting *à la* Bessen and Maskin, introducing an innovative experimental design requiring real effort on the part of subjects. The players are asked to actually 'innovate' producing words from given letters, and face the choice to *copyright* or *copyleft* their words.

Preliminary results show that copyleft is more likely to emerge when royalty fees are relatively high, and when the extendability, modularity and manipulability of inventions is enhanced.

## 1   Introduction

The standard economic argument in favour of patents and copyright states that, since knowledge is a public good with a high fixed cost of production and a relatively small cost of imitation, the state needs to grant for limited time a monopoly on the invention to allow the inventor to recover his costs, thus giving him an incentive to invest in the first place. While a long-standing debate in economics exists over the exact nature of the cost-benefit tradeoff at the base of intellectual property regulations from society's point of view, it has been widely assumed that an innovator would copyright/patent his innovation whenever given the chance to do so. This assumption seems to need further inquiry as we witness the rise of FOSS.

It has recently been suggested in the economics of innovation literature that the very nature of software - encoded, algorithmic knowledge - could generate incentives to disclose the creation of one's labour and ingenuity. It has been shown

P. Ågerfalk et al. (Eds.): OSS 2010, IFIP AICT 319, pp. 53–72, 2010.

by [1] that when innovations are sequential and research efforts are complementary, in a dynamic setting it is optimal for innovators to fully disclose their work, since every inventor gains from the marginal future contributions of others to his own innovation, made possible by the open nature of his contribution.

The converse argument, i.e. that in dynamic, sequential settings introducing intellectual property can slow down innovation, has been made several times in economics [10], and has recently been the focus of the literature about the so-called *tragedy of the anticommons*. The name derives from the famous tragedy of the commons paper by [4]: while in the commons the absence of exclusive usage rights generated overutilisation and waste, in the anticommons the presence of overlapping and fragmented exclusion rights generates underuse and waste. The imposition of patents on research results generates an anticommons whenever every right holder independently sets a price for the license without taking into account negative externalities, with the result of general stagnation of downstream innovation. This mechanism has been outlined theoretically [2,9], analysed experimentally [3] and proved to be at work in the field of biomedical research [6].

In this paper we present the preliminary results of a new real-effort experimental design that enables to explore jointly, in a controlled setting, the intellectual property choices of subjects in a dynamic sequential innovation game and the aggregate consequences of individual behaviour in terms of anticommons.

## 2   The Model in Bessen and Maskin [1]

When innovations are sequential, each innovation is both an output and an input for further innovations developed by different inventors; the true value of the first invention might only be revealed when the original idea is extended, in directions possibly unforeseen by the first innovator.

In this context, copyright has ambiguous effects, from both the society and individual points of view, as Intellectual Property is likely to affect not only the revenue, but also the costs of the innovator. Copyright generates an expected stream of revenues for the copyright holder, thus providing incentives to put effort into the (costly) innovating activity. At the same time, follow-on innovators face the cost of licensing and increasing transaction costs if bundling of many innovations patented by different inventors is needed, thus generating a negative effect on the number of follow-on innovations.

A possible - if radical - solution to the anticommons problems in R&D would be for the innovator to release its discovery under copyeft licences ('private-collective' innovation mode, [11]). Nonetheless, while it is straightforward to see why a 'private-collective' solution could spread once initiated - innovators can freely use and improve upon increasing amounts of knowledge - it is harder to see why a self-interested innovator should voluntarily forgo any direct appropriation of revenue from his discovery, releasing it free to the public.

According to [1], the very sequential nature of innovations can provide incentives to start sharing and to forgo direct appropriation through IP. In their theoretical paper patents are shown to be a valid policy tool - enhancing social welfare as well as providing incentives to innovate to all firms involved, thus increasing the likelihood of discovery - in a static model; when the same model is enlarged to include dynamics and sequential and complementary innovations, though, patents have a counterintuitive effect, and are shown, in line with the anticommons literature, to become *hurdles* to innovation rather than *spurs*, the social welfare being higher *without* patents than with them. More to our point, Bessen and Maskin show that if profit dissipation due to increased competition is low - i.e. if the availability of further innovations expands the market, increasing opportunities for further R&D on the one side and increasing the number of interested customers on the other - the firms themselves have an incentive *not to* patent their discoveries even if a patent system is available at no cost to them.

The present paper presents preliminary results of a new experimental design developed to test in the lab the argument of [1].

## 3   Experimental Design

The experiment has been designed to create a dynamic, interdependent setting in which subjects make choices in a collectively and dinamically co-created landscape. To simulate innovative activity, the experiment follows a real effort protocol; the players have to actually innovate - over a set of given rules - using both economic (experimental money) and cognitive (creative effort) resources.

The real effort task chosen is a sort of *Scrabble* game. The idea of using creation of words to mimic innovative activity in investigating FOSS has been pioneered by Lang et al. [7], who design a double-auction market for words, inducing demand but leaving players free to make supply decisions. The design here presented differs greatly from theirs, their interest being in market design, while the main focus here is on sequential innovations.

The design chosen creates a situation in all respects similar to the [1] model, but it is not a formal equivalent of the latter; rather, it is a transposition of the assumptions and workings of the model into an intuitive but controlled setting.

### 3.1   The *Copyleft* Game

1. *Game structure.* The game is played by **6** players. Players play in turns until there are no letters left in the letter set, with random assignment of starting positions. The players aim to maximise their monetary payoff from the game. Players get payoffs from two sources: 'use value' of the words they produced or extended, and '(net) royalty fee revenue'. The game interface is shown in Figure 3.
2. *Letter set.* The letter set is composed of **200** letters (see Figure 1). Frequencies of letters are the same as in standard *Scrabble*. Every letter comes with an attached payoff this letter will give the player when inserted into a produced or extended word.

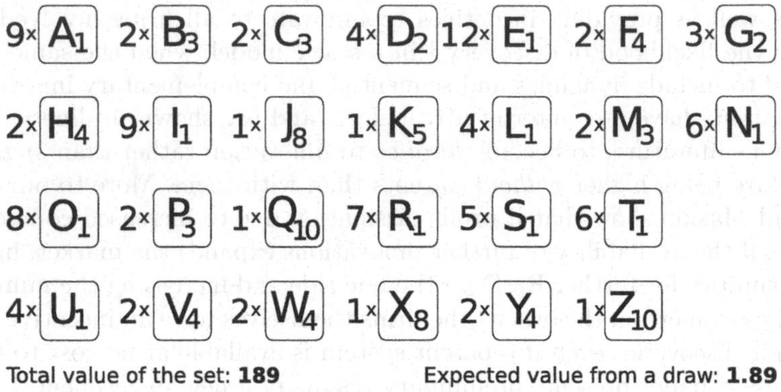

9× $A_1$    2× $B_3$    2× $C_3$    4× $D_2$    12× $E_1$    2× $F_4$    3× $G_2$

2× $H_4$    9× $I_1$    1× $J_8$    1× $K_5$    4× $L_1$    2× $M_3$    6× $N_1$

8× $O_1$    2× $P_3$    1× $Q_{10}$    7× $R_1$    5× $S_1$    6× $T_1$

4× $U_1$    2× $V_4$    2× $W_4$    1× $X_8$    2× $Y_4$    1× $Z_{10}$

Total value of the set: **189**          Expected value from a draw: **1.89**

**Fig. 1.** Letter set used for the game

3. *Turns.* In any turn, a player can make at most three actions: buy a random letter (for a fixed price); either produce a word or extend an existing word; decide whether to copyright or copyleft the produced word. Players can choose to make just one or two decisions, or to pass. Turn structure is summarised in Figure 2.

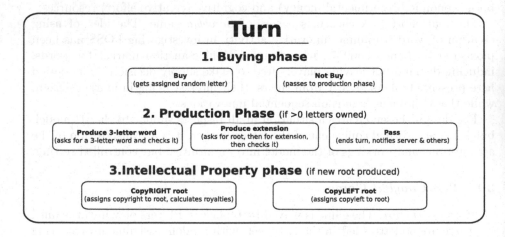

**Fig. 2.** Actions available to the players in every turn

4. *Buy phase.* When it is their turn, players can buy a random letter for a fixed price, out of their show-up fee. If a player buys, a letter is drawn from the letter set (without substitution) and assigned to the player. The price is set at 2 experimental points, which is slightly above the expected value from the set, 1.89.

5. *Words phase.* Players collect payoffs by composing or extending words. A spelling check is performed on produced words. At every turn, every player

can perform only one action in the words phase, either producte a new or extend an existing word.

— *Producing words.* Players may produce from scratch only three-letter words, called *roots*. This rule mimics the fact that an innovation must be more-or-less working when released. Production of a word generates a 'use value' payoff that is the sum of the values of the letters used to form the word. This payoff is incurred only once, the moment the root is produced. Produced words are common knowledge; the use of the root as input for extensions is instead subject to the IP choice of the creator.

— *Extending words.* Players may extend existing words, by one letter at a time, inserting this letter in any position within the word or at its edges. Anagrams are not permitted. Words have no length constraints apart from the ones implicit in the English language. Example extension paths for the root *car* are *car* → *care* → *scare* → *scared*, or *car* → *card* → *cards*. Extending a word gives as payoff the sum of all the letters composing the extended word but might be subject to copyright fees to be paid to the word original creator.

6. *Spellchecker.* To avoid language skills biases, the players are provided with an interactive free-access spellchecker embedded in the main game interface.

7. *Word availability.* Extended words do not replace existing roots, but exist in parallel to them. A root can be used by different extenders to generate different word trails.

8. *Intellectual Property Phase.* Players that have produced new roots must decide whether to copyright or copyleft the word or extension. The IP choice cannot be undone in later stages and lasts throughout the game.

(a) *Copyrighting.* Copyright can be obtained free of charge, and gives the copyright holder the exclusive right over the use of his root/extension. This implies that no other player can produce the same root/extension and that the copyright holder will receive a royalty fee every time some player uses it for an extension. The fee is fixed, and is proportional to the value contributed to the word. A root which value is $v$ generates a fee of $\alpha v$, in which $0 < \alpha < 1$ is a fixed known parameter. Every extension falls under the same rule, and if the value added to the word is $w$, an extension generates a right to obtain a royalty fee of $\alpha w$. The $n^{th}$ extender of a word of value $v + w$ pays $\alpha(v + w)$ in fees, distributed automatically to other players according to the value contributed. Free copyright (an irrealistic assumption) has been introduced to strengthen the external validity of the emergence of copyleft in the lab.

(b) *Copylefting.* Copyleft is free of charge. It gives the copyleft owner no exclusive rights, but it endows users with a large set of use, redistribution and modification rights. Extenders can use the word as an input for free, but they are not allowed to copyright their extension. When extending a word, a copyleft extender earns all of the use value of the word, without having to pay any fee, but not getting any further fee from extenders either.

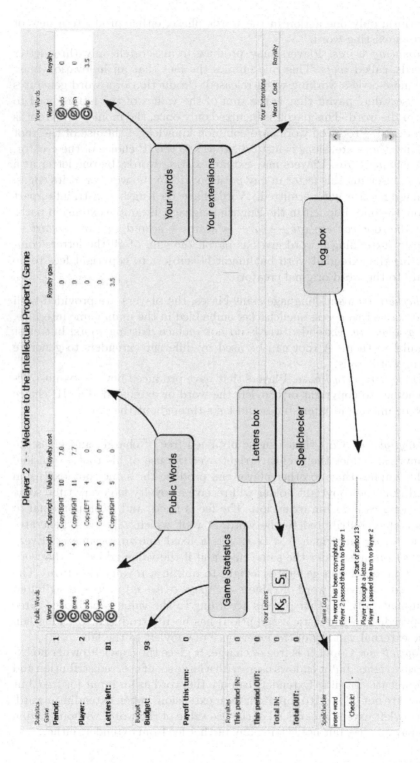

**Fig. 3.** The main interface for the Copyleft Game

The game described above has been designed to share as many features as possible with the model in [1], but it is not a perfect match. Every deviation from the model was made on the safe side, lowering the likelihood that the Bessen and Maskin outcome would emerge.

## 4   Experimental Results

The pilot experiment was held in October 2009 in the EXEC lab, University of York, UK. The software was developed in Python [8]. The experiment involved in total 23 English native speakers. Four sessions of 6 players were run, divided into two treatments: in the *low-fee* treatment the royalty parameter was set to $\alpha = 0.7$, while in the *high-fee* treatment it was set to $\alpha = 0.9$. We expected the *low-fee* session to show less copyleft activity and anticommons problems than the *high-fee* sessions.

Copyleft emerged in three out of four sessions (the two *high-fee* and one of the *low-fee*), dominating one; anticommons features (underuse of resources, low cooperation, lower payoffs) emerged in the two *high-fee* sessions as expected.

### 4.1   General Results

The sessions differed from one another in many respects. Two main problems for data analysis emerged: first, Session 4 featured only 5 players instead of the standard 6; second, in Session 3 the players found a clever way of altering the game rules to their advantage, and hence played a completely different game. Safe comparisons can be made between Session 1 (*low-fee*) and Session 2 (*high-fee*). Results from Session 3 will be separately discussed (Section 4.4).

The experiment showed significant emergence of copyleft in the *high-fee* treatments, and a number of recurring features.

1. The number of roots created is 30 in all sessions. What differentiates the sessions is the number and nature of extensions; the behaviour of follow-on innovators determines social welfare and payoffs.
2. Payoffs are significantly higher in the *low-fee* Session 1 w.r.t. the *high-fee* Session 2 (Mann-Whitney U, p-value = 0.041, K-S p-value = 0.1429) and less so w.r.t. Session 4 (Mann-Whitney U, p-value = 0.129, K-S p-value = 0.5909).
3. The *high-fee* treatment sessions resulted in a lower number of words created, and in a higher waste of productive resources (unused letters), resulting in lower average payoffs. In *high-fee* sessions players tended to extend their own roots and to avoid other player's roots, to access which they had to pay royalties; this behaviour created quite isolated word trails and lower extension opportunities.
4. Copyrighted roots are much more likely to be created than copylefted roots. Players did not create any (but one on the very last period) copylefted root in Session 1 (*low-fee*), but did create 5 and 1 copylefted roots in Sessions 2 and 4 (*high-fee*).

**Table 1.** Summary data for the four experimental sessions

| | Low-fee | | High-fee | |
| --- | --- | --- | --- | --- |
| | Session1 (6) | Session3 (6) | Session2 (6) | Session4 (5) |
| Payoff: average (£) | 17.76 | 26 | 13.95 | 15.7 |
| Standard Deviation | 2.6 | 4.8 | 1.95 | 2.52 |
| Total number of words | 112 | 154 | 84 | 91 |
| % copyrighted | 99.1% | 29.2% | 73.8% | 90.1% |
| Total net value created | 471 | 968 | 237 | 288 |
| Royalty turnout | 268.8 | 71.4 | 119.7 | 71.1 |
| Royalties/value | 32.2% | 5.3% | 21.6% | 11.6% |
| Unused letters | 26 | 16 | 46 | 47 |
| Number of copyrighted roots | 30 | 11 | 30 | 30 |
| Extensions per root | 2.7 | 3.1 | 1.06 | 1.73 |
| Relative extensions per root | 0.19 | 0.27 | 0.1 | 0.12 |
| Number of copylefted roots | 1 | 4 | 5 | 1 |
| Extensions per root | 0 | 26.25 | 3.4 | 8 |
| Relative extensions per root | 0 | 0.4 | 0.42 | 0.34 |
| Average word length | 4.18 | 4.53 | 3.76 | 3.89 |
| Copyrighted words | 4.18 | 3.73 | 3.58 | 3.8 |
| Copylefted words | 3 | 4.86 | 4.27 | 4.67 |
| Average word value | 7.45 | 8.63 | 6.6 | 6.76 |
| copyrighted words | 7.46 | 6.84 | 6.22 | 6.62 |
| copylefted words | 6 | 9.37 | 7.68 | 8.11 |

5. Copylefted roots are three to seven times more likely to be extended than copyrighted roots. Once copylefted words exist, the incentives to extend them are higher, because the foregone royalty from one's extension is more than counterbalanced by the fact that no royalties are paid on the root.

6. Copylefted roots generate a two to four times higher exploration of extension possibilities (see 'Relative Extensions' rows in Table 1). The 'average relative extensions per root' index was computed as the average of the ratio of the actual number of extension of each root and the theoretically possible number of extensions of each root allowed by the English language.

7. Copylefted words are on average longer and have higher value than copyrighted words. This is because copylefted words are extended more and new roots tend to be copyrighted but not extended: when faced with the choice of copyrighting or copylefting a root, the player most of the time decides to copyright it; but this decreases the likelihood of follow-on innovations to be built on top of that root, and hence slows down the pace (extensions per root), and reduces the exhaustivity (relative extensions per root) of further inventions.

8. The amount of royalties paid or received is higher in *low-fee* sessions. When fees are low, players tend to seize the best opportunities available to them irrespectively of the fact that there is a royalty to be paid: this results in a high number of words, lower letter waste, and a high amount of royalties exchanged. When royalties are higher (*high-fee* treatment) players tend to restrict their innovative efforts to the 'free' roots, either copylefted or owned by the extender, and hence many opportunities are missed, a higher number of letters is left unused and a lower amount of royalties flows in the system.

9. Players used the spellchecker extensively - on average more than 100 times - in all sessions. There is no significant correlation between the use of the spellchecker and the result in terms of payoff for the player.

A treatment effect can be argued to exist. Copyleft emerged and saw sustained contributions in *high-fee* sessions, while it did not emerge in *low-fee* session; anticommons features appeared in (*high-fee*) sessions, in which payoffs were lower, number of unused letters higher, and royalty flows reduced.

The paradoxical effect that better patent protection results in a lower amount of innovations, one of the central points in [1] is reproduced in the lab; moreover, players recognise the mild incentives to copyleft, and provide some (though not many) copylefted roots; once a copylefted root exists, players reap much more benefits from it, extending it further and deeper, than from copyrighted roots. There is not enough evidence, though, to support the stronger argument in Bessen and Maskin's paper that in a sequential settings the firms prefer not to patent their innovations, relying on subsequent innovations rather than on royalties from upstream contributions: if that were the case, we would have recorded a surge in copylefted *roots* as the game progressed. What happened instead was that in all sessions players preferred to release the *roots* created as copyright, even when this generated a suboptimal anticommons situation; nonetheless, players preferred to extend existing copylefted roots much more than they extended copyrighted roots, and hence copylefted words appeared and accounted for up to a quarter of all words created.

## 4.2   Session 1 (*low-fee*): Copyright and *business as usual*

Six subjects participated in Session 1, a *low-fee* treatment. In the session 112 words, all copyrighted but one, were produced: on average, every player produced 18.7 words, 3.6 words per period. The only copyleftd word was created in last turn by the last player, when the choice was of no importance any more.

The session describes a 'business as usual' situation. Players copyrighted every root created and pursued the most profitable extensions without taking into account the royalty fee. 81.4% of the extensions came from extending a root created by other players, and hence incurring the royalty fee; conversely, only 18.6% of extensions were built upon own words (Figure 4). Since a player is likely to produce one sixth of the roots, this figure implies that, on average, the players extended any root available and had no biases favouring their own creations.

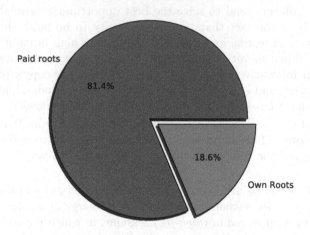

**Fig. 4.** Distribution of extensions by root type, session 1

Royalties make up an important share of players' payoffs. In the session the net added value created was 471 experimental points (worth 47.1£); royalties exchanged were worth 268.8 points, i.e. 57% of the value added.

The effect of royalty flows on individual payoffs can be appreciated in Figure 5: some players produced a lot of added value, but paid high amounts in royalties

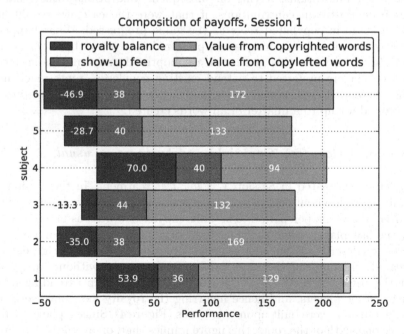

**Fig. 5.** Breakdown of payoffs, session 1

to do so, while other players produced fewer - but more likely to be extended - roots and extensions, and received a consistent amount of royalties. The two strategies are somewhat complementary: some inventors prefer to create many, if marginal, innovations; others focus on seminal, if fewer, innovations.

The session showed an intellectual property system working well: copyright does not get in the way and royalty fees are considered low enough not to block follow-on innovations. Copyright generates considerable incentives to innovate rewarding the creators of the best innovations (in this case, the most extendable) through royalties and the hard-working 'marginal' inventors through use-value of their marginal inventions. No 'tragedy of the anticommons' appears, as downstream innovators are not blocked in their endeavour by excessively high royalties, upstream inventors can trust that their roots will be used, and royalties flow from the marginal inventors, extending existing technologies, to the basic inventors, creating promising roots. In an ideal situation of perfect information, with no transaction nor legal costs associated with IP and with low, fixed and ex-ante known royalty fees, the copyright system works and delivers incentives to different types of innovators.

## 4.3   Session 2 (*high-fee*): Copyleft and Anticommons

Six players took part in Session 2, a *high-fee* treatment. The players left 23 letters unbought and 23 more unused. The players created 84 words, of which 22 copylefted: 14 words per player and 2.5 per period. The session showed both the emergence of copyleft and of anticommons gridlock: players in Session 2 created less words and enjoyed lower final payoffs w.r.t. Session 1 (see Table 1); moreover, the letter waste was double (46 *vs* 23 unused letters), the net value added half (237 *vs* 471), the average length (3.76) and value (6.6) of the words produced was substantially lower than in Session 1 (4.18 and 7.45).

73.8% of the words created were copyrighted, 26.2% copylefted. The players created 32 copyrighted extensions from 30 roots (1.07 per root), and 17 copylefted extensions from 5 roots (3.4 per root, three times as much). Copylefted words are furthermore longer and more valuable (see Figure 6). The best efforts of the players were directed to extend copylefted words. The players showed a somewhat schizophrenic behaviour: when creating new roots, players opted for copyright, hoping to enjoy a royalty stream from extenders, but, when extending, preferred to extend either their own or copylefted roots (Figure 8, right). This resulted in a mild but significant 'tragedy of the anticommons'; the players could have broke out of it either by adopting a 'business as usual' strategy as seen in Session 1 (but the high royalty fee seems to have made this less likely) or by starting to create copylefted roots. Having failed to do both, the players ended up producing less words and earning lower payoffs.

Players started copylefting roots at the very beginning, and switched to copyright after the first 'defections', replicating the usual 'decay of contributions' phenomenon of Public Good games (Figure 8, left). After period 8 all newly created roots were copyrighted; nonetheless, copylefted roots continued to be

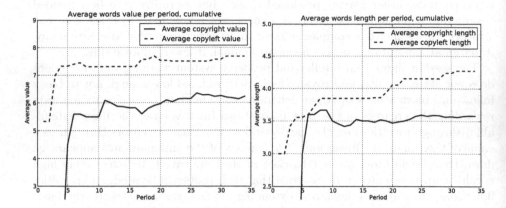

**Fig. 6.** Average word value and length per period per type, session 2

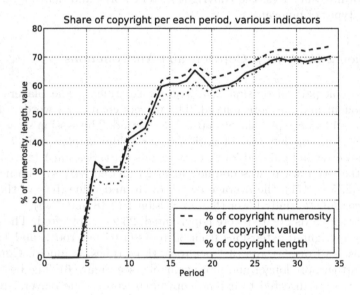

**Fig. 7.** Share of copyrighted words, length and value per period, session 2

extended during all the rest of the game, possibly to avoid paying royalty fees to the other players.

The players tended to extend other player's roots much less than in the *low-fee* treatment: only 51% of the extensions came from a root owned by another player, down from 81.4% in Session 1 (Figure 8, right). This resulted in many opportunities for extension being lost.

In Session 2 anticommons effects appeared, breaking the flow of innovations and resulting in lower value added and lower payoffs. Awarding a higher royalty to innovators generates the perverse effect of lowering the amount of innovation

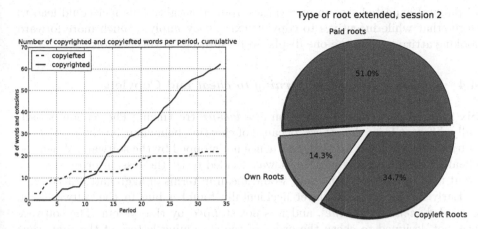

**Fig. 8.** Session 2: Words created by period and by type (left) and extensions by root type (right)

created in the system, at the same time not benefiting the inventors. Upstream patents, coupled with high royalty fee, discourage follow-on inventors to innovate on top of someone else's basic research on the one side, and to invest *tout-court* on the other hand (unused letters). Contrary to expectations, though, players fail to see the incentive to get rid of the patent system altogether, as argued by [1]: they extend copylefted roots but do not forgo the (low, given the behaviour

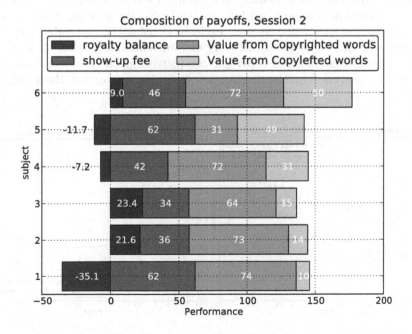

**Fig. 9.** Breakdown of payoffs, session 2

of players) expected stream of royalties from copyright. This fact could lead to argue that while incentives to copyleft exist, they imply a much more forward looking attitude than the one displayed by subjects in this session.

### 4.4   Session 3 (*low-fee*): *learning to cheat* and Copyleft

Six subjects participated in Session 3, a *low-fee* treatment. The session is radically different from the others, because of cheating behaviour on the part of the subjects that was not foreseen and hence not stopped by the software. When the cheating behaviour was noticed, it was decided to let the session proceed to see what this 'innovative' behaviour would mean in terms of aggregate statistics.

Early in the game, a player accidentally found a bug in the software: she extended *'new'* into *'wine'*, and was not stopped by the system. The software was not designed to check the order of the remaining letters of the root, thus leaving the door ajar for a particular kind of anagram, generated by adding one letter in any position and displacing the other letters at will. Actually created examples of this anagrams are given by *lady → delay, preen → opener*. All other types of extensions were correctly turned down by the software, including simple displacement (with no added letter, e.g. *not → ton*) and inclusion of more than one letter (e.g. *sit → shift*).

This new rule implies higher tinkering possibilites and higher payoffs, but at much higher cognitive costs, and it allows for a highly enhanced extendability of words, thus giving less incentives to create new roots.

After the first player successfully anagrammed a root to form an extension, the other players noticed and, all but one, endorsed the new practice. Around period 20, five out of six players were exclusively using the new rule. A further consequence of this endogenous discovery of value-enhancing rules was that the players eventually abandoned copyright completely: given the much enhanced extendability, players found it much more profitable to copyleft the word and

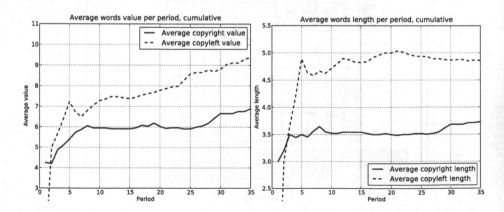

**Fig. 10.** Average word value and length per period per type, session 3

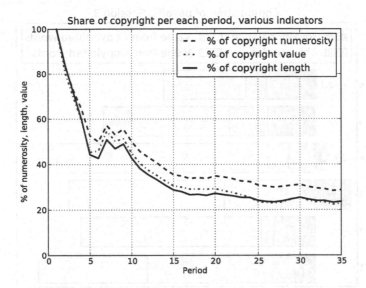

**Fig. 11.** Share of copyrighted words, length and value per period, session 3

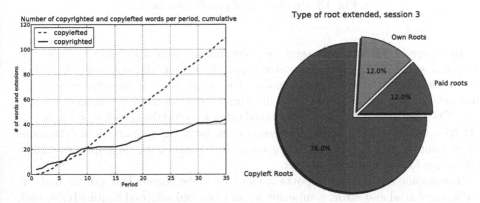

**Fig. 12.** Session 3: Words created by period and by type (left) and extensions by root type (right)

then build on the contributions of the other players than to rely on royalty fees that were unlikely to be received anyway.

The results of the new rules were striking. In the session 154 words were created, 25.6 per player and 4.5 per period; the net value added reached the level of 968, twice the level of Session 1, and four times the level of Session2; the royalty flow accounted for a mere 7.4% of it, and the unused letters were just 16, the lowest value of the experiment. The average value and length of copylefted words increased during the experiment much more than those of copyrighted words (see Figure 10), and the percentage of copyrighted words dropped steadily, ending at 29.2% (see Figure 11).

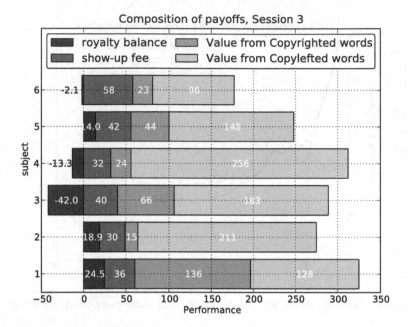

**Fig. 13.** Breakdown of payoffs, session 3

In Figure 12, left, it can be seen the point, around period 10, when the new rule began to spread. The creation of new copyrighted roots did not stop altogether, but continued at a much lower pace. The breakdown of payoffs records the importance of copyleft for the very high final payoff enjoyed by subjects (Figure 13). The number of extensions generated using a copyrighted root (see Figure 12, right) dropped to 12%, as did the use of own roots; three-quarters of extensions were built on top of the existing 4 copylefted roots, with an impressive 26.25 extension for each copylefted root.

The session showed two important facts: first, the players learned, from observation and trial and error, a superior set of rules and adopted it quickly; second, when extendability is enhanced the incentives to copyleft are much higher, and the role of copyright is more limited.

### 4.5   Session 4 (*high-fee*): Copyleft and Anticommons

Only five subjects participated in Session 4, a *high-fee* treatment. The players created 91 words, of which 9 copylefted; 18 words per player and 2.67 per period. The session had many results in common with Session 2, the other *high-fee* session: the players were affected by the high royalty fee and preferred to extend their own roots rather than pay royalties to the other players (Figure 14, right); this resulted in a substantial letter waste (47 unused letters), low payoffs (average 15.7£), a low number of words produced (even though higher than in Session 2), low average word length (3.89) and value (6.76).

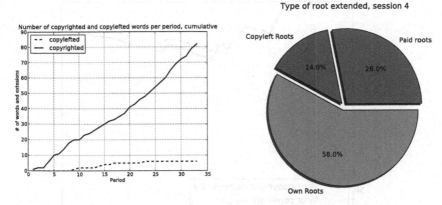

**Fig. 14.** Session 4: Words created by period and by type (left) and extensions by root type (right)

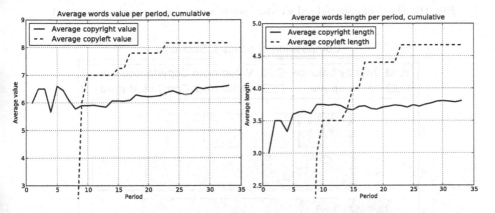

**Fig. 15.** Average word value and length per period per type, session 4

In this session copylefted roots were extended deeper and further than copyrighted ones, but the players failed to find a way out of the anticommons: all roots but one were copyrighted, despite the fact that players were not likely to extend other player's roots (just 28% of extensions came from a paid root). The 'isolation' of players was the highest of all sessions, with 58% of extensions created from one's own set of roots (Figure 14, right).

As in all other sessions, the copylefted words showed a higher average value and length than the copyrighted ones (see Figure 15); here they represented just 10% of the overall value, though, and had a minor impact on payoffs (Figure 17). Royalties also played a minor role: the total net value added was 288, with royalty flows at a low 71.1, 24.7% of the added value. This was due to the high reliance of players on own roots rather than on cross-fertilisation of word trails with other players.

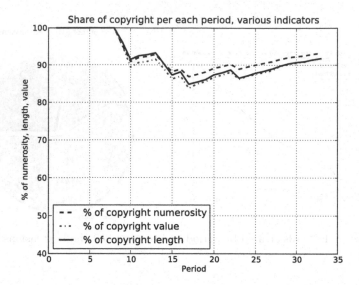

**Fig. 16.** Share of copyrighted words, length and value per period, session 4

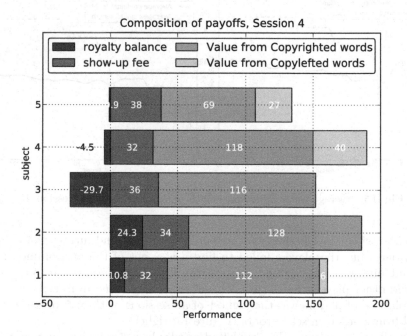

**Fig. 17.** Breakdown of payoffs, session 4

Session 4 confirms that when royalty fees are high, players fail to build innovation on top of other players' roots, reducing the number of profitable opportunities. Copyleft is a possible way out of this gridlock situation, but players fail to recognise this fact.

# 5   Conclusions

This paper sets up a dynamic real-effort experimental game to test the argument of [1] that in a dynamic setting featuring sequential and complementary innovations the innovators themselves would, at certain conditions, prefer not to impose copyright on their creation and instead welcome imitation.

Even if the statistical validity of such a small experiment is admittedly very low, preliminary results show that both anticommons problems and copyleft are more likely to emerge when royalty fees are relatively high. Increasing intellectual property rights leads to less innovation and to potential gridlock on the one hand, and to firms giving up copyright protection and embracing 'open' innovations. Copyleft emerged and saw sustained contributions in *high-fee* sessions, while it did not emerge in the *low-fee* session; anticommons features appeared in *high-fee* sessions, resulting in less innovations created, lower payoffs, higher number of wasted resources (on the form of unused letters), and reduced royalty flows, w.r.t. *low-fee* sessions.

The paradoxical effect that better patent protection (higher fees) results in a lower amount of innovations is reproduced in the lab. There is not enough evidence, though, to support the argument that in a sequential settings the firms prefer not to patent their innovations, relying on subsequent innovations rather than on royalties from upstream contributions.

The experiment is also potentially able to account for the gradual engagement in the production of Open Source software of more and more corporate giants. In the experiment, while it is hard to find someone willing to start a copyleft trail, the incentives to contribute to the copyleft commons with extensions once such an alternative exists are high. This leads, in time, to the building of a sizeable body of copyleft roots and extesions. As this grows, more and more subjects - even the ones that would strongly prefer copyright - have incentives to extend the copyleft words, because they take out more and more value from the commons. The copyleft rule forces them to give away any rights on their marginal contributions; but, as the size of the copyleft commons increases, the cost-benefit balance of contributing to it shifts gradually to the benefit side. The copyleft clauses are able to create a growing commons; no matter why the first contributors decided to forgo their copyright fees, the existence of a free alternative is an attractor for a larger and larger number of developers.

The preliminary experimental evidence calls for further exploration of this experimental design, both deepening - i.e. with more replications of the same treatments, to gain statistical significance - and widening the research agneda. The word-game design could be used as a platform for testing the effect of different intellectual property policies or different technological settings on the amount of innovative activity, licensing choices, social welfare, commons and anticommons problems. The experimental design would allow for instance to test the consequences on innovation levels and quality of varying the length and breadth of patent protection, or varying the menu of licenses available, or allowing patent pools.

# References

1. Bessen, J., Maskin, E.: Sequential innovation, patents, and imitation. Economics Working Papers 0025, Institute for Advanced Study, School of Social Science (Mar 2006), http://ideas.repec.org/p/ads/wpaper/0025.html
2. Buchanan, J.M., Yoon, Y.J.: Symmetric tragedies: Commons and anticommons. Journal of Law & Economics 43(1), 1–13 (2000), http://ideas.repec.org/a/ucp/jlawec/v43y2000i1p1-13.html
3. Depoorter, B., Vanneste, S.: Putting humpty dumpty back together: Experimental evidence of anticommons tragedies. SSRN eLibrary (2004)
4. Hardin, G.: The tragedy of the commons. Science 162(3859), 1243–1248 (1968), http://www.sciencemag.org/cgi/content/abstract/162/3859/1243
5. Heller, M.A.: The tragedy of the anticommons: Property in the transition from marx to markets. Harvard Law Review 111, 621–688 (1998)
6. Heller, M.A., Eisenberg, R.S.: Can patents deter innovation? the anticommons in biomedical research. Science 280(5364), 698–701 (1998), http://www.sciencemag.org/cgi/content/abstract/280/5364/698
7. Lang, K.R., Shang, R.D., Vragov, R.: Designing markets for co-production of digital culture goods. Decision Support Systems 48, 33–45 (2009)
8. van Rossum, G.: Python reference manual. CWI Report (May 1995)
9. Schulz, N., Parisi, F., Depoorter, B.: Fragmentation in property: Towards a general model. Journal of Institutional and Theoretical Economics 158, 594–613 (2002)
10. Scotchmer, S.: Standing on the shoulders of giants: Cumulative research and the patent law. Journal of Economic Perspectives 5(1), 29–41 (Winter 1991), http://ideas.repec.org/a/aea/jecper/v5y1991i1p29-41.html
11. Von Hippel, E., Von Krogh, G.: Open source software and the 'private-collective' innovation model: Issues for organization science. Organization Science 14, 209–223 (2003)

# Voting for Bugs in Firefox:
# A Voice for Mom and Dad?

Jean-Michel Dalle[1] and Matthijs den Besten[2]

[1] Université Pierre et Marie Curie, Paris
jean-michel.dalle@upmc.fr
[2] Ecole Polytechnique, Paris
matthijs.den-besten@polytechnique.edu

**Abstract.** In this paper, we present preliminary evidence suggesting that the voting mechanism implemented by the open-source Firefox community is a means to provide a supplementary voice to mainstream users. This evidence is drawn from a sample of bug-reports and from information on voters both found within the bug-tracking system (Bugzilla) for Firefox. Although voting is known to be a relatively common feature within the governance structure of many open-source communities, our paper suggests that it also plays a role as a bridge between the mainstream users in the periphery of the community and developers at the core: voters who do not participate in other activities within the community, the more peripheral, tend to vote for the more user-oriented Firefox module; moreover, bugs declared and first patched by members of the periphery and bug rather solved in "I" mode tend to receive more votes; meanwhile, more votes are associated with an increased involvement of core members of the community in the provision of patches, quite possibly as a consequence of the increased efforts and attention that the highly voted bugs attract from the core.

## 1 Introduction

Firefox is an open source project that explicitly tries to cater to the needs of a mainstream audience. Judging by its market share, as estimated for instance by the firm StatCounter (http://gs.statcounter.com), Firefox is succeeding. What might explain this success? To a large extent it is due to the leadership of people like Blake Ross, Dave Hyatt and Asa Dotzler and to their apparently correct judgement in deciding which features and bugs deserve the highest priority for development. In addition, however, we expect that the success is due to explicit mechanisms that have been put in place in order to make sure that the needs of mainstream users are assessed and addressed correctly. Voting for bugs in the Bugzilla bug tracking system is one such mechanism.

Voting for bugs is of course not unique to Firefox. It is a feature of the Bugzilla bug tracking system that has been activated by many of the projects who use it. Moreover, it is generally assumed that some sort of voting is a standard element of the open source software development model (see, e.g., [11]). Yet, apart from

P. Ågerfalk et al. (Eds.): OSS 2010, IFIP AICT 319, pp. 73–84, 2010.
© IFIP International Federation for Information Processing 2010

the governance of Apache [10] and Debian [12], there are to our knowledge surprisingly few *explicit* analyses or even descriptions of voting procedures in free/libre open source software communities.

In her analysis of the emergence of voting procedures in Debian, O'Mahony focuses on the greater efficiency and transparency that is provided by voting compared to consensus based collective decision making usually dominant in small groups and presents the introduction of voting as a reaction to the increases in scale and scope of the Debian enterprise. Voting can however be useful even in small groups: see for example the seminal paper on collective problem identification and planning by Delbecq and Van De Ven [4], who propose a model of an effective group process in which the first phase of problem exploration is concluded with a vote. This, they note, serves to make the results of the exploration explicit and create pressure for change on the people who will be responsible for resolving the problem. Here voting is not just about representation, it is also about "getting heard".

In this context it might be useful to ponder the framework that Hirschman [5] developed on the means by which patrons can influence their organizations. Patrons have basically two options: either they leave the organization and buy or produce with a competitor ("exit"), or they express their concerns more explicitly through channels that might be provided by the organization ("voice"). In open source software development, the primary mechanism for "exit" is forking [9] and the primary mechanism for "voice" is to "show code" by proposing patches [1]. For mainstream users such as "mom and dad", however, neither forking nor patching is a realistic option, as they typically do not have the skills to do either. Since Firefox precisely wants to be a product for mom and dad [6], other options have to be found: "exit" can be implemented by switching to other browsers like Internet Explorer, Safari, Opera, and Chrome. For "voice", voting might be an appropriate answer.

In what follows, we present circumstantial evidence to support our conclusion that voting for bugs in Firefox is a means to provide a voice to mainstream users. This evidence is drawn from a sample of bug-reports maintained by Mozilla's Bugzilla and from explicit information on voters found at the same site. We present results regarding voters and bugs in sections 4 and 5, respectively. We present some more background information on our research in section 2 and describe our sampling strategy in section 3.

## 2   Background

This paper is number four in a succession of papers that we have presented at OSS conferences. In the first paper, presented in Limerick in 2007 [2], we analyzed bug reports from Mozilla and found that there are some bugs that take exceedingly long to resolve and that part of the reason for the existence of these "superbugs", as we named them, could be related to the insufficient provision of contextual elements such as screenshots in the bug threads. The second paper, presented in Milan in 2008 [3], analyzed bug reports that are associated with the Firefox branch and exploited the existence of the so-called "CanConfirm" privilege — the privilege to declare bugs as "NEW" given that the initial status of bugs is "UNCONFIRMED" by default — to distinguish between core and peripheral participants within bug resolution processes and inquired whether various variables could influence the speed at which NEW and UNCONFIRMED bugs were patched. In the third and latest paper, presented in Skövde in 2009 [8], we used text mining techniques to arrive at a further characterization of participants within the core and the periphery of the Firefox community, stressing notably the fact that members of the core tend to use to pronoun "We" disproportionately while members of the periphery, conversely, seem to prefer to use "I".

By focusing on voters this paper tries to exploring an additional layer of the onion model of the Firefox community. To echo the language we used in our 2008 paper, while the likes of Blake Ross and Dave Hyatt clearly belong to the inner core, the periphery is probably mostly populated by "alpha-geek" users — that is, exactly by those people who Ross and Hyatt professed to ignore in Firefox development [6]. "Mom and dad" are not there. If at all, they are more likely to be among the *outer* periphery of people who are simply voting and contributing very little otherwise.

This line of inquiry obviously owes a lot to the pioneering work by Mockus et al. [10] and also to subsequent work by Ripoche [13], who established bug reports as an object of study. Together with others, e.g. [14], we continue to exploit this extremely rich source of information. Conceptually we have been inspired by the analysis of MacCormack at al. [7], who argue that it was necessary to make the existing code that was left by Netscape more modular before Mozilla could attract patches from the periphery. We wonder here whether votes could constitute another element which could help to explain how Firefox could turn the unwieldy open source project that Mozilla had become into the sleek browser for the mainstream market that we have now.

## 3 Sampling Method

For our analysis we constructed two types of data-sets. The first type relates to bugs, which may or may not have attracted votes, and concerns the history of the bug resolution process as it is recorded by Bugzilla as well information about the bugs that can be found in the logs of the CVS code repository. The second type concerns information that is stored in Bugzilla about the activities of the people who have voted for one or more bug. For each type we have obtained several sets and sub-sets of data based on three criteria: first of all, we are interested in bugs for which we can assume that they have had an impact on the Firefox code-base in the sense that they are mentioned in the commit-comments and the CVS repository; secondly, we are interested in bugs that have received votes; and finally, we are interested in bugs that Bugzilla associates with the Firefox project. Similarly, we are interested in people who voted for bugs which eventually found their way into the CVS; we are also interested in the other bugs they voted for; and we are interested in the people who voted with them

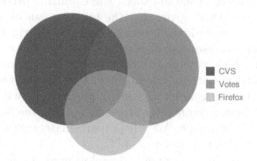

CVS
Votes
Firefox

**Fig. 1.** Sampling of bugs: 37408 appear in the code base (CVS); among them 3787 have attracted at least one vote (Votes ∩ CVS) and 418 among these are associated with the Firefox project (Firefox ∩ Votes ∩ CVS)

on Firefox bugs. Figure 1 is a Venn diagram which illustrates this admittedly somewhat complex constellation of sets. Our main interest is in the intersection of bugs from CVS that have been voted on and are also officially associated with Firefox as well as in the people who have voted for this particular set of bugs. Other than that we also have some interest in the other subsets formed by the intersection of the three criteria in order to be able to compare and contrast with what we find in our main set.

In practice we followed a procedure similar to snowball sampling in order to obtain our sets and subsets of bugs and voters. We started with the bugs that we found in the logs of our copy of the Mozilla's CVS archive concerning the code for Firefox up until version 2.0. Only a small subset of these bugs, some 10%, has ever attracted a vote. Associated with those bugs is a list of all people who have cast a vote (people can also retract there vote later; in that case they fall through our net). Our next step was to retrieve all those lists and compile a list of CVS voters constituted by union of all sets of voters contained in these lists — we found 11826 of them. The list of voters that is attached to the bug-reports in Bugzilla also contains, for each voter, a link to a voter-page that lists all the bugs the voter in question has voted for split according to the project with which Bugzilla associates the bug. As far as we can tell, this "project" field is a new data element in the bug-reports that was not yet displayed in the original bug-reports that constitute the CVS-set of bugs. That is why we do not have such project-data for most of the bugs. Nevertheless via the voter-pages we can determine the project for those bugs for which at least one vote was cast. Contrary to what one might assume just a fraction of the bugs that we associate with code in the Firefox-branch of the Mozilla CVS are associated with Firefox by Bugzilla as well. Partly, this may be due to misattribution from our part, but mostly this seems to be a reflection of the fact that Firefox shares code with other applications and the fact that it is based on code that was developed earlier for other projects. In addition to the project affiliation of CVS bugs the voter-pages also give the project affiliation for bugs that did not make it into that set — that

**Table 1.** Summary indication of source of the main sets of bugs and voters

|  | **Bugs** | **Voters** |
|---|---|---|
| **CVS** | Bugs associated with Firefox code before version 2.0. | Union of lists of voters on bugs in set defined in previous column. |
| **Votes** | All bugs that received one or more votes by voters on the CVS bugs. | As above, ceteris paribus. |
| **Firefox** | All the bugs associated with Firefox in the listing of bugs per voter. | " " |

is, bugs that did get votes but where not mentions in the commit-comments we looked at. The union of all these bugs constitutes the voter-set of bugs. We did not retrieve the complete Firefox-set of bugs, but we do not which part of the voter-set overlaps with it and that another part of the CVS-set should do as well. Finally, our set of Firefox-voters is by looking at the voter-lists of all bugs in the voter-set that are associated with Firefox through Bugzilla and taking the union of the sets of voters listed in those lists. Table 1 gives a short summary of the manner in which we obtained our sets described above.

Another and, for now, final detail about the data preparation concerns the way we link voter-identities with other activities by the same people in the bug resolution process. For this we rely on the fact that up until recently the voters as well as participants to the bug forum discussions, bug-reporters, bug-assignees, etcetera, were identified by their email address. Hence we could assign various bug-activities to the voters by matching these email addresses. This method works fine for the bugs that we focus on for this study. However, studies concerned with the most recent bugs would not be able to apply this method as Bugzilla has moved to enhanced identity management and does no longer provide the full email address for voters.

## 4 Characterizing Voters

A first dimension along which voters can be distinguished from non-voters, reported in Table 2, is with respect to both groups' status in the community.

**Table 2.** Number of voters by status for participants in the resolution of bugs which have received at least one vote, are mentioned in the CVS, and are associated with the Firefox project ($n = 2968$; $\chi^2 = 366$, $df = 3$, $p$-value = $5.243e^{-79}$)

|  | Status | | | | |
|---|---|---|---|---|---|
|  | Unconfirmed | New | Both | Other | Total |
| Vote | 184 | 25 | 64 | 1408 | 1681 |
| No Vote | 286 | 139 | 182 | 680 | 1287 |

Table 2 is a contingency table comparing community status and voting activity of participants in the bug resolution process for the intersection of 418 bugs with the following properties: they appear in the CVS; they have attracted votes; and they are associated with Firefox. As participants, we consider people who have either contributed at least one comment to at least one discussion thread or who have cast at least one vote for these bugs. Among the 418 bugs of interest, this definition yields a total of 2968 participants, 1681 of whom have cast a vote and 1755 of whom have written at least one message, which implies that 1213 participants have voted but never written a message. A first conclusion that can be drawn from these numbers, then, is that most voters do not engage in other activities. If we go one step further and check participation on a bug-by-bug basis, this finding becomes even more pronounced: typically, voters who are active in the bugs in the set tend to engage in this activity on bugs for which they did not vote.

In order to gauge the status of participants we rely on the CanConfirm privilege mentioned earlier. In particular, we check for each participant whether he or she has ever declared bugs contained in the set of 37408 bugs that appeared in the CVS logs. For those participants who did declare bugs we look at the initial status of those bugs. If all the bugs that a participant declared start with status NEW the participant is considered to belong to the core; if all the bugs start with status UNCONFIRMED the participant belongs to the periphery; if some bugs start with UNCONFIRMED and others with NEW the participant is considered to be a freshly joined or a freshly expelled member of core; finally if the participant has not declared any bug, he or she is classified as member of the outer-periphery, here denoted as "Other".

Given all this, the main conclusion from table 2 is that the people who cast a vote are mostly, yet not exclusively, outsiders. Interestingly, this finding also holds for the status of the people who voted for any one of the bugs that appeared in the CVS logs. Of the 11850 voters among these 3787 bugs 10269 have never declared a bug while only 283 participants have had all the bugs they declared accepted as NEW right-away. Furthermore, when we consider the top 1000 most active voters, 688 among them can still be classified as outsiders while only 36 belong to the core.

Figure 2 gives some indications about the voting activities of the people who have cast a vote for at least one of the 418 bugs in the latter sample. It shows the distribution of the number of bugs to which people have cast their vote as split according to projects and status. Actual outsiders, with status Other, tend to declare "pure" Firefox bugs while contributors with another status disperse their votes much more.

Figure 3 tries to shed some light on the timing of bug activity. Votes do not come with timestamps in the Bugzilla records, but we know that bug-ids are assigned sequentially. Figure 3 shows how many bugs had received votes before a given bug id against the sequence number that is used to identify them and this gives a rough indication about the distribution of voting over time. On the left is shown this distribution for bugs with votes and that appear in the

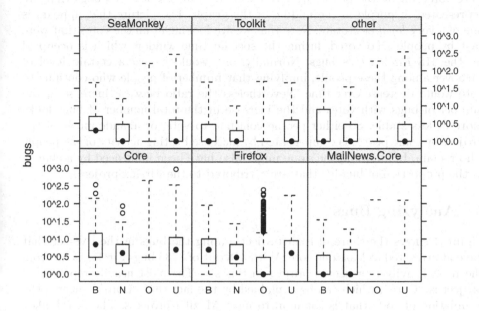

**Fig. 2.** Distribution of number of bug declared by status per project. B = both; N = new; O = other; U = unconfirmed.

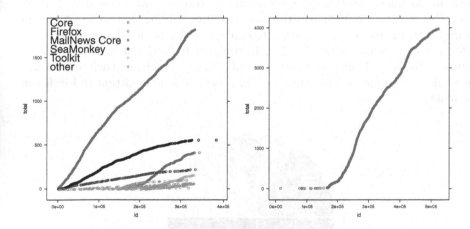

**Fig. 3.** Cumulative sum of bugs per project against bug-id; in the plot on the left specifically for Firefox (see text)

CVS logs: typically, a steady proportion of bugs attract votes; not all projects are active at the same time; and some, for instance SeaMonkey, are gradually abandoned while others take off. The right graph focuses on bugs associated with the Firefox project more in particular. This graph includes the bugs that were voted on by the people who voted for the 418 Firefox CVS bugs that are represented in purple in the middle of the graph. The picture that appears is one of a very *loyal* electorate. Note that we are looking at all the votes that were cast by people who voted during the specific time window which is occupied by the 418 Firefox CVS bugs. Normally, one would expect a certain level of turnover among these people, implying that number of people who continue to vote would decrease over time. Nevertheless the ratio between increase in the number of bugs with votes and the increase in the total number of bugs looks more or less stable. So either people continue to vote, or abandonment of the project by some people is compensated by increased voting activity of the people who remain. In addition, the slope might also have been influenced by a change in the proportion of bug-ids that are attributed to the Firefox project.

## 5  Analyzing Bugs

Figure 4 shows the share of each project among the bugs in the CVS-set that have at attracted at least one vote. With 418 out of 3785 bugs Firefox is far from the most heavily represented project in this set. The most heavily represented project is "Core", which can be explained by the fact that "Core" concerns the foundation of code that is common to most Mozilla projects. The SeaMonkey project "is a community effort to develop the SeaMonkey all-in-one internet application suite" (http://www.seamonkey-project.org/). In a sense, it pursues a strategy that Firefox explicitly chose not to follow and it is a little ironic that there are so many SeaMonkey bugs mentioned in commits related to Firefox code. This may be a reflection of the fact that these siblings have a lot of code in common. "MailNews Core" is harder to explain as this project is concerned with code for email clients, which Firefox, in contrast to SeaMonkey, choose not to include. "Toolkit", finally, includes cross-platform components such as "Gecko" the rendering engine for web-pages — which is of course important to Firefox in particular.

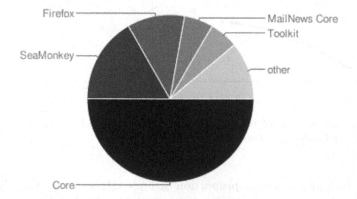

**Fig. 4.** Pie chart showing the share of Mozilla projects among the bugs that received bugs in our sample

In order to compare the bugs that attract votes with those that don't, Table 3 gives the estimations of a generalized linear model, assuming a Poisson distribution for the variable totalVotes, which is a numeric variable whose value represents the total number of votes that a bug has attracted. The estimates are based on data from a subset of about ten thousand cases, bugs for which we don't have complete information, very old bugs with a low bug-number and other outliers having been removed. Please refer to [3] for a detailed description of these variables in so far as they are not explained below.

The number of votes is statistically correlated with NumAttention, which relates to the number of people who have put themselves on the CC-list associated

**Table 3.** Generalized linear model with Poisson distribution for dependent variable totalVotes based on 10763 observations

| Parameter | Est. | Std.Er | [Wald | 95% ] | $\chi^2$ | $Pr > \chi^2$ |
|---|---|---|---|---|---|---|
| Intercept | 0.3231 | 0.1697 | -0.0096 | 0.6558 | 3.62 | 0.057 |
| numberOfEditsByBugReporter | 0.0061 | 0.0029 | 0.0005 | 0.0118 | 4.57 | 0.0326 |
| numberOfEditsByLastAssignee | -0.0025 | 0.0014 | -0.0052 | 0.0003 | 3.13 | 0.0771 |
| NoSgRestriction | 0.5972 | 0.1046 | 0.3922 | 0.8022 | 32.6 | <.0001 |
| numberOfTimesAssigned | 0.1064 | 0.006 | 0.0946 | 0.1182 | 311.23 | <.0001 |
| bugWasReopened | 0.3264 | 0.0264 | 0.3781 | 0.2747 | 153.14 | <.0001 |
| nauth | 0.1799 | 0.021 | 0.1387 | 0.2212 | 73.19 | <.0001 |
| patches | 0.1217 | 0.0032 | 0.1154 | 0.1281 | 1410.15 | <.0001 |
| attach_patch | 0.1065 | 0.0049 | 0.097 | 0.116 | 481.08 | <.0001 |
| com_com | -0.5699 | 0.0152 | -0.5997 | -0.54 | 1400.64 | <.0001 |
| comauthrate | -0.2283 | 0.0154 | -0.2584 | -0.1981 | 220.65 | <.0001 |
| severity | 0.0341 | 0.0131 | 0.0085 | 0.0597 | 6.82 | 0.009 |
| priority | 0.0053 | 0.0097 | -0.0138 | 0.0244 | 0.3 | 0.5848 |
| priorityNotIncreased | -0.133 | 0.0698 | -0.2698 | 0.0039 | 3.63 | 0.0568 |
| priorityNotDecreased | 0.1907 | 0.0461 | 0.1005 | 0.281 | 17.15 | <.0001 |
| severityNotIncreased | -0.321 | 0.0295 | -0.3788 | -0.2632 | 118.44 | <.0001 |
| severityNotDecreased | -0.4144 | 0.0403 | -0.4934 | -0.3354 | 105.71 | <.0001 |
| nfile | -0.0068 | 0.0009 | -0.0086 | -0.0051 | 58.35 | <.0001 |
| dependson | 0.0216 | 0.0075 | 0.007 | 0.0363 | 8.39 | 0.0038 |
| blocked | 0.0561 | 0.0037 | 0.0489 | 0.0633 | 232.89 | <.0001 |
| NumAttention | 1.0415 | 0.0164 | 1.0093 | 1.0737 | 4018.84 | <.0001 |
| version2 1.7 | 1.325 | 0.1105 | 1.1084 | 1.5416 | 143.73 | <.0001 |
| version2 Trunk | -0.4801 | 0.0285 | -0.536 | -0.4242 | 283.24 | <.0001 |
| Ltpsassign | 0.0822 | 0.0053 | 0.0718 | 0.0926 | 239.27 | <.0001 |
| DuplicatesBeforeLastResolved | 0.1741 | 0.0044 | 0.1654 | 0.1828 | 1531.72 | <.0001 |
| UX | 0.3312 | 0.0454 | 0.2421 | 0.4202 | 53.12 | <.0001 |
| I/we | 0.0537 | 0.0014 | 0.0509 | 0.0565 | 1383.95 | <.0001 |
| os3 NonWin | 0.1279 | 0.0299 | 0.0693 | 0.1864 | 18.31 | <.0001 |
| OC | 0.1415 | 0.0278 | 0.0871 | 0.196 | 25.94 | <.0001 |
| CO | 0.424 | 0.0361 | 0.3533 | 0.4947 | 138.19 | <.0001 |
| OO | 0.8045 | 0.0366 | 0.7327 | 0.8763 | 482.33 | <.0001 |
| Scale | 1 | 0 | 1 | 1 | | |

with the bug so that they can follow the progress on the bug resolution, i.e. with the attention received by a bug.

More generaly, more votes tend to be associated with "problems" in solving a bug and notably to neglect: patches, which counts the number of different patches needed to solve a bug; duplicatesBeforeLastResolved, which indicates repeated declarations of the same bug; Ltpsassign, the log of the time need for the bug to be assigned to someone; numberOfTimesAssigned, the number of times a bug has been assigned; or else bugWasReopened, a self-explanatory dummy variable, are all variables of this kind. Conversely, high values for com_com and com_authrate, which are indicative for a high level of activity and a lot of commitment on a bug, are correlated with a lower number of votes: votes would be less needed when there is enough commitment on a given bug.

Interestingly then, several variables generally relevant for the patching and triage of bugs are not or only weakly significant here: numberOfEditsByBugReporter and numberOfEditsByLastAssignee, two variables that reflect the level of activity of the people who are most directly involved with the resolution of a bug; severity, which reflects the current community estimate of the severity of a bug; priority, the assessment by the community of the importance of a bug; and dependson, the number of other bugs which have been found to depend on a given bug.

Finally, and closer to our interest in this article, I/We, which represents the ratio between the number of times that the personal pronouns I and We appear in the bug thread, is highly significant: bugs patched in "I-mode" tend to recieve more votes, either because people vote for their own bugs or because bug patching in I-mode is associated with the involvement of peripheral members of the community. The significance of OC, CO, and esspecially of OO — dummy variables indicating that the bug reporter and the first "patcher" for that bug stem from periphery and core (in case of OC), core and periphery (CO), or both from the periphery (OO), respectively — tends to support the view that the involvement of the periphery in patching a bug would be statistically associated with more votes, even while controlling by attention, commitment, neglect, and various other problems affecting bug patching.

In this last respect, Figure 5 is the result of an attempt to delve deeper into the relationship between patching and voting in the context of the relationship between periphery and core. Most of the patches are proposed by people from the core of the Firefox community. In many cases the first patch that is proposed is accepted as the solution for a bug. There are however a few cases in which multiple patches are proposed before a final patch is accepted as the solution for a bug. Figure 5 focuses on bugs whose first patch was proposed by a member of the periphery. We compute how many of these bugs do not receive any subsequent patches by members from the core, relative to bugs that have received zero votes, one to five votes, six-to-ten votes, and so further, respectively. What Figure 5 shows is that there is a clear increase in the level of participation of core members of the community when the number of votes increases. A possible interpretation of this could be that votes are used by the periphery to attract the attention of the core.

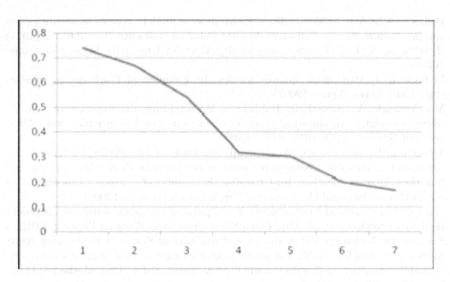

**Fig. 5.** Proportion of bugs that received patches from the periphery only among all bugs whose first patch was proposed by a member of the periphery (y-axis) against number of votes as average over a 5-vote range (i.e. $1 = 0$ votes, $2 = 1 - 5$ votes, $3 = 6 - 10$ votes etc) on the x-axis

## 6  Conclusion

When Blake Ross and Dave Hyatt initiated Firefox, they established themselves as benevolent dictators fulfilling the *volonté générale*, to borrow from Rousseau, while emphatically reserving the right to ignore the *volonté de tous*. However, and contrary to acounts of voting in open source communities which tend to present it as a method to arrive at a fair representation of the will of *the* community, they allowed voting as as a channel through which *voices from outside the community* could be heard. Consequently, we found that most votes originate in the outer periphery. Bugs that attract most votes tend to be bugs that are relatively neglected or bugs where the periphery is heavily involved. Hence one can surmise that the votes were cast in order to attract the attention from the core. It would be interesting to see whether there are additional mechanisms apart from the votes that help the core to focus its attention.

## References

1. Cox, A.: Cathedrals, bazaars and the town council (1998),
   http://www.linux.org.uk/Papers_CathPaper.cs
2. Dalle, J.-M., den Besten, M.: Different bug fixing regimes? A preliminary case for superbugs. In: Proceedings of the Third International Conference on Open Source Systems, Limerick, Ireland (June 2007)
3. Dalle, J.-M., den Besten, M., Masmoudi, H.: Channelling Firefox developers: Mom and dad aren't happy yet. In: Proceedings of the Fourth International Conference on Open Source Systems, Milan (September 2008)

4. Delbecq, A.L., Van de Ven, A.H.: A group process model for problem identification and program planning. Journal of Applied Behavioral Science 7(4), 466–492 (1971)
5. Hirschman, A.O.: Exit, voice, and loyalty. Harvard University Press, Cambridge (1970)
6. Livingston, J.: Blake Ross; creator, Firefox. In: Founders at Work: Stories of Startups' Early Days, Apress (2007)
7. MacCormack, A., Rusnak, J., Baldwin, C.Y.: Exploring the structure of complex software designs: An empirical study of open source and proprietary code. Management Science 52(7), 1015–1030 (2006)
8. Masmoudi, H., den Besten, M., de Loupy, C., Dalle, J.-M.: Peeling the onion: The words and actions that distinguish core from periphery in Firefox bug reports, and how they interact together. In: Crowston, K., Boldyreff, C. (eds.) Proceedings of the Fifth International Conference on Open Source Systems (2009)
9. Garcia, J.M., Edward Steinmueller, W.: Applying the open source development model to knowledge work. INK Open Source Research Working Paper 2, SPRU - Science and Technolgy Policy Research, University of Sussex, UK (January 2003)
10. Mockus, A., Fielding, R.T., Herbsleb, J.D.: Two case studies of open source software development: Apache and mozilla. ACM Trans. Softw. Eng. Methodol. 11(3), 309–346 (2002)
11. O'Mahony, S.: The governance of open source initiatives: what does it mean to be community managed? Journal of Management and Governance 11(2), 139–150 (2007)
12. O'Mahony, S., Ferraro, F.: The emergence of governance in an open source community. Academy of Management Journal 50(5), 1079–1106 (2007)
13. Ripoche, G.: Sur les traces de Bugzilla. PhD thesis, Université Paris XI (2006)
14. van Liere, D.W.: How shallow is a bug? Technical report, Rotmon School of Management, University of Toronto, November 16 (2009)

# The Nagios Community: An Extended Quantitative Analysis

Jonas Gamalielsson, Björn Lundell, and Brian Lings

University of Skövde, Sweden
{jonas.gamalielsson,bjorn.lundell,brian.lings}@his.se
http://www.his.se

**Abstract.** The health of an Open Source ecosystem is an important decision factor when considering the adoption of an Open Source software or when monitoring a seeded Open Source project. In this paper we assess the ecosystem health using approaches involving domain analysis and social network analysis of mailing lists for the Nagios project. We elaborate approaches for how involvement of different roles can be analysed through quantitative analysis, specifically focusing on core developers and professional providers. Our contribution is a step towards a deeper understanding of professional involvement in professional Open Source ecosystems.

## 1 Introduction

Before an organisation adopts an Open Source project it is important to evaluate its community in order to make sure that it is healthy and that the project is likely to be maintained for a long time (van der Linden et al. 2009), and it can also be the case that a seeded Open Source project needs to be monitored. One important means in such an evaluation is to quantitatively assess the health of an Open Source community (Crowston and Howison 2006). A number of studies have investigated large, well known Open Source projects through quantitative analysis, including the Linux kernel (Moon and Sproull 2000), Apache (Mockus et al. 2002), Mozilla (Mockus et al. 2002), Gnome (German 2004) and KDE (Lopez-Fernandez 2006). Several of these studies focus on social network analysis from different kinds of data sources such as CVS/SVN (Martinez-Romo et al. 2008), bug reports (Crowston and Howison 2005) and mailing lists (Kamei et al. 2008).

We have chosen to analyse social networks derived from the mailing lists of an Open Source project. Such an analysis can reveal how active members in a community are interacting with each other. In earlier work (Gamalielsson et al. 2009) we adopted the approach of Kamei et al. (2008), who studied the community of the Apache web server project, in a study of the Nagios project (www.nagios.org). Kamei et al. (2008) studied the user and developer mailing lists for a period of six months; three months before and after a major release of Apache. In our earlier work, we chose to study a period of six months from January 2009 to June 2009. The aim was to investigate what extent core developers act as mediators within and between the sub-communities emerging around the user- and developer mailing lists. In this work

P. Ågerfalk et al. (Eds.): OSS 2010, IFIP AICT 319, pp. 85–96, 2010.

we extend the earlier work by studying 22 six-month periods from January 2004 to September 2009, and also suggesting other kinds of analyses.

Important roles in a typical Open Source project are users, developers, core developers and project leaders (Crowston and Howison 2006). The importance of core developers applies to any Open Source project as it is well established in the literature that core developers "contribute most of the code and oversee the design and evolution of the project." (Crowston et al. 2006). The influence of Nagios core developers was explored in the earlier work (Gamalielsson et al. 2009), and also in this work. Another role to consider, especially with the advent of OSS 2.0 (Fitzgerald 2006), is the provider role. According to Fitzgerald (2006), OSS 2.0 product support is characterised by "customers willing to pay for a professional 'whole-product' approach". The provider role is also emphasised in the OSS stakeholder triangle proposed by Lundell et al. (2009). The model conceptualises an healthy Open Source ecosystem involving different kinds of professional actors. In such ecosystems there are mutualistic relationships between all roles (developer, user and provider). Concerning the provider aspect, it is evident that there are still misconceptions about support of Open Source software (Lundell 2009), something which was identified as a myth ten years earlier by Tim O'Reilly (O'Reilly 1999). This shows that it is important to recognise the professional involvement in Open Source communities and that companies can provide professional support. Of particular interest to our work is that companies can provide different types of support to an Open Source project, including "the participation to online forums in order to keep the community alive by answering to users and customers" (Capra et al. 2009).

Earlier research has had limited focus on professional involvement in Open Source communities (Capra et al. 2009). One notable exception is the study of the Maemo platform (Ghosh 2006), which reported the extent to which different companies contributed to the project.

In this study we have access to informative data in the form of email addresses of mailing list contributors. We also have access to explicit lists of core developers and providers at the website of the project. Given this information, the purpose of our work is to elaborate approaches for studying how these groups of individuals contribute to the community around the mailing lists. As a motivation, it has been suggested that an organisation planning to professionally engage with Open Source needs to assess "the health of an OS product's ecosystem" as part of their development of an Open Source strategy (Watson et al. 2005).

## 2   The Nagios Project

Nagios is a tool for monitoring IT infrastructure that has been used in many professional organisations and mission critical systems. For example, Toland et al. (2007) found "the Nagios availability tool to be effective in the proactive support of mission-critical radiology and other clinical imaging systems.". It has received several awards over the years, e.g. InfoWorld's Best of Open Source Award for both 2008 and 2009. In a recent survey at "thegeekstuff.com" (September 2009), Nagios was voted as by far the most popular Open Source monitoring tool. Nagios has a large base of users, with many active contributors to the project mailing lists. The core developers of Nagios are explicitly listed at the Nagios website (www.nagios.org/development/ teams/core, accessed on

**Table 1.** Events in the more recent version history of Nagios

| Event | Date |
|---|---|
| V2.0 | Feb 2006 |
| V2.2 | Apr 2006 |
| V2.3, V2.4, V1.4 | May 2006 |
| V2.6 | Nov 2006 |
| V2.8, V3.0a1 | Mar 2007 |
| V2.9 | Apr 2007 |
| V3.0b1 | Jul 2007 |
| V2.10 | Oct 2007 |
| V3.0rc1 | Dec 2007 |
| V2.11, V3.0 | Mar 2008 |
| V3.1.0 | Jan 2009 |
| V3.0.2, V2.12 | May 2009 |
| V3.2.0 | Aug 2009 |

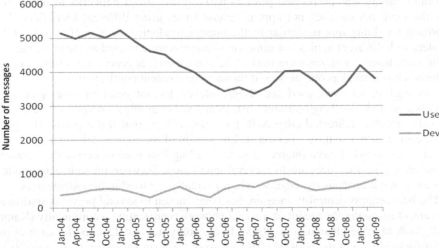

**Fig. 1.** Number of messages for different and six-month time windows starting at the month shown at the x-axis. "Use" denotes the user mailing list for Nagios, and "Dev" denotes the developer mailing list.

21 December, 2009). Furthermore, Nagios has a number of recognised professional providers. There are two preferred service providers (OP5, and Wuerth Phoenix) and 15 additional service partners for Nagios listed at the project website (www.nagios.org/support/servicepartners, accessed on 21 December, 2009). The explicit lists of core developers and providers are utilised in the result analysis.

Table 1 shows the more recent version history of Nagios, which was derived from the website of the project (www.nagios.org/news, accessed on 21 December, 2009). The first commits on Nagios were performed as early as in the end of 2001.

In order to illustrate the active community of Nagios, Figure 1 shows number of messages for 22 different and overlapping time windows for the user- and developer mailing lists of Nagios. It is evident that there is much more activity in the user list. In fact, the level of activity in these two mailing lists of Nagios is in the same range as the activity of the corresponding mailing lists for the Apache webserver project for a six month period (Gamalielsson et al. 2009).

## 3   Research Approach

Data was collected from the GMANE (gmane.org) archives of the SourceForge mailing lists for Nagios for the period from January 2004 to September 2009. The GMANE export interface was used to retrieve the raw mbox files for the project. For a message, the "message-id" field was used to get a unique message identifier. The "in-reply-to" field of a message was used, which contains the identifier for the message it was a reply to. Each message also has a "from" field from which the name and email address of the sender were derived. Finally, the "date" field was used to obtain the time and date of the message.

Data cleaning was performed prior to deriving the social networks to make sure that the same person does not appear several times using different identifiers. The approach for doing this is similar to the approach adopted by Kamei et al. (2008). Senders with different names but same email address were treated as the same person. If the same name (or slight variations of the same name) appears with different email addresses it is treated as one person, if the message content confirms this.

Two mailing lists were used: the "Nagios-devel" list, intended for development related issues; and the "Nagios-users" list, intended for problems related to the use of the software. An undirected edge A-B in the network is created if a person B replies to a message earlier sent by a person A. The weight of the edge is defined as the number of times A and B have interacted in the mailing lists used to derive the network, where each edge may account for several interactions between the connected nodes. However, in our analysis the edge weight is currently not used in the calculations.

The betweenness centrality measure has been chosen in several previous studies as a means of assessing the health of a social network representing a community (Kamei et al. 2008, Martinez-Romo et al. 2008). In our study it was calculated for each of the nodes (individuals) in the network. This measure takes on values in the interval [0,1], and quantifies the ability of a node to act as a mediator in the network (Kamei et al. 2008). More precisely, betweenness centrality reflects the number of shortest paths that pass through a specific node. The loss of a node with high betweenness centrality may therefore disconnect parts of the network that the node "glues" together. Betweenness centrality is described in mathematical detail in Kamei et al. (2008).

## 4   Results

In this section we present different approaches for establishing ecosystem health that may be of a particular interest to a potential adopter in a professional Open Source context. In so doing, we specifically illustrate professional stakeholders in the Nagios community.

Figure 2 shows an early social network derived from six months of email correspondence (before release of Nagios V2.0, from April to September 2004) in the

**Fig. 2.** Social network derived from Nagios developer mailing list Apr-Sep 2004

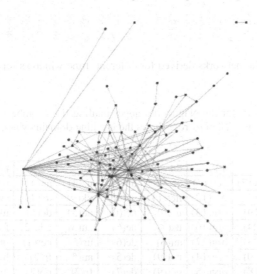

**Fig. 3.** Social network derived from Nagios developer mailing list Apr-Sep 2009

Nagios developer mailing list, and Figure 3 shows the corresponding network for the most recent six month period (April to September 2009). It can be observed that for both periods, the network has a distributed topology indicating no single point of "failure". The networks for the user mailing list are also distributed, but have more nodes. The Nagios networks (Figure 2 and 3) have a shape similar to the corresponding networks for the Apache webserver project (Kamei et al. 2008).

Simple metrics showing the size of an active community is the number of nodes and edges in the social network. By calculating these metrics for different time windows it is possible to study the community dynamics. Figure 4 illustrates this for the Nagios developer mailing list. Here we can observe a positive trend with a long term growth in both number of nodes and edges. As expected, there is a clear correlation between number of nodes and edges (Pearson correlation of 0.92).

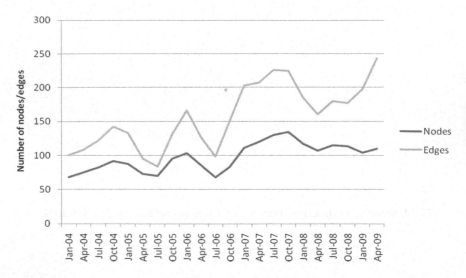

**Fig. 4.** Statistics for social networks derived for different time windows for the Nagios developer mailing list

**Table 2.** Top nine domains for the Nagios developer mailing list. Numbers in brackets represent percentage of messages. The last row shows the top nine domains when accumulated over time.

| Win | #1 | #2 | #3 | #4 | #5 | #6 | #7 | #8 | #9 |
|---|---|---|---|---|---|---|---|---|---|
| Jan-04 | com(29) | org(17) | net(14) | se(8) | uk(8) | us(4) | au(4) | ru(3) | de(2) |
| Apr-04 | com(28) | se(13) | org(12) | net(11) | us(9) | uk(9) | de(4) | au(3) | ru(3) |
| Jul-04 | com(33) | se(19) | org(13) | us(9) | net(6) | au(5) | de(5) | uk(2) | br(2) |
| Oct-04 | com(30) | se(24) | org(16) | us(7) | de(6) | au(5) | edu(2) | fr(1) | pt(1) |
| Jan-05 | com(27) | se(23) | org(17) | au(6) | de(6) | us(5) | net(4) | pt(2) | edu(2) |
| Apr-05 | com(34) | se(20) | org(14) | net(9) | dc(5) | au(5) | fr(2) | hu(2) | no(2) |
| Jul-05 | com(37) | se(18) | org(15) | net(9) | de(7) | fr(2) | no(2) | hu(2) | gov(1) |
| Oct-05 | com(23) | org(17) | se(17) | de(13) | net(10) | br(4) | info(2) | uk(1) | tm(1) |
| Jan-06 | com(22) | org(19) | net(12) | se(11) | de(11) | br(5) | au(2) | nl(2) | no(1) |
| Apr-06 | com(29) | org(20) | net(11) | de(9) | se(6) | edu(4) | au(4) | br(3) | nl(3) |
| Jul-06 | com(39) | org(16) | de(10) | net(10) | se(7) | edu(4) | us(2) | hu(2) | nl(2) |
| Oct-06 | com(34) | org(22) | de(15) | net(8) | se(6) | edu(3) | ca(2) | at(2) | hu(2) |
| Jan-07 | com(32) | org(21) | de(18) | net(7) | se(4) | edu(3) | ca(3) | at(2) | br(1) |
| Apr-07 | com(35) | de(18) | org(17) | se(6) | net(6) | ca(2) | edu(2) | br(2) | at(2) |
| Jul-07 | com(29) | org(21) | se(14) | de(14) | net(4) | dk(2) | edu(2) | ru(2) | ca(1) |
| Oct-07 | com(27) | org(19) | de(16) | se(12) | net(4) | dk(3) | edu(3) | uk(2) | fr(2) |
| Jan-08 | com(38) | de(17) | org(12) | se(6) | fr(4) | net(3) | ca(3) | uk(3) | dk(3) |
| Apr-08 | com(43) | de(16) | se(10) | org(7) | ca(5) | net(4) | fr(4) | br(2) | uk(2) |
| Jul-08 | com(37) | de(19) | se(14) | org(8) | ca(5) | net(3) | fr(3) | br(2) | edu(1) |
| Oct-08 | com(38) | de(18) | se(10) | org(9) | ca(7) | br(5) | fr(3) | net(2) | edu(1) |
| Jan-09 | com(41) | se(14) | de(13) | org(11) | ca(6) | br(5) | net(3) | fr(3) | edu(1) |
| Apr-09 | com(44) | org(15) | se(14) | de(10) | net(4) | ca(4) | br(3) | fr(2) | at(1) |
| Acc. | com(34) | org(15) | se(12) | de(12) | net(6) | ca(2) | br(2) | edu(2) | fr(2) |

As earlier mentioned, the mbox files used to derive the social networks contain email addresses for the contributors. This data makes it possible to analyse the domain origins of messages, and assess the involvement of different groups of individuals. As an example, Table 2 shows the top nine email domains for contributed messages in the developer mailing list over time. Each row represents a six-month time window starting at the month specified in the leftmost column. It can be observed in this case that the "com" domain is most dominant, followed by "org" and "se" when studying all messages accumulated from January 2004 to September 2009.

As an example of how the top domains can be analysed in more detail, Table 3 shows the top five subdomains for the top five domains accumulated for all windows over time. It can be noted that for several of the top domains there are some very dominant subdomains like "nagios" for the "org" domain and "op5" for the "se" domain.

**Table 3.** Top five subdomains for the accumulated top five top domains in the Nagios developer mailing list

| Top dom | #1 | #2 | #3 | #4 | #5 |
|---|---|---|---|---|---|
| com (34) | gmail(8) | opservices(2) | zango(2) | altinity(1) | ena(1) |
| org(15) | nagios(11) | ldschurch(2) | clewett(<1) | gmane(<1) | debian(<1) |
| se(12) | op5(12) | forumsql(<1) | cendio(<1) | dokad(<1) | iis(<1) |
| de(12) | process-zero(4) | netways(1) | gmx(<1) | consol(<1) | ederdrom(<1) |
| net(6) | seanius(1) | lordsfam(<1) | gmx(<1) | netshel(<1) | elan(<1) |

Similarly, it is possible to derive domain trees using the email addresses in the messages for a mailing list. This can be particularly informative if there are several levels of subdomains in an email address.

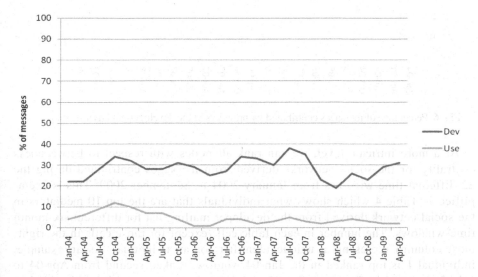

**Fig. 5.** Percentage of messages contributed by the core developers for the developer- and user mailing lists of Nagios

The availability of mailing list data containing domain information makes it is possible to assess involvement of a certain group of stakeholders in an OSS ecosystem. As an example of this, Figure 5 shows the percentage of messages contributed over time by the group of explicitly listed core developers in the Nagios project. The results show that core developers in this project communicate more on development related issues than on user related issues, and that they with some fluctuation on average contribute to about one third of the development related correspondence.

Another specific group of stakeholders are the providers. The percentage of messages contributed by providers for Nagios is illustrated in Figure 6. It is evident that the providers contribute more to the developer mailing list, and it can also be noted that the shape of the curves is quite similar (Pearson correlation of 0.59). This is partly due to the fact that one individual of the most influential provider (OP5), who contributes to both lists, is the most active contributor in the set of providers.

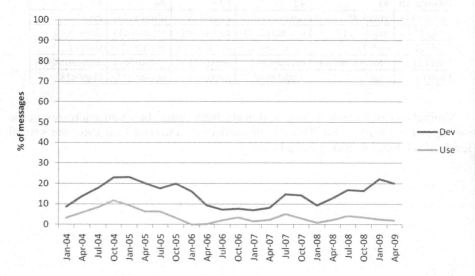

**Fig. 6.** Percentage of messages contributed by providers in the developer- and user mailing lists

At a more intricate level, we can rank all nodes with respect to betweenness centrality for the social networks derived from messages contributed during the 22 different time windows from January 2004 to September 2009. This is exemplified in Table 4 which shows what individuals that are the top 10 mediators in the social network derived from the developer mailing list for different six-month time windows. The number in each table cell (except for those cells in the rightmost column) represents a unique identifier for an individual. As an example, individual 1 is top ranked in the Jan-04 window, ranked second from Apr-04 to Jul-05, and ranked first from Oct-05 to Jan-08. It can be argued that this kind of table reflects the sustainability of the community that actively participates in mailing lists. The rightmost column shows the number of new mediators in the top 10 of a window, which have not been in the top 10 before in the table. One general observation is that there is a constant addition of new mediators over time and that there are only a few individuals that contribute over all time windows (mediator 1 and 3). Mediators at lower ranks (towards 10) also tend to exhibit lower endurance over time. For example, mediator 12 is ranked as the 8th most important mediator for the two time windows starting in April 2004 and January 2005, and is thereafter not amongst the ten most important mediators.

**Table 4.** Mediator index at ranks 1 to 10 for the developer mailing list

| Win | #1 | #2 | #3 | #4 | #5 | #6 | #7 | #8 | #9 | #10 | $N_{new}$ |
|---|---|---|---|---|---|---|---|---|---|---|---|
| Jan-04 | 1 | 2 | 3 | 4 | 5 | 6 | 7 | 8 | 9 | 10 | 10 |
| Apr-04 | 3 | 1 | 6 | 10 | 11 | 5 | 4 | 12 | 13 | 14 | 4 |
| Jul-04 | 3 | 1 | 11 | 15 | 6 | 10 | 5 | 16 | 13 | 17 | 3 |
| Oct-04 | 3 | 1 | 11 | 15 | 16 | 18 | 17 | 19 | 5 | 20 | 3 |
| Jan-05 | 3 | 1 | 7 | 11 | 18 | 20 | 15 | 12 | 5 | 21 | 1 |
| Apr-05 | 3 | 1 | 7 | 5 | 20 | 22 | 23 | 18 | 24 | 21 | 3 |
| Jul-05 | 3 | 1 | 25 | 22 | 23 | 26 | 27 | 28 | 24 | 29 | 5 |
| Oct-05 | 1 | 3 | 20 | 30 | 31 | 32 | 33 | 26 | 34 | 29 | 5 |
| Jan-06 | 1 | 3 | 20 | 11 | 35 | 31 | 33 | 36 | 37 | 32 | 3 |
| Apr-06 | 1 | 3 | 38 | 39 | 40 | 36 | 11 | 41 | 31 | 42 | 5 |
| Jul-06 | 1 | 43 | 31 | 24 | 3 | 40 | 44 | 45 | 46 | 38 | 4 |
| Oct-06 | 1 | 3 | 44 | 47 | 45 | 46 | 37 | 33 | 48 | 49 | 3 |
| Jan-07 | 1 | 33 | 3 | 45 | 44 | 47 | 50 | 37 | 51 | 31 | 2 |
| Apr-07 | 1 | 33 | 3 | 45 | 52 | 50 | 48 | 53 | 54 | 31 | 3 |
| Jul-07 | 1 | 3 | 33 | 52 | 45 | 55 | 24 | 56 | 57 | 53 | 3 |
| Oct-07 | 1 | 3 | 33 | 45 | 55 | 24 | 58 | 59 | 44 | 60 | 3 |
| Jan-08 | 1 | 3 | 33 | 45 | 61 | 24 | 62 | 31 | 63 | 59 | 3 |
| Apr-08 | 3 | 1 | 33 | 45 | 47 | 61 | 44 | 16 | 31 | 64 | 1 |
| Jul-08 | 3 | 33 | 1 | 45 | 64 | 44 | 16 | 58 | 31 | 65 | 1 |
| Oct-08 | 3 | 45 | 33 | 1 | 61 | 66 | 31 | 65 | 58 | 67 | 2 |
| Jan-09 | 3 | 33 | 45 | 1 | 65 | 68 | 61 | 69 | 70 | 63 | 3 |
| Apr-09 | 3 | 1 | 33 | 71 | 65 | 44 | 72 | 73 | 74 | 16 | 4 |

When studying a particular group of stakeholders and the ability to act as a mediator, it may be better to study the ranking of each of the individuals in the group. Table 5 illustrates this for the six core developers in the developer mailing list over time. Like in Table 4, this kind of information gives an indication of sustainability, but here with respect to a particular group of interest. In this particular case it can for example be noted that CD1 and CD2 have been very important mediators in the developer mailing list community since it started, and that CD4 has become equally important since January 2007. Results of this kind can also be derived for the provider list.

**Table 5.** Mediator rank for the six core developers (CD1-CD6) in the developer mailing list. N denotes total number of contributors in a specific time window.

| Win | N | CD1 | CD2 | CD3 | CD4 | CD5 | CD6 |
|---|---|---|---|---|---|---|---|
| Jan-04 | 69 | 1 | 3 | 57 | - | - | - |
| Apr-04 | 76 | 2 | 1 | 66 | - | - | - |
| Jul-04 | 83 | 2 | 1 | - | - | - | - |
| Oct-04 | 93 | 2 | 1 | - | - | - | - |
| Jan-05 | 88 | 2 | 1 | - | - | - | - |
| Apr-05 | 74 | 2 | 1 | - | - | - | - |
| Jul-05 | 71 | 2 | 1 | - | - | - | - |
| Oct-05 | 96 | 1 | 2 | 25 | 7 | - | - |
| Jan-06 | 104 | 1 | 2 | 25 | 7 | - | - |
| Apr-06 | 87 | 1 | 2 | 14 | 18 | - | - |
| Jul-06 | 69 | 1 | 5 | 7 | 16 | - | - |
| Oct-06 | 84 | 1 | 2 | 3 | 8 | - | - |
| Jan-07 | 112 | 1 | 3 | 5 | 2 | 9 | - |
| Apr-07 | 121 | 1 | 3 | 13 | 2 | 26 | - |
| Jul-07 | 131 | 1 | 2 | 39 | 3 | 126 | - |
| Oct-07 | 135 | 1 | 2 | 9 | 3 | - | 125 |
| Jan-08 | 118 | 1 | 2 | 11 | 3 | - | 114 |
| Apr-08 | 108 | 2 | 1 | 7 | 3 | 104 | - |
| Jul-08 | 116 | 3 | 1 | 6 | 2 | 108 | - |
| Oct-08 | 114 | 4 | 1 | 17 | 3 | 106 | - |
| Jan-09 | 105 | 4 | 1 | 16 | 2 | 41 | 25 |
| Apr-09 | 111 | 2 | 1 | 6 | 3 | 35 | 20 |

## 5   Conclusion and Discussion

In this paper we have presented an extended quantitative analysis of the Nagios community by performing domain- and social network analysis of mailing lists. It can be concluded from our study that there is an increasing ability to develop useful approaches and metrics to cover the broader aspects of the OSS ecosystem. This means that we have shown that not only the user and developer role can be analysed, but also the provider role. This is an important contribution that can offer a more comprehensive understanding of professional involvement in professional Open Source.

Specifically, the way domain analysis has been performed in the context of mailing lists contains elements of novelty. We have proposed approaches for showing how a specific group of stakeholders contribute to a mailing list community. In particular, we have contributed approaches for analysing the sustainability of a mailing list community or a specific group of stakeholders in a community in terms of mediation of information when studying social networks derived from mailing list data over time.

One limitation of the study is that no account was taken of the content of email responses, so that each response was equally weighted. A clearer indication of the nature and value of responses would add value to such analyses, but would require techniques beyond the scope of this study.

Since quantitative research on OSS to a large extent is data driven, the results can only be as informative as the available data sources permit. OSS projects that would like their projects to be taken up more broadly should consider providing as informative data as possible. In fact, the community as a whole would benefit from a continuous joint effort to make available informative data sources in order to be able to cover broader aspects of the OSS ecosystem.

In healthy Open Source communities people are active and responsive to questions during the life cycle of a software system. It is important to consider such indicators of health prior to any organisational adoption or during the seeding of a community. The kind of analyses elaborated in this paper serve as important means for establishing the health of an Open Source community.

We are currently working on developing approaches for obtaining valuable information about professional involvement in Open Source communities. Such information is essential for any professional organisation wishing to better understand the broader Open Source ecosystem.

## Acknowledgement

The authors would like to thank their colleagues in the OSA-project for their encouragement and support. The OSA-project is financially supported by the Knowledge Foundation ("KK-stiftelsen").

## References

Capra, E., Francalanci, C., Merlo, F.: A Survey on Firms' Participation in Open Source Community Projects. In: Proceedings of the Fourth Conference on Open Source Systems (OSS 2009), pp. 225–236 (2009)

Crowston, K., Howison, J.: The social structure of Free and Open Source software development. First Monday 10(2) (2005)

Crowston, K., Howison, J.: Assessing the Health of Open Source Communities. IEEE Computer 39(5), 89–91 (2006)

Crowston, K., Wei, K., Li, Q., Howison, J.: Core and Periphery in free/Libre and Open Source software team communications. In: Proceedings of the 39th Hawaii International Conference on System Sciences, p. 118.1(2006)

Fitzgerald, B.: The Transformation of Open Source Software. MIS Quarterly 30(3) (2006)

Gamalielsson, J., Lundell, B., Lings, B.: Social Network Analysis of the Nagios project. In: Open Source Workshop (OSW 2009), Skövde, Sweden, October 15-16 (2009)

German, D.: The GNOME project: a case study of open source global software development. Journal of Software Process: Improvement and Practice 8(4), 201–215 (2004)

Ghosh, R.A.: Study on the: Economic impact of open source software on innovation and the competitiveness of the Information and Communication Technologies (ICT) sector in the EU (2006), http://ec.europa.eu/enterprise/sectors/ict/files/2006-11-20-flossimpact_en.pdf (accessed December 23, 2009)

Kamei, Y., Matsumoto, S., Maeshima, H., Onishi, Y., Ohira, M., Matsumoto, K.: Analysis of Coordination Between Developers and Users in the Apache Community. In: Proceedings of the Fourth Conference on Open Source Systems (OSS 2008), pp. 81–92 (2008) van der Linden, F., Lundell, B., Marttiin, P.: Commodification of Industrial Software: A Case for Open Source. IEEE Software 26(4), 77–83 (2009)

Lopez-Fernandez, L., Robles, G., Gonzalez-Barahona, J.M., Herraiz, I.: Applying Social Network Analysis Techniques to Community-driven Libre Software Projects. International Journal of Information Technology and Web Engineering 1, 27–48 (2006)

Lundell, B.: Being Open about the ten Open Source myths: Myth #2 is still alive! Lightning talk presented at OpenMind 2009, Tampere, Finland, September 30 (2009)

Lundell, B., Forssten, B., Gamalielsson, J., Gustavsson, H., Karlsson, R., Lennerholt, C., Lings, B., Mattsson, A., Olsson, E.: Exploring Health within OSS Ecosystems. In: First International Workshop on Building Sustainable Open Source Communities (OSCOMM 2009), Skövde, Sweden, June 6 (2009)

Martinez-Romo, J., Robles, G., Ortuño-Perez, M., Gonzalez-Barahona, J.M.: Using Social Network Analysis Techniques to Study Collaboration between a FLOSS Community and a Company. In: Proceedings of the Fourth Conference on Open Source Systems (OSS 2008), pp. 171–186 (2008)

Mockus, A., Fielding, R.T., Herbsleb, J.D.: Two case studies of open source software development: Apache and Mozilla. ACM Transactions on Software Engineering and Methodology 11(3), 309–346 (2002)

Moon, Y.J., Sproull, L.: Essence of distributed work: The case of the Linux kernel. First Monday 5(11) (2000)

O'Reilly, T. Ten myths about open source software. Transcript of talk given to a group of Fortune 500 executives (October 1999), http://www.oreillynet.com/lpt/a/2019 (accessed December 22, 2009)

Toland, C., Meenan, C., Warnock, M., Nagy, P.: Proactively Monitoring Departmental Clinical IT Systems with an Open Source Availability System. Journal of Digital Imaging 20, 119–124 (2007)

Watson, R.T., Wynn, D., Boudreau, M.-C.: Jboss: The Evolution of Professional Open Source Software. MIS Quarterly Executive 4(3), 329–341 (2005)

# Collaborative Development for the XO Laptop: CODEX 2

Andrew Garbett, Karl Lieser, and Cornelia Boldyreff

Centre of Research on Open Source Software – CROSS
University of Lincoln, UK

**Abstract.** At the University of Lincoln, undergraduate students are given the opportunity to take part in an Undergraduate Research Opportunities Scheme (UROS), which allows students to not only contribute to a particular field of research but also to enrich their knowledge and understanding of the chosen research topic. The Centre of Research for Open Source Software (CROSS) within the School of Computer Science at the University of Lincoln offered UROS students the opportunity to research into the Collaborative Development for the XO Laptop (CODEX). The aim of this project is to provide an easily accessible Open Source platform within which students are able to develop activities for the One Laptop Per Child (OLPC) XO-1 laptop, as well as create Open Source applications and contribute to the OS Community. Under the supervision of Professor Cornelia Boldyreff, two students (Andrew Garbett and Karl Lieser) were tasked with continuing the initial research and development undertaken by James Munro on the original CODEX project in the previous year. The resulting CODEX 2 project has managed to produce sufficient tutorial materials for new students to begin development for the XO laptop and more specifically for its Sugar interface utilising a development environment that met the requirements of the project and thus resulting in a successful research outcome.

## 1 Introduction

The One Laptop Per Child (OLPC) scheme aims to "create educational opportunities for the world's poorest children by providing each child with a rugged, low-cost, low-power, connected laptop with content and software designed for collaborative, joyful, self empowered learning. When children have access to this type of tool they get engaged in their own education. They learn, share, create, and collaborate. They become connected to each other, to the world and to a brighter future." [14] In order to provide the learning experiences for the users of the XO 1 laptop, developers are required to contribute their own time and effort to create useful, educational and enjoyable applications; and Computer Science students have been identified as an excellent source of volunteer developer effort.

The first One Laptop Per Child XO 1 laptop used an Open Source distribution of the Linux operating system known as Fedora [11] which was coupled with Sugar [15], a Graphical User Interface, with the aim to be as multilingual as possible through the use of illustrative graphical icons rather than text. In doing so the XO

P. Ågerfalk et al. (Eds.): OSS 2010, IFIP AICT 319, pp. 97–104, 2010.

User Interface does not conform to modern operating system desktop norms; most notably, Sugar does not contain a taskbar or 'start menu' but rather different levels at which the user is able to interact with not only their own applications but also collaboratively with the applications of other Sugar users on-line. The interface uses activity based operations rather than a tree structured collection of documents and applications and can be seen as quite esoteric at first. However, Sugar developers insist that the interface encourages all its users to learn through experiences rather than prior knowledge of similar products [15]. Therefore developers hope that likeminded individuals will create activities that will continue the learning process and aid to a child's educational experience.

In order to encourage further application development for the XO 1 laptop and Sugar, the Centre of Research for Open Source Software (CROSS) at the University of Lincoln has accepted the challenge and has been working to offer student developers within the university the opportunity to contribute to the worldwide Open Source community that has grown around the OLPC project. The availability of this offer is provided by the university's Undergraduate Research Opportunities Scheme (UROS) where undergraduate students are provided with the opportunity to gain an understanding of the processes and activities entailed when performing a research project. The UROS research scheme is normally undertaken by students who will be commencing their third year of study and carried out over the summer period under the supervision of a member of academic staff at the University. The students are usually "embedded" with an existing research group and work alongside other researchers in the group.

The CROSS was granted funding to begin a student-led project known as Collaborative Development for the XO Laptop or "CODEX" [5] and it was initialised with the aim of bringing an easily accessible development environment to the students at the University of Lincoln in the hope that they would also become a part of and contribute to the OS community.

The initial research was undertaken by James Munro [1] at the University of Lincoln as part of UROS, under the supervision of Professor Cornelia Boldyreff. The CODEX project produced a Live CD that contained Sugar development software as well as a wiki [6] which contained tutorial information about the setup processes involved and other related resources including a report to the community at the Fifth International Conference on Open Source Systems held during June 2009 [4].

The success of the CODEX project allowed for a continuation project, aptly named "CODEX 2", where students in the following year could contribute by carrying forth the original stated aims and building upon the work that had previously been completed. The CODEX 2 project was undertaken by Andrew Garbett and Karl Lieser, with supervision provided again by Professor Cornelia Boldyreff.

The following discussion describes the aims of the CODEX 2 project (Section 2), the evolution of objectives that occurred as it was carried out, and the accomplishments that were made (Section 3). A comparison of the accomplishments is provided thereafter (Section 4) with further work being indicated (Section 5) and finally a conclusion provided (Section 6).

## 2  Project Aims

The principle aims of CODEX 2 project, as initially outlined in the project brief which was provided to students, were to follow on from the objectives of the previous CODEX 1 project. These objectives were to investigate the OLPC and Sugar projects in order to develop methods by which student developers and the outside community alike, could contribute towards developing activities, i.e. applications for the Sugar interface. These methods would be supported by producing a suitable development environment, accompanying documentation and tutorial content that could be used in aid of activity development. Throughout both CODEX projects, the overall aims have remained the same and are as follows:

1. Identify a suitable development environment for Sugar activities
2. Modify and develop a development environment for student use
3. Produce tutorial content and documentation as a student aid
4. Develop example activities using the environment

Within the CODEX 2 project, there were a number of initial sub-goals identified in order to build on the previous work completed in CODEX 1. These were to update the Live CD which had already been produced, via providing a more up to date software distribution and more suitable means for publication via a LiveUSB replacing the LiveCD (Aims 1 & 2); and to use the software provided on the LiveUSB to develop further activities for Sugar and then produce tutorial content and documentation on how to do so (Aims 3 & 4).

## 3  Project Process and Accomplishments

As proposed and abstracted from a brief provided to the students, the CODEX 2 project was envisaged to involve three main phases, these being:

- Investigation phase, whereby background research on all areas of the project could be accomplished, including OLPC, XO Sugar Software and results achieved in previous CODEX project
- Development phase, allowing the CODEX LiveCD to be updated into a more enhanced version on a more suitable medium, LiveUSB
- Trial and Evaluation phase, where by applications could be developed using the deliverable from the previous phase, and production and updating of the current tutorial guide and documentation could take place

Many of the initial presumptions from the brief, which further detailed the outlined continuation of the original CODEX project, were soon found to be unwarranted once the research commenced. The key change that is noted from the main phases of the project above was the discovery of work relating to the development of software which already allowed implementation of LiveUSB type mediums; these being "Sugar-on-a-Stick" and "Fedora Edu Sig" which are discussed later.

Details of how the research actually progressed was recorded in the blogs that were kept by the two CODEX 2 student researchers. This was accomplished through the use of Lincoln University Blogging system [2] [3]. These have provided a public record of the research and development of the project, thus producing a useful resource for future participants in the CODEX project.

Collaborative software technology has also been used in order to consolidate results from CODEX 1 and CODEX 2, and to manage information during the production of the tutorial content and further documentation. These came in the form of the project Wiki [9] which allowed for the information produced within the blogs and via research to be filtered. Additionally an online repository Git Hub [12] was used for storage of documentation and produced software; such as activities that could be managed and stored.

Before the start of the CODEX 2 project, the student researchers' knowledge in the area of open source operating systems and subject areas of the project was limited, and so during the start of the research, time was spent investigating and learning within four main areas:

1.  Understanding and using Linux; in particular Ubuntu and Fedora
2.  Understanding and learning to script in the Python Programming Language
3.  Testing the required software for emulation and porting operating systems to USB devices
4.  Investigating CODEX 1, OLPC & Sugar projects, and other background research

In addition to the initial research into the above areas, both researchers made contact with the online community, thus creating working ties whereby additional help and support could be obtained in the later stages of the research. It soon became clear after the initial investigations began that the OLPC and Sugar projects had made substantial progress from the previous summer's research reported on the original CODEX 1 project, and that much more information and online community presence was now available.

A number of discoveries were made during research and contact with the online community which meant that there were already materials available to begin developing a LiveUSB. The first discovery was the release of "Sugar on a Stick" [15], which officially allowed the online community to download and create a version of the Sugar software on a USB storage device.

The second important discovery was finding that Sugar had been included in the Fedora Linux distribution which allowed users to directly download Sugar through the Fedora Package Management System. Finally the Sugar Project had made substantial progress with their online presence, providing an abundance of information on all aspects of Activity and Sugar Interface Development, and availability of "Sugar JHBuild", a repository of the latest Sugar Interface source code.

These discoveries meant that some of the initial CODEX 1 project outputs had now become obsolete; such as the Ubuntu LiveCD. It could be replaced by either "Sugar on a stick" or a Fedora installation with Sugar installed, which could also be ported to a USB drive with ease. The research focus now shifted due to the option of obtaining a built version of Sugar or its latest released source code online, coupled with the availability of an abundance of tutorial content in helping to develop for Sugar. Attention was now being directed to a related and complementary project, as proposed by

**Fig. 1.** Fedora Edu Sig Logo

some of the online Fedora / Sugar development community, called "Fedora Edu Sig" [10]. The project logo is displayed in Fig.1.

The "Fedora Edu Sig" project is an educational "spin" of the Fedora project where software is being developed, which includes Sugar pre-installed for easy use to port to USB devices (Fig.2.); along with developing tutorial information on how to download and install. These goals fell in line with the objectives outlined in the CODEX projects; and so the remaining time was spent helping the current developers of the "Fedora Edu Sig" to build, port and test the software in preparation for its final release.

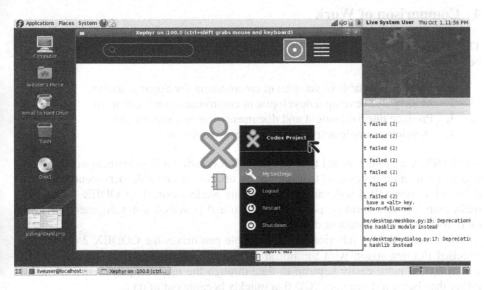

**Fig. 2.** Edu Spin Emulating Sugar

Although much of the information on the areas covered thus far is available online, time was needed to locate and compile information into a meaningful resource. There was no single central on-line source that could point student researchers and developers who had objectives such as the CODEX 2 project in the right direction. Although tutorial content for activity developers was located, some of the current tutorial content already developed needed more explanation for entry level student developers, and thus it was proposed that a primary output of the CODEX 2 project would be the development of a knowledge base or "gateway" that student developers could use to find all relevant links. This would be accompanied with the production of short tutorials that would complement existing information to allow beginner users, such as new student researchers, to understand and start developing in a shorter time frame. These new outputs now form the base of the content in the CODEX 2 section of the project Wiki, and additional content in Git Hub, with each development being recorded in the project blogs by each student researcher.

The CODEX 2 project team were invited to attend the Summer 09 Open Source Schools Nottingham 'Unconference' [13] which allowed the student researchers to not only participate in talks, but also to present materials, raise awareness and demonstrate their knowledge learnt about the OLPC, Sugar Interface and Sugar on a Stick projects by helping in the Sugar (OLPC) presentation. Both student researchers participated in an impromptu video interview by other students at the Unconference [8]. The student researchers also produced a project poster for distribution at the Unconference [7].

This attendance gave the student researchers an opportunity to engage in dissemination activities normally associated with academic research and gain feedback from a wider potential user community in schools.

## 4    Comparison of Work

Throughout the CODEX project lifecycle the aims have remained the same and have been as follows:

1. Identify a suitable development environment for Sugar activities
2. Modify and develop a development environment for student use
3. Produce tutorial content and documentation as a student aid
4. Develop example activities using the environment

The CODEX 2 project aimed to take the idea of creating a development environment and make it more accessible to students by offering a bootable persistent USB pen drive with preinstalled software. Much like its predecessor, the CODEX 2 project's main output took a form of portable medium and provided a development environment for the potential student developers.

However, with new addition of the bootable pen drives the CODEX 2 project has excelled in comparison with the initial project. This is due to the 'Edu-spin' USB which can be more easily kept up to date through the use of a persistent USB drive rather than being a 'burn once' CD that quickly became out of date.

Other benefits of the persistent storage include the ability to store the student user's files and being able to transport them with the development tools available. This is arguably the biggest improvement yet as it allows truly portable development from almost any computer. Another aspect of availability is the ability to install the environment on different sized pen drives; as well as the ease at which the user is able to do so, this allows developers to download additional software and install it within the operating environment. As almost all modern computers used by students have USB ports, the development of the persistent pen drive environment has meant that the students are able to utilise the Sugar development environment with most computers rather than having to ensure that their machine had a CD drive to boot up the original CODEX 1 disk.

A disadvantage of having USB driven development is that there are some mother-boards that do not allow the user to boot from USB, although this may be overcome through the use of "CD Loaders" which recognise the USB device for the system. It has also been found impossible to boot from USB on some networked computers where the administrator has locked the BIOS settings thus disallowing the user to se-lect 'Boot from USB', although this too may be overcome through communication with network administrators. A further concern with the Edu-spin approach is that the environment had been cut down significantly with regards to non-essential software and driver files. This stripping of non-essential files may interfere with some student user's software and hardware requirements and thus reduce the effectiveness of the USB by not including certain code libraries and drivers. Although USB drives are limited in size, there is a good improvement on the storage space available when compared to the original Live CD produced in the CODEX 1 project.

## 5  Further Work

Further development for the CODEX project may include a deployment plan in order to provide the students of The University of Lincoln with all the resources required to obtain the development environment. This may include the physical aspect of de-ployment such as looking into the distribution of the necessary software from a cen-tral or multiple locations. Other aspects concerning the deployment are areas such as the rebuilding and updating of the development environment ensuring that software, bookmarks and tutorial materials are relevant.

Since USB drives are expected to increase in size, users will soon be able to store many more files on their drives. With this being the case, a future investigation could consider any other software and files that are required to extend the effectiveness of the current distribution.

## 6  Conclusions

The CODEX 2 project has been successful in meeting its project aims; and most im-portantly, the student researchers undertaking the UROS project have gained an in-sight into collaborative development within the wider Open Source community. Alongside this, the students have also been given the opportunity to work with Open Source tools and environments, and have had the chance to learn new skills such as python development and wiki editing.

The students also attended an Open Source conference and had the chance to listen to new and exciting ideas from Open Source developers from the open source and education communities. Whilst undertaking the CODEX 2 project, all research mate-rials and ideas have been recorded on two separate blogs which show the progress of the project.

The project managed to produce an abundance of tutorial materials for new students to use when they begin developing activities for the XO laptop as well as providing students with an Open Source development environment which can be in-stalled upon a persistent USB pen drive. The CODEX project will definitely provide other students with interesting challenges and topics to research in the near future.

# References

[1] Blog JMunro (2008), http://james.blogs.lincoln.ac.uk/?s=CODEX [Checked September 2009]

[2] Blog KLieser (2009), http://CODEX2project.blogs.lincoln.ac.uk/ [Checked September 2009]

[3] Blog AGarbett (2009), http://andygarbett.blogs.lincoln.ac.uk/ [Checked September 2009]

[4] Boldyreff, et al.: Undergraduate Research Opportunities in OSS. In: Open Source Ecosystems: Diverse Communities Interacting, vol. 299, pp. 340–350. Springer, Boston (2009), http://www.springerlink.com/content/b54858r345437548 (Checked September 2009)

[5] CODEX Project (2008), http://learninglab.lincoln.ac.uk/wiki/Collaborative_Developm ent_for_the_XO-1_laptop_%28CODEX%29 (Checked September 2009)

[6] CODEX Wiki (2008), http://learninglab.lincoln.ac.uk/wiki/OLPC_XO-1#2008_CODEX_1_Project (Checked September 2009)

[7] CODEX 2 Poster (2009), http://tinyurl.com/CODEX2-Poster (Checked September 2009)

[8] CODEX 2 Video (2009), http://vle.hamblecollege.co.uk/course/view.php?id=1137 (Checked September 2009)

[9] CODEX 2 Wiki (2009), http://learninglab.lincoln.ac.uk/wiki/OLPC_XO-1#2009_CODEX_2_Project (Checked September 2009)

[10] Edu. Sig. (2009), https://fedoraproject.org/wiki/SIGs/Education (Checked September 2009)

[11] Fedora Project (2009), http://fedoraproject.org/ [Checked September 2009]

[12] GitHub Repo. (2009), http://github.com/Kodex/CODEX-tutorial-content (Checked September 2009)

[13] OSS Unconference (2009), http://opensourceschools.org.uk/unconference-programme.html (Checked September 2009)

[14] OLPC (2009), http://laptop.org/en/vision/index.shtml (Checked September 2009)

[15] Sugar Project (2009), http://www.sugarlabs.org/ (Checked September 2009)

# Risks and Risk Mitigation in Open Source Software Adoption: Bridging the Gap between Literature and Practice

Øyvind Hauge, Daniela Soares Cruzes, Reidar Conradi, Ketil Sandanger Velle, and Tron André Skarpenes

Norwegian University of Science and Technology
{oyvh,dcruzes,conradi,ketilsan,skarpenes}@idi.ntnu.no

**Abstract.** The possible benefits of open source software (OSS) have led organizations into adopting a variety of OSS products. However, the risks related to such an adoption, and how to reduce these risks, are not well understood. Based on data from interviews, a questionnaire, and workshops, this paper reports ongoing work in a multi-national telecom company. The paper has three main contributions. First, it identifies and discusses several risks related to OSS adoption. Second, it identifies steps for reducing several of these risks. Third, it shows how research can be used to increase the visibility of, and involve the employees in, ongoing OSS efforts.

## 1 Introduction

The promise of reduced costs, increased flexibility, and independence from vendors of proprietary products has convinced organizations worldwide into deploying open source software (OSS) products in their production environments and integrating OSS components into their software systems [15,16,19,20]. While a couple of studies have looked at the benefits and drawbacks of such OSS adoption [2,24,35], few have discussed steps for dealing with related risks.

Our primary goal is to identify relevant risks and risk mitigation steps for organizations that adopt OSS products. The secondary goal of the study presented here is to explore the opportunities for increasing organizations' adoption of OSS. This includes identifying potential benefits of an increased OSS adoption. The main research questions investigated in this study are:

RQ1. What are the perceived benefits of an increased adoption of OSS products?

RQ2. What are the perceived risks of such an adoption?

RQ3. What steps may organizations take to reduce these risks?

The study presented in this paper was partially conducted at Telenor, a large international telecommunications company. Telenor's Norwegian IT division has already adopted some OSS products, but it is currently looking into increasing its adoption. However, to avoid the possible pitfalls of OSS adoption, Telenor IT wanted to identify (1) the benefits s and risks which are relevant to their context and (2) how to deal with potential risks. To support Telenor in finding the answers to these questions, we

P. Ågerfalk et al. (Eds.): OSS 2010, IFIP AICT 319, pp. 105–118, 2010.

have conducted a study consisting of semi-structured interviews, a questionnaire with 86 responses, and three workshops.

## 2  Related Literature

OSS can be adopted in different ways. In [19] we show that OSS can be adopted through deploying OSS products, using OSS CASE tools, integrating OSS components, participating in the development of OSS products, providing OSS products, or through using OSS development practices. Grand et al. [18] present a four level model for resource allocation to OSS. In a company perspective, the four levels are (1) company as a user of OSS software, (2) OSS software as complementary asset, (3) OSS software as a design choice, and (4) OSS compatible business mode. This paper focuses on the **deployment of OSS products** (like operating systems, database servers, application servers etc) within in a company at **level 1 or 2** in Grand et al.'s model for resource allocation. The following sections are mainly based on a systematic literature review focusing on OSS adoption [19].

### 2.1  Possible Benefits of OSS

The literature discusses several possible benefits (**B**) of OSS adoption. However, some of these benefits may be perceived as drawbacks as well [35]. Cost cuts (B1) are, for instance, frequently mentioned as a benefit of OSS adoption, while hidden costs (R1) are mentioned as a risk.

**Cost cuts (B1):** OSS has been claimed to enable costs cuts through for instance reduced license fees, hardware requirement, scaling costs, etc. [2,14,24,33].

**Independence from vendors of proprietary products (B2):** The adopter of OSS may also get increased freedom from vendor lock-in and increased influence on providers of both proprietary and OSS products [2,5,6,23,24].

**Simplified procurement and license management (B3):** The majority of OSS products tend to come with only a few different licenses and without licensing fees. This may simplify the procurement of the software and the licensing of derivative products [23,28].

**Software reuse (B4):** Through adopting software products that are developed, tested, and used by others, we may gain the benefits of software reuse. This includes extra/new functionality, increased R&D and innovation, improved quality (e.g. reliability, security, performance, defect density) and increased productivity [2,5,6,14,25,34]. OSS may also contribute to increased standardization [1,21] or to establishing de-facto standards if no standards exist [23].

**High availability (B5):** OSS products are most often easily available together with source code and trustworthy information about the products' true state [22,23,35].

**Community support (B6):** This openness may lead to increased collaboration between community members [2,24]. The community might also provide free maintenance and upgrades of the software together with user support [5,24,33].

## 2.2  Potential Risks of OSS Adoption

There are also risks (**R**) related to adopting OSS but not all organizations consider them, as there are organizations adopting OSS without performing any cost/benefit analysis [35]. There are no papers that explicit focus on potential risks of OSS adoption, but the literature mentions several possible drawbacks of OSS adoption.

**Hidden costs (R1):** Adoption of OSS products is not without costs: It may be time-consuming to evaluate them [31]. Adoption may involve user training and configuration [24,31]. We might need to spend resources on community participation [23]. Many organizations would need premium professional support [14,35].

**Lack of products (R2):** While there are many OSS products available, there may still be a lack of products with specific functionality [6,24]. The quality of these products can also be questionable [14,35]. OSS products may furthermore suffer from limited standardization and compatibility with document formats or with versions of other software products [2,24,25].

**Lack of providers, expertise, and support (R3):** Despite the significant adoption of OSS, there may still be a lack of expertise and support for specific products [2,24,31,35]. The lack of professional providers may also introduce unclear liability and uncertainty about the longevity of OSS project as OSS projects may lack roadmaps and documentation [2,24]. Holck et al. hypothesized that this lack of traditional vendor-customer relationship could stop the adoption of OSS [22].

**Customization needs (R4):** It may be necessary to customize the OSS products to fit them into the context in which they are going to be used [1]. When changing an OSS product we may get a maintenance responsibility [36] as these changes must be updated when more recent versions of the software are adopted. When these situations arise, the adopter must decide to follow the new releases or ensure backward compatibility with his own changes [23].

**Licensing issues (R5):** The variety of OSS licenses available is confusing, as there is a lack of guidance on how to interpret them [31]. When adopting OSS and when integrating it into derivative software systems, it may be challenging to combine code under an OSS licenses with proprietary licenses and APIs [23].

## 2.3  Risk Mitigation in OSS Adoption

As there are few publications discussing risk mitigation (**RM**), this section describes literature on success criteria and enabling versus inhibiting factors of OSS adoption. Some of these issues may contribute to reduce the risks of OSS adoption.

**Employee attitude, awareness, and skills (RM1):** A positive employee attitude towards OSS and the OSS ideology can enforce the adoption of it [4,5,16,25,34]. The adoption of OSS should also be made visible, so that end users have an awareness of the technology adoption [6]. Finally, if employees have the necessary skills and experience with OSS, the probability of a successful adoption will increase [6,16,32].

**Management support (RM2):** Management support is also important for OSS adoption [6,14,16]. Management should provide resources for driving the adoption, and a clear plan for analysis, testing, and pilot projects [6, 13, 16].

**Access to support (RM3):** The quality of the OSS components [29] is an important factor for a successful adoption, and for many adopters it is also necessary to have access to professional support [6,13].

**Success stories (RM4):** It is furthermore an advantage if the products have been successfully adopted by other companies [16]. Lack of such success stories or lack of other users could easily complicate the OSS adoption [2,4].

**No lock-in (RM5):** An organization may have a hard time adopting OSS if they are locked in by industry-wide purchasing agreements and standards for IT or already have a coherent stable IT infrastructure based on proprietary or legacy technology [6,16]. The costs related to moving away from such a lock-in situation can significantly impede the adoption of OSS [4,17,35].

## 3   Context and Research Method

Telenor is currently among the ten largest mobile operators in the world Telenor Norway IT (hereafter Telenor IT) is the information technology and software support division under Telenor's Norwegian branch. It has about 380 employees that together with several external partners develop, maintain, and support more than 500 IT systems. Open Source 2010 is an Telenor IT project aiming at exploring the opportunities of adoption of OSS products like databases, application servers etc.

Telenor IT's motivation for conducting this study was threefold. First, Telenor IT wanted to increase the awareness of OSS and its Open Source 2010 project within the organization. Second, it wanted to get feedback from and involve the employees in the project's work. Third, it wanted to weaken the grip providers of proprietary products have on Telenor, and reduce Telenor's expenses on licensing and support.

This study consisted of four main steps. As part of a **systematic literature review** on OSS in organizations, we reviewed the literature for evidence on the perceived benefits and drawbacks of OSS [19]. Some of the output from this review was used as an input for Section 2. Next, we conducted **semi-structured interviews** with four employees. The respondents had different positions (developer, system specialist, chief engineer, and manager) within different parts of Telenor IT (mobile, landlines, data warehouse & business intelligence, and customer relation management). They had been with the company from two to nine years. The interviews were carried out through 30 minutes face-to-face sessions that were recorded and later transcribed. Based on these interviews, we developed a **questionnaire**. The questionnaire was pretested by colleagues at the university and nine employees from different parts of Telenor IT. The questionnaire was written in Norwegian, and the final version contained 42 open and closed questions (using 5-point Likert scales). However, we will focus mainly on the questions below:

Q1. Which advantages and disadvantages do you see with the use of OSS in Telenor IT? (See Table 1);

Q2. For which reasons do you think Telenor IT should select OSS instead of proprietary products and vise versa? (See Table 2);

Q3. Why should Telenor IT increase its use of OSS? (Open);

Q4. Which risks do you see with increased use of OSS in Telenor IT? (Open);

Q5. If the use of OSS in Telenor IT should be increased, what should do Telenor do to facilitate this? (See Table 3);

Q6. Where would an increased use of OSS be appropriate? (Open).

The questionnaire was conducted with a sample of 140 employees from Telenor IT that were handpicked by our local contact. This sampling technique was used to get a representative sample of employees from all relevant parts of the organization while avoiding employees who were not involved in development and/or support of Telenor's software systems. In total 86 respondents completed the survey, giving a response rate of over 60%. The analysis of the data consisted of descriptive statistics, statistical tests, and grouping of about 200 comments from the open questions.

After the analysis, we held three **workshops**. First, we presented the results from the questionnaire to several employees from various parts of the organization. Second, three project members and three employees with experience from different operating environments participated in a discussion around (1) benefits, (2) risks, and (3) approaches related to increasing the organization's adoption of OSS. These three sessions were performed as "KJ sessions" [3], where each of the workshop participants used post-it notes to write down their concerns and put these notes on a white board. In total 152 post-it notes were collected. Then the participants re-arranged related notes into groups as a collaborative effort. These groups of related issues were then discussed. Finally, we presented these results during a third dissemination workshop, open to all employees at Telenor IT.

# 4  Results

Results presented in this section were grouped according to the research questions (RQ) stated in the introduction. In the following, we summarize the main findings related to these RQs, while keeping a focus on the results most relevant to Telenor.

RQ1. Based on the interviews, questions (Q1, Q2, Q3), and the workshop, we have identified the main perceived benefits (**BT**) of OSS adoption;

RQ2. Based on the interviews, question (Q4), and the workshop, we have identified several potential risks (**RT**) related to adoption of OSS;

RQ3. Mainly through the workshop and the interviews, but also questions (Q5, Q6), we have identified steps for (1) facilitating the adoption of OSS and (2) steps for mitigating (**RMT**) some of the risks related to it.

## 4.1  RQ1: Potential Benefits of OSS

**Reduced costs (BT1):** Cost reduction is the most cited advantage of OSS adoption. Table 1 shows that the respondents to the questionnaire agreed (Q1.1). Several respondents stressed the value of reducing the expenses on support agreements and claimed that OSS could contribute to this. One respondent suggested that they could simplify the administration of (proprietary) software licenses. Moreover, Table 2 shows that the respondents expected both development and maintenance costs to be lower with OSS (Q2.1 and Q2.2). Finally, if Telenor could standardize on one OSS platform, the IT department could increase its productivity and reduce costs from running on a more homogeneous and cheaper hardware platform.

**Independence from vendors of proprietary products, and the ability to apply pressure on providers (BT2)** was frequently discussed by interviewees, workshop participants and many of the responses to Q3 (see also Q1.2 and Q1.3). They highlighted in particular the ability to use OSS to apply pressure on their vendors in order to make them lower their license and support fees. As one respondent wrote "[when using OSS, one] *may chose to pay for support if you actually need it (often one does not need it)*" (Q3).

**Attractive and future-oriented technology as a motivational factor for the employees (BT3):** Several popular technologies are offered as OSS, and the interviewees mentioned that using OSS could improve the Telenor brand (Q2.4), be a source of motivation for current employees (Q1.4), and be a way to attract skilled employees. The ability to work with new and open technology was also perceived as being fun by the workshop participants. In fact, quite a lot of attention was drawn to this issue. OSS technology was also considered to be the future for several areas. For instance, one workshop participant wrote that "*OSS is future oriented and it enables access to competency*". A respondent in the questionnaire wrote that "*OSS is becoming the industry standard in many areas*" (Q3).

**Ease of use through access to information and the source code (BT4):** The respondents suggested that OSS technology was easier to use because of the high availability of the software, its source code, and related information (see also Q1.5, Q2.3, and Q2.5). One workshop participant wrote that because of this availability "*[it] is easier to make prototypes and to evaluate the software*". A respondent in the questionnaire wrote: "*it is better to modify what is meant to be modified rather than buying a final package and doing extra development around it [the package]*" (Q3). The workshop participants furthermore believed that the flexibility and openness of OSS could give them better and more innovative solutions. Easy access to technology, development tools, together with the technical support, documentation, and other resources, could further reduce the effort needed to develop and maintain their systems.

Table 1. Potential advantages and disadvantages with OSS in Telenor IT (Q1)

| ID | Statement | Mean | STD |
|----|-----------|------|-----|
| Q1.1 | Reduced licenses costs | 4.56 | 0.86 |
| Q1.2 | Independence from providers | 4.48 | 0.88 |
| Q1.3 | Ability to apply pressure on providers | 4.41 | 0.93 |
| Q1.4 | Motivational factor for the employees | 4.16 | 0.99 |
| Q1.5 | Access to read and modify source code | 4.10 | 1.01 |
| Q1.6 | Confidence and experiences with provider | 3.26 | 1.18 |
| Q1.7 | Existing contracts with providers | 3.19 | 1.28 |

Having access to the communities behind the OSS products was seen as an advantage, not only to get support, but also to influence the development of the products. One responded: *"OSS products are quite often having active communities with dedicated users who are more than willing to help"* (Q3). OSS communities were considered to be more accessible than vendors of proprietary products.

**Table 2.** Reasons for selecting OSS versus proprietary software (Q2)

| ID | Statement | Mean | STD |
|----|-----------|------|-----|
| Q2.1 | Reduced maintenance costs. | 4.15 | 1.15 |
| Q2.2 | Reduced development costs. | 4.05 | 1.13 |
| Q2.3 | Possibility to run pilot-tests (alpha/beta tests) before release. | 3.94 | 1.2 |
| Q2.4 | Improve Telenor's brand and reputation. | 3.76 | 1.17 |
| Q2.5 | Adaptability to existing systems. | 3.68 | 1.32 |
| Q2.6 | Development time. | 3.64 | 1.13 |
| Q2.7 | Influence on provider (add new or changed functionality). | 3.64 | 1.43 |
| Q2.8 | Availability of external expertise and experience. | 3.48 | 1.35 |
| Q2.9 | Availability of support during development. | 3.32 | 1.33 |
| Q2.10 | Available information (manuals etc.). | 3.24 | 1.43 |
| Q2.11 | Functional requirements (adequate functionality) | 3.19 | 1.17 |
| Q2.12 | Non-functional requirements (quality, reliability, security, scalability, performance, usability etc. | 2.95 | 1.26 |
| Q2.13 | Availability of support in production | 2.87 | 1.38 |

## 4.2 RQ2: Potential Risks and Drawbacks

**Lack of support and expertise (RT1):** The lack of a professional provider is not necessarily a problem. However, the lack of support and expert advice, in particular for complex problems, was considered as one of the major challenges with OSS. One of the interviewees feared that they would need to increase their internal resources quite dramatically. Telenor requires professional support 24/7. However, providers of such support are not necessarily available for all OSS products. One workshop participant pointed this out and wrote that *"[there are] few/no international support organizations (for instance when you need 24/7 operation)"*. Moreover, since the diffusion of OSS products is not always as large as their proprietary equivalents, the workshop participants feared that it could be difficult to get hold of both expert consultants and highly skilled employees.

**Hard to select the right OSS product (RT2):** The respondents expressed an uncertainty related to whether there existed OSS equivalents for some of the largest and most advanced systems they had. The respondents moreover feared that existing OSS products were immature and would miss key functions. One respondent wrote that *"there are in some cases no OSS products, or no OSS products which are good enough, for solving certain problems"* (Q4). The products may also lack support from a viable community and they may therefore have an uncertain future. Adopting such immature or unsupported products can introduce significant costs further down the road, and it was therefore considered important to find the right products.

**Change and hidden costs (RT3):** OSS products would in most cases be acquired and maintained somewhat differently than proprietary products. Most OSS products are available over the Internet and do not have the same number of providers pushing and supporting the products. These changes may improve the way the organization works but any change introduces challenges, uncertainty, and at least some costs. A workshop participant wrote that *"[Telenor] has to find and relate to new partners"*, something which would include both change and cost. The respondents were uncertain whether the cost savings from reduced licensing and support fees would outweigh the cost related to switching technology and changing the way they worked, as some of them described the total cost of adopting OSS as *"foggy"*. One respondent wrote that *"replacing familiar technology"* (Q4) could be a potential risk. Replacing existing technology would also make current expertise less valuable.

**Unclear liability and responsibility (RT4):** As of today Telenor's partners have relatively clearly defined responsibilities. Changing these relationships was considered an important challenge. One responded that it could lead to *"unclear distribution of roles between provider - customer [Telenor]"* (Q4). Most OSS products lack a clear (professional) vendor and the respondents feared ending up in situations with unclear liability, where they were unable to influence the provider, and where they would not get sufficient support. One respondent wrote *"[we have] no provider to make responsible in situations with critical errors"* (Q4). Such situations could put a significant strain on Telenor's internal resources.

**Uncontrolled adoption and modification (RT5):** Changes, or potential anarchy, related to the acquisition of software was discussed to some length in the workshop. This is because (1) there are a lot of easily available OSS products (in many different versions), (2) there is a lot of hype around many of these products, and (3) they are very easy to modify. Some participants feared that this could lead to uncontrolled adoption and modification of new OSS products. This would give Telenor a diverse and expensive to maintain a software portfolio. One workshop participant wrote that he feared that *"one [Telenor employees] selects products because they are OSS, not because they solve our problems"*. A respondent in the questionnaire feared what he called *"product anarchy"* meaning that the selected a lot of products without really making sure that they were the right ones.

### 4.3　RQ3: Mitigating the Risks Related to OSS Adoption

**Place responsibility, dedicate resources, and ensure support (RMT1):** To make sure that Telenor IT has the necessary resources to develop, support, and operate OSS based systems, it was considered important to place the responsibility for the adopted products between internal resources and external partners. This was highlighted by several participants in our study. One of them wrote that *"[Telenor must] coordinate with development, internal operations, and external [service] providers"*. This could involve increasing the internal resources or allocating employees to, not only support OSS solutions, but also to developing new solutions and monitoring the OSS community. It may also involve dealing with new partners, or driving existing partners into adopting new technology. The participants in the study expressed particular concerns about ensuring support for the really difficult problems.

**Start pilot projects (RMT2):** The respondents agreed that it was important not to rush the adoption of OSS, but promoted instead a cautious, stepwise approach to OSS. According to them, Telenor IT had to gain experience with one project at the time through identifying projects where OSS would be give real benefit. These pilot projects could then be used to illustrate the potential and true benefits of OSS within the organization. The respondents acknowledged that pilot projects were important not only to illustrate the potential of OSS products, but also to have a more moderate learning curve and limit the consequences of problems. One workshop participant wrote that *"[Telenor should] incrementally introduce OSS and consider new/revise measures based on our own experience"*.

**Increase awareness and make the OSS initiative visible (RMT3):** The first thing which could be done, is making the organization's current and planned use of OSS visible to, not only its employees and management, but also its partners (see Table 3). In the workshop one participant wrote that *"[Telenor IT must] make the concrete advantages visible"*. By identifying successful cases of OSS adoption and making these visible, they may create a positive attitude towards OSS and show that it is a viable option for the future. Moreover, it was considered important to explain why Telenor IT is planning to increase its adoption of OSS.

**Include OSS in strategies supported by top management (RMT4):** Finally, the adoption of OSS should not be left up to chance and the individual employees' taste. A workshop participant wrote that *"[Telenor] should not allow the system or project select freely [it should rather] be part of a strategic technological decision"*. To ensure that the OSS adoption was planned, it should be part of a strategy where (top) management, developers, operations, and support were involved in the decision making process. It was furthermore considered important to assess the benefits versus the costs in each specific case. Management support was considered important because Telenor IT mainly used OSS products in risk-free development environments. The consequences of failure in production environments is obviously higher, and it was therefore perceived important to ensure the support of management.

**Table 3.** Possible steps for increasing the adoption of OSS (Q3)

| ID | Statement | Mean | STD |
|----|-----------|------|-----|
| Q3.1 | Start one/several pilot projects to show possible effects of OSS | 4.54 | 0.85 |
| Q3.2 | Make the OSS initiative visible for all employees | 4.48 | 0.63 |
| Q3.3 | Make visible the OSS already present in the organization | 4.44 | 0.85 |
| Q3.4 | Top management commitment to the OSS initiative | 4.42 | 0.95 |
| Q3.5 | Make someone responsible for monitoring selected OSS domains | 4.15 | 0.93 |
| Q3.6 | Improve both internal and external knowledge management (e.g. with a Wiki, message boards, mailing lists, blogs or similar) | 4.14 | 0.94 |
| Q3.7 | Hire new employees with OSS experience | 3.96 | 0.99 |
| Q3.8 | Hire external consultants with updated expertise | 3.06 | 1.19 |
| Q3.9 | Restructure the business model of Telenor IT | 2.74 | 1.12 |

**Table 4.** Possible risks and steps for reducing these risks

| Potential risks of adopting OSS products | Possible risk reduction steps | |
|---|---|---|
| | From the Telenor case | From the literature |
| OSS products may lack (professional) support. There may be limited access to expertise, and situations involving unclear liability and division of responsibility may occur. (**R3+RT1+RT4**) | - Place responsibility at an early stage (**RMT1**)<br>- Make sure that your service providers support OSS products (find new ones or ask existing ones to extend their service offering) (**RMT1**)<br>- Increase/dedicate internal resources to OSS (**RMT1**)<br>- Increase employee skills (hire new or train existing) (**RM1+RMT1**) | - Encourage local "OSS champions" [16] |
| Hidden costs related to adopting OSS, replacing existing technology, and changing current processes. (**R1+RT3**) | - Conduct risk assessments<br>- Execute pilot studies and a planned stepwise adoption (**RM2+RMT2**)<br>- Adopt (only) products which show a clear added-value and have a proven track record (**RM4&5**) | - Evaluate the total costs of ownership of OSS products in your own context [35] |
| Hard to select the right product due to (1) lack of products or products with matching functionality and/or quality, and (2) the amount of products and information available. (**R2+RT2**) | - Adopt only mature products which give clear benefits (**RMT4**)<br>- Dedicate personnel to monitoring the OSS community and selecting OSS products (**RMT1**) | Research suggests several methods for selecting OSS products like for instance [7,9,30] |
| Uncontrolled adoption and modification, due to the high availability of OSS products, their low purchase price, and the access to these products' source code. (**RT5**) | - Have a plan/strategy behind adopting the various OSS products (**RMT4**)<br>- Adopt products which show a clear added-value and have a proven track record (**RM4&5**)<br>- Standardize on a limited set of technologies/products (**RMT4**)<br>- Begin with a few products (e.g. operating systems, databases, and server applications)<br>- Require that new products should run on OSS platforms when writing call for tenders and requirements specifications<br>- Keep track of the adopted software<br>- Create guidelines for adoption<br>- Dedicate personnel with responsibilities for OSS adoption (review and monitoring) (**RMT1**)<br>- Conduct risk assessments<br>- Involve management, development, operation, support, (and external partners). (**RMT1**) | - Define a strategy for maintenance and modifications [35]<br>- Set up a central software repository for adopted products [10] |

## 5  Risks and Risk Mitigation Strategies

Our empirical results confirm many of the findings from the literature review presented in Section 2. Through the literature review and our study we have identified several risks related to the adoption of OSS products. Table 4 shows an aggregation of the results from this study and from the literature, in a first step towards a risk mitigation approach in OSS adoption. Most of these are already presented in Section 2 or 4. The table is divided in three main columns. The first column lists the main risks identified in the study. The second column describes possible steps for mitigating these risks. This is once more based in our study and the papers that implicitly or explicitly discuss these steps. The third column describes other possible steps that were only identified in the literature.

Besides the results shown in Table 4, we identified some general steps for reducing the risks related to adoption of OSS products such as: (1) increasing the employees' skills (hire new or train existing) (**RM1**), (2) increasing the employees' attitude towards, and awareness of, current adoption of OSS and ongoing OSS initiatives (**RM1**), (3) ensuring top management commitment to the OSS initiative (**RM2**), and (4) avoiding going from a proprietary to an OSS lock-in (**RM5**).

The literature also mentions licensing and customization of the OSS products as po-tential risks related to OSS adoption. These risks were not discussed in the table or in our results. First, Telenor IT's Open Source 2010 project did not consider licensing issues to be a problem, particularly since Telenor is not going to distribute its software. Issues related to releasing the source code were therefore not relevant. However, it was suggested to seek legal advice to approve a set of OSS licenses, and adopt only prod-ucts with these licenses. Second, customization needs was not given much attention. One possible explanation could be that Telenor IT focused on software like operating systems, database servers, and application servers. These products constitute a "soft-ware infrastructure" and are mainly configured and deployed. Customization problems is perhaps more relevant for other kinds of software products or components.

## 6  Limitations of This Study

The sampling for the questionnaire was conducted by our contact person at Telenor. This may pose a possible threat to the validity of our results, since the sample and respondents may have more experience with OSS than the rest of the organization. However, our contact has long experience from the company, we got a high response rate, and the respondents reflect the organization at large. Moreover, when we presented the results at the workshops the audience was allowed to participate, and we did not get any feedback indicating that the results were wrong.

The study benefits from data triangulation through the use of interviews, a questionnaire, and workshops. However, we conduct only four interviews of 30 minutes each. The study would benefit from further, more in-depth interviews.

There are many different OSS products available, and these products do not share the same properties. The same holds for proprietary products. Asking about benefits, risks, and steps for reducing risks related to an increased OSS adoption is therefore somewhat problematic. We must have in mind that the answers reflect the individual respondent's perception of OSS and proprietary products. To get more precise data one would need to compare individual OSS products against specific proprietary products.

## 7 Conclusion and Future Work

Based on an extensive literature review and a study from a telecom company (Telenor IT Norway), we have **identified several risks** related to the deployment of OSS products. However, the paper's main contributions are the **identified steps for reducing these risks**. In addition, we establish a link between our results from a company and results the literature in Section 5. There are limitations associated with the findings from this paper. Nevertheless, we believe the results of this study are a first step towards focusing the research, on risks of OSS adoption, on more measurable approaches for such evaluation. Finally, our study focuses on bridging the gap between OSS research and practice by focusing on topics highly relevant to practitioners. The study is furthermore an example of how researchers and practitioners may benefit from closer collaboration.

As future work we intend to follow the process of adoption of OSS at this company to further investigate and measure the real effect of the adoption of OSS. A particular focus will be directed towards the relationship between the Telenor IT's internal development and support, and their partners. We also acknowledge that many of the risks and mitigation steps described in this paper are similar to the ones described in the literature of adoption/diffusion of general information technology e.g. [12, 26]. This research could also lend research on OSS adoption valuable support (see e.g. [13]). We intend to do more research in order to investigate these issues, so we can focus the OSS research on the issues that are mostly related to the OSS adoption, and not part of the general issues related to general adoption/diffusion of information technology.

## References

1. Adams, P., Boldyreff, C., Nutter, D., Rank, S.: Adaptive Reuse of Libre Software Systems for Supporting On-line Collaboration. In: Feller, J., Fitzgerald, B., Hissam, S.A., Lakhani, K.R., Scacchi, W. (eds.) Open Source Application Spaces: Proceedings of the Fifth Workshop on Open Source Software Engineering (WOSSE 2005), pp. 1–4. ACM, New York (2005)
2. Ågerfalk, P.J., Deverell, A., Fitzgerald, B., Morgan, L.: Assessing the Role of Open Source Software in the European Secondary Software Sector: A Voice from Industry. In: Scotto, Succi (eds.) [26], pp. 82–87
3. Birk, A., Dingsøyr, T., Stålhane, T.: Postmortem: Never Leave a Project without It. IEEE Software 19(3), 43–45 (2002)
4. Bonaccorsi, A., Giannangeli, S., Rossi, C.: Entry Strategies Under Competing Standards: Hybrid Business Models in the Open Source Software Industry. Management Science 52(7), 1085–1098 (2006)

5. Bonaccorsi, A., Rossi, C.: Comparing motivations of individual programmers and firms to take part in the open source movement: From community to business. Knowledge, Technology, and Policy 18(4), 40–64 (2006)
6. Brink, D., Roos, L., Weller, J., Van Belle, J.-P.: Critical Success Factors for Migrating to OSS-on-the-Desktop: Common Themes across Three South African Case. In: Damiani, et al. (eds.) [8], pp. 287–293
7. Cruz, D., Wieland, T., Ziegler, A.: Evaluation Criteria for Free/Open Source Software Products Based on Project Analysis. Software Process: Improvement and Practice 11(2), 107–122 (2006)
8. Damiani, E., Fitzgerald, B., Scacchi, W., Scotto, M. (eds.): Proceedings of the 2nd IFIP Working Group 2.13 International Conference on Open Source Software (OSS 2006) - Open Source Systems. IFIP International Federation for Information Processing, vol. 203. Springer, Heidelberg (2006)
9. del Bianco, V., Lavazza, L., Morasca, S., Taibi, D.: Quality of Open Source Software: The QualiPSo Trustworthiness Model. In: Boldyreff, C., Crowston, K., Lundell, B., Wasserman, A.I. (eds.) Proceedings of the 5th IFIP Working Group 2.13 International Conference on Open Source Systems (OSS2009) - Open Source Ecosystems: Diverse Communities, June 3-6. IFIP International Federation for Information Processing, vol. 299, pp. 199–212. Springer, Heidelberg (2009)
10. Dinkelacker, J., Garg, P.K., Miller, R., Nelson, D.: Progressive Open Source. In: Tracz, W., Magee, J., Young, M. (eds.) Proceedings of the 24th International Conference on Software Engineering (ICSE 2002), Orlando, Florida, May 19-25, pp. 177–184. ACM, New York (2002)
11. Feller, J., Fitzgerald, B., Scacchi, W., Sillitti, A. (eds.): Proceedings of the 3rd IFIP Working Group 2.13 International Conference on Open Source Software (OSS 2007) - Open Source Development, Adoption and Innovation, Limerick, Ireland, June 11-14. IFIP International Federation for Information Processing, vol. 234. Springer, Heidelberg (2007)
12. Fichman, R.G.: Information Technology Diffusion: A Review of Empirical Research. In: DeGross, J.I., Becker, J.D., Elam, J.J. (eds.) Proceedings of the Thirteenth International Conference on Information Systems (ICIS '92), Dallas, USA, Minneapolis, MN, December 13-16, pp. 195–206. University of Minnesota (1992)
13. Fitzgerald, B.: Open Source Software Adoption: Anatomy of Success and Failure. International Journal of Open Source Software & Processes 1(1), 1–23 (2009)
14. Fitzgerald, B., Kenny, T.: Developing an Information Systems Infrastructure with Open Source Software. IEEE Software 21(1), 50–55 (2004)
15. Ghosh, R.A.: Study on the Economic Impact of Open Source Software on Innovation and the Competiveness of the Information and Communication Technologies (ICT) Sector in the EU. Technical report, UNU-MERIT (2006)
16. Glynn, E., Fitzgerald, B., Exton, C.: Commercial Adoption of Open Source Software: An Empirical Study. In: Verner, J., Travassos, G.H. (eds.) Proceedings of International Symposium on Empirical Software Engineering (ISESE 2005), Noosa Heads, Australia, November 17th-18th, pp. 225–234. IEEE Computer Society, Los Alamitos (2005)
17. Goode, S.: Something for nothing: management rejection of open source software in Australia's top firms. Information & Management 42(5), 669–681 (2005)
18. Grand, S., von Krogh, G., Leonard, D., Swap, W.: Resource allocation beyond firm boundaries: A multi-level model for Open Source innovation. Long Range Planning 37(6), 591–610 (2004)
19. Hauge, Ø., Ayala, C.P., Conradi, R.: Open Source Software in Organizations - A Systematic Literature Review. Submitted to Information and Software Technology

20. Hauge, Ø., Sørensen, C.-F., Conradi, R.: Adoption of Open Source in the Software Industry. In: Russo, B., Damiani, E., Hissam, S.A., Lundell, B., Succi, G. (eds.) Proceedings of the 4th IFIP Working Group 2.13 International Conferences on Open Source Software (OSS2008) - Open Source Development Communities and Quality, Milano, Italy, September 7-10. IFIP International Federation for Information Processing, vol. 275, pp. 211–222. Springer, Heidelberg (2008)

21. Hauge, Ø., Sørensen, C.-F., Røsdal, A.: Surveying Industrial Roles in Open Source Software Development. In: Feller, et al (eds.) [11], pp. 259–264

22. Holck, J., Larsen, M.H., Pedersen, M.K.: Managerial and Technical Barriers to the Adoption of Open Source Software. In: Franch, X., Port, D. (eds.) ICCBSS 2005. LNCS, vol. 3412, pp. 289–300. Springer, Heidelberg (2005)

23. Jaaksi, A.: Experiences on Product Development with Open Source Software. In: Feller, et al. (eds.) [11], pp. 85–96

24. Morgan, L., Finnegan, P.: Benefits and Drawbacks of Open Source Software: An Exploratory Study of Secondary Software Firms. In: Feller, et al. (eds.) [11], pp. 307–312

25. Ozel, B., Jovanovic, U., Oba, B., van Leeuwen, M.: Perceptions on F/OSS Adoption. In: Feller, et al. (eds.) [12], pp. 319–324

26. Rogers, E.M.: Diffusion of Innovations, 5th edn. Free Press, New York (2003)

27. Scotto, M., Succi, G. (eds.): Proceedings of The First International Conference on Open Source Systems (OSS 2005), Genova, Italy, July 11th-15th (2005)

28. Serrano, N., Calzada, S., Sarriegui, J.M., Ciordia, I.: From Proprietary to Open Source Tools in Information Systems Development. IEEE Software 21(1), 56–58 (2004)

29. Sohn, S.Y., Mok, M.S.: A strategic analysis for successful open source software utilization based on a structural equation model. Journal of Systems and Software 81(6), 1014–1024 (2008)

30. Taibi, D., Lavazza, L., Morasca, S.: OpenBQR: a framework for the assessment of OSS. In: Feller, et al. (eds.) [11], pp. 173–186

31. Tiangco, F., Stockwell, A., Sapsford, J., Rainer, A.: Open-source software in an occupational health application: the case of Heales Medical Ltd. In: Scotto, Succi (eds.) [26], pp. 130–134

32. Ven, K., Van Nuffel, D., Verelst, J.: The Introduction of OpenOffice.org in the Brussels Public Administration. In: Damiani, et al. (eds.) [8], pp. 123–134

33. Ven, K., Verelst, J.: The Organizational Adoption of Open Source Server Software by Belgian Organizations. In: Damiani et al. [8], pp. 111–122

34. Ven, K., Verelst, J.: The Impact of Ideology on the Organizational Adoption of Open Source Software. Journal of Database Management 19(2), 58–72 (2008)

35. Ven, K., Verelst, J., Mannaert, H.: Should You Adopt Open Source Software? IEEE Software 25(3), 54–59 (2008)

36. Ven, K., Mannaert, H.: Challenges and strategies in the use of Open Source Software by Independent Software Vendors. Information and Software Technology 50(9-10), 991–1002 (2008)

# Usability Innovations in OSS Development – Examining User Innovations in an OSS Usability Discussion Forum

Netta Iivari

University of Oulu, Department of Information Processing Science
P.O. Box 3000, FIN-90014 University of Oulu, Finland
netta.iivari@oulu.fi

**Abstract.** This paper examines the emergence and evolution of user innovations in Open Source Software (OSS) development, with focus on usability innovations. Existing literature on user innovation and usability is reviewed, after which usability innovation is empirically explored in OSS development. The interpretive case study shows that usability innovations emerge and evolve in OSS development. They emerge after a user recognizes a need, after which she invents a fix to meet the need, thereafter needing a developer to realize the fix in the OSS. Afterwards, the user experiments with the solution and may provide feedback, which again may lead to the developer adjusting the OSS accordingly. The process is characterized as a collaborative negotiation process among the users and developers. The results also reveal that the usability innovations may be need, opportunity or creativity based, and connected to improving efficiency, effectiveness or satisfaction. Implications both for theory and practice are discussed.

## 1 Introduction

This paper examines how end user innovations emerge and evolve in Open Source Software (OSS) development, limiting the focus to end users' *usability* innovations. The concept of usability innovation refers to innovations that are trying to contribute to OSS usability, i.e. to 'the extent to which a product can be used by specified users to achieve specified goals with effectiveness, efficiency and satisfaction in a specified context of use' [15]. Therefore, the focus is on innovations that try to enable the specified users to achieve their specified goals in the specified context of use more effectively, more efficiently, or generally in a more satisfactory way.

Innovation, on the other hand, refers to "an idea, practice, or object that is perceived as new by an individual or other unit of adoption [15: 12]. Innovation necessitates not only the invention part, but also its implementation and adoption [8]. End user innovation is defined as construction of innovations by end users, instead of manufacturers (or OSS developers) producing the innovations for the users [22]. The term end user implies that the innovation is to be produced by people without skills or interest to develop the solution themselves. Therefore, this kind of user innovation always necessitates the developers joining in the implementation phase. In OSS development the developers

P. Ågerfalk et al. (Eds.): OSS 2010, IFIP AICT 319, pp. 119–129, 2010.

tend to be also users of the software, and their innovation behaviour has already been examined [6], which will not be done in this paper. Researchers have argued that the user population of OSS solutions is constantly growing, including a growing number of users who do not have technical competence or interest to develop the OSS solution [6], [17], [23]. These users' innovation behaviour will be examined in this paper. It has also been argued that usability of the OSS solutions has not traditionally been a major concern among the developers, who have been more interested in improving the functionality of the software than in improving its usability, but since the non-developer users are entering the scene, also improvement of usability is becoming a legitimate concern [1], [17], [20].

OSS development is distributed and the OSS projects rely heavily on Internet tools supporting their work, i.e. on mailing lists, discussion forums, chat and bug reporting and version control systems [6], [19], [23]. In this paper the focus will be on end user innovation, which is observed during communication among OSS users and between OSS users and OSS developers in an OSS usability discussion forum that has been established by an OSS project to increase the software's usability and therefore provides a unique opportunity to study this phenomenon.

The next section reviews user innovation and usability literatures and relates them to the OSS development context. The third section discusses the interpretive case study that has been carried out, particularly focusing on a usability discussion forum of an OSS project, the fourth section outlining the main empirical results. The fifth section discusses their implications and limitations, and identifies paths for future work.

## 2    User and Usability Innovation

### 2.1    User Innovation

OSS development has been praised as an arena fostering user innovations (e.g. [3], [6]). Franke and von Hippel [6] have carried out an empirical study on user innovation in OSS development, arguing that user innovation is needed since the users' needs are highly heterogeneous. However, also they show that there are clear differences in the skill levels of the OSS users, some of them being able to modify the software, while others not, the skilled users being more satisfied with the software than the less skilled users. [6] Therefore, it can be assumed that the possibility to innovate leads to increased user satisfaction, but the results indicate that this may be very challenging for the non-developer users. This paper will examine how, if at all, non-developer user innovation happens in OSS development.

Even though non-developer user innovation has not been examined in OSS development, it has been studied in other contexts. Researchers have analysed it e.g. in sports communities and in the research and development context (see e.g. [7], [10], [16]). Hyysalo [10] discusses micro-innovations as self-made modifications and adaptations to the existing products. Micro-innovations involve people making "various small alterations" both to the products and to their practices and usage patterns "to make their activity more enjoyable" and products "work better" [10: 250]. Also other researchers [16], [21] have considered user innovations contributing either or both to the development in the user domain and in the technical domain, and the non-developer users have been conceptualised as co-developers in the process that necessitates also the developers joining in [16]. Also in this paper the non-developer user innovations can address either or both the user and the technical domain, the non-developer users needing developer assistance in realizing the innovation in the OSS.

From the viewpoint of the emergence of user innovations, the following phases have already been identified: user recognizes a need, solves the problem by invention, builds a prototype and, finally, establishes the prototype's value in use [22]. Floyd and colleagues [5] have examined user-driven technological innovation, which, according to them, starts from user's everyday life and experiences, during which the user encounters problems, to which she rapidly and easily produces fixes through utilizing technology, after which she reflects on the solution, i.e. on how it fits the overall systems, and adjusts the system as a result. The cycle starts over when the user, again, encounters problems in her everyday life. This type of a process can support the creativity, needs and desires of the users, and it is characterized as rapid, participatory and collaborative [5]. OSS solutions enable users to explore how technology could be utilized in their work practices and how their work practices could be conceptualized anew with technology [5].

## 2.2 Usability Innovation

Usability innovations have not been discussed in depth in the literature. However, there exists a huge amount of literature addressing the development and evaluation of usability. The widely cited usability definition presented already establishes that usability can be developed only after one understands and specifies who the intended users are and what their goals in using the software are and in what kind of context of use the software will be used [14]. For gaining this understanding, different kinds of methods are suggested, e.g. interviews, observation and surveys. Thereafter, one needs to carefully redesign the users' future tasks that are to be carried out to achieve the goals. Only after this the computer-based solution should be considered. Finally, essential is also to evaluate these design solutions, as early as possible and continually during the development life cycle [14], [11].

Typically these activities are expected to be carried out by usability (or user-centered design, usability engineering etc.) specialists [11]. Even though not explicitly mentioning usability innovations, the literature seems to assume that these specialists produce the innovations, after being in contact with the users. Part of the literature views this as a structured engineering process while other part maintains that it is a creative, artistic process, but generally in the both literatures it seems to be assumed that there needs to be some understanding of the users and their work practices, some understanding of the capabilities and limitations of technology and HCI background [11], the innovations being generated based on the combination of these understandings. This paper, instead, examines the innovations produced by the end users themselves, when given the chance to do that.

However, it needs to be underlined that usability in this paper is an etic concept introduced by the researcher. Another approach could be to analyze the multitude of meanings attached to usability by the OSS developers and users, but this has been done already, revealing that surprising and rather technical aspects are attached to usability in OSS development [1], [12]. Instead of showing this again, this paper adopts the established, etic usability definition from the literature and limits the focus to the innovations that adhere to this definition.

Existing OSS research has already outlined many issues, which may cause problems for usability innovation in OSS development. Typically, the developers do not have knowledge about the non-developer users, their goals and their contexts of use, and they do not necessarily have any interest in learning about them either. In addition, communicating usability problems to the development has proven to be difficult for the non-developer users. [2], [17], [20], [24] It has been argued that usability bugs are very complex to fix and difficult to explain textually [20]. Furthermore, usability specialists would be very useful in improving OSS usability, but typically there is a lack of usability specialists in OSS projects [2], [17], [24].

## 3 Research Design

This research relies on the qualitative research tradition [4]. One case is analyzed: an OSS project developing a media application for end user without necessarily any technical knowledge or programming skills. The project is interested in their users and their feedback regarding the solution. The *usability discussion forum* asks the users of the OSS to take part in improving the program. Altogether, over 1600 messages and nearly 400 topics have emerged in this discussion forum. Nearly 600 message senders have contributed to the discussion forum. The project is a small but active one: there are 9 developers listed for the project, the development status being 5 (production/stable).

The analysis focused on the discussions in the usability discussion forum, which is assumed to be the place where user innovation occurs in this OSS project. The research data consists of the usability discussion forum messages (altogether 1600), all of which were printed out for the analysis purposes. First, the researcher familiarized her with the discussion forum and the message senders. All posts of the usability discussion forum were printed and read thorough. The focus was in identifying messages that could be connected to non-developer users' usability innovation, i.e. on messages that were suggesting something considered new [8], [18] by the community as well as dealing with improving usability, i.e. proposing improvements related to how users can achieve their goals more effectively, efficiently or in a more satisfactory way [15]. After the examination, it became evident that there were a lot messages without any replies. However, there were also some topics that had gained popularity in the sense of the number of replies as well as some of the discussions indicating that changes have resulted in the OSS due to these discussions. The focus was further limited to these discussions.

They were analyzed from the viewpoint of non-developer users' usability innovation. The literature presented in section 2 was used as a sensitizing device. In addition, a textual approach was utilized as a theoretical tool. It postulates that during development, the developers always 'configure the users' [9]; i.e. the developers (knowingly or not) delineate the future users and their work practices already while producing the software text [9]. OSS development arena enables the users to gain access and voice in the process through mailing lists, discussion forums, chat and such, which do not necessitate competence in technical development [13]. The users even if they were not capable to directly affect the source code, can participate in the discussions taking place in the OSS projects' websites, during which user innovation might occur. The innovating users "constitute a specific group of users that adopt specific, informed ways of not just reading but also introducing new scripts, by inscribing characteristics of their specific use situation into the product" [21: 186].

# 4 Empirical Insights

In this OSS project non-developer user innovations improving effectiveness, efficiency and satisfaction in users' goal achievement were all evident. Especially innovations improving how users achieve their goals more *efficiently* were outlined. Some of them were implemented also fast by the developers, as in the following example. A user asks:

"Better Keyboard/Shortcut integration: I'm asking for Shortcuts to show/hide the side bar, to change between the different sidebar tabs or a way to focus [an element]. To edit [information] which is not in the first column, a mouse is needed." (User)

A developer advises the user that the user can use tab and no mouse is needed. The user replies, after some experimentation.

"I never thought about using the tab in the edit mode - thanks for this information. Now I found some odd behaviour: the tab key circles through all possible entries including the non-visible one. So in some situations I have to press the tab key three or more times" (User)

Another developer joins the discussion the same day and announces that a new editor has been implemented and asks the user to test the solution.

The discussion reveals that the developer accepts the idea proposed by the user, i.e. he lets the user to establish certain aspects related to the future use practices (efficiency in editing), afterwards realizing those in the OSS text.

Another user also proposes improvement to the efficiency:

"What about a keyboard shortcut for selecting the next tag in the same column?" (User)

A developer suggests to the user to use tab, to which the user replies:

"I have in fact tried tab. Maybe I'm an unusual user; I'd like to jump to the next row, same column. Anyhow, in a recurring action like tag editing possibly it could be more useful with more control over navigation - e.g. the ability to move back and forwards, up and down" (User)

The developer agrees with the user and promises to try to enable it. Therefore, one can again conclude that the developer lets the user to establish certain aspects related to the future use practices (efficiency in editing). Related to realizing those in the OSS text, he promises to do as much as he can, the existing technological solution, however, restricting his possibilities.

Both of these examples illustrate that creating something 'new' involves a collaborative negotiation process among users and developers.

Also *effectiveness* in achieving goals is brought up, e.g. in the following relating to user goal of organizing his files:

"[Another user wrote]: "I would like a way to view [an element] like in [another application]."") Another vote for it. (…) When I have my list sorted [according to a field/field] I cannot just toggle them because the entries are greyed out. This is not convenient. Please permit it and use only the primary entry in this case." (User)

The same day a developer replies that the suggestion was smart and it has already been implemented. A noteworthy observation is also the voting for the suggestion of another user with which the message starts. In this case the developer does not implement it, however. Imitation of other applications is not appreciated in this OSS project, as can be concluded from numerous messages criticizing it in the discussion forum.

The user innovations implemented can be all labelled as small scale, incremental [8] micro-innovations [10] based on the users' needs. In these discussions, the users refer to their needs or problems in the 'configuration of the user' they expect to be met or removed through the solutions they propose. On the other hand, some discussions make it clear that not only 'needed' but also 'cool' and 'eye candy' features are to be included in this OSS, causing not only improvements in effectiveness or efficiency in goal achievement, but clearly also in *satisfaction* in doing so. For example, a user argues for tag guessing and for a tab with favourite and last files that both would be 'cool', and another user argues for using certain kinds of visualizations in the application that he calls 'eye-candy', maintaining that they would make use more fun but also introduce 'cool' usability improvements. Based on the discussions, one can conclude that the users are not only motivated by efficiency or effectiveness improvements, but also by 'cool, 'fun' and 'eye candy', which, nevertheless, also contribute to the users' goal achievement.

However, in the discussions it is not only the 'configuration of the user' in the OSS to be refined according to the users' requests, but the innovation concerns also the users' use practices that are conceptualised anew (see [5]), the change being informed or inspired by the existing 'configuration of the user'. In the community there has been a lot of discussion related to enabling accurate ranking. The OSS incorporates ranking but the users argue that it is not done correctly and they wish to be able to have more power related to that. An example message from a user is shown below:

"I've been thinking a lot about how to do automatic ranking given implicit info on [user's use patterns], and there's really no good way. The problem is that the user's mood changes often. The entire scoring system can be totally broken based on the user's mood. [The system] would have to take the moods in to account to score correctly. You could achieve this with statistics on e.g. whether you were in front of your computer and how long you were there, plus what you did and didn't. Then make a decision how your "mood" changes your taste. There are two solutions for that: 1) Use stats from an entire community instead of making all new users train the system every time. 2) Give the users more guidance in the training process. A simple low tech way would be to provide a slider that tells how much skipping affects the score." (User)

Therefore, the user wishes to be able to achieve his goal related to accurate ranking of media files and proposes few solutions that in this case address both the technical solution and the 'configuration of the user' in the OSS. However, a developer informs the user that the idea has been to achieve another goal, altogether: the idea has been to let the OSS to set the scores based on the longer time span usage habits, but the users are arguing for the possibility to give ratings themselves. Nevertheless, the developer informs that due to the users' requests the developers have implemented the rating functionality as well. However, he also argues that his own habits have changed due to the scoring information that indicates to the users what their long time favourites are. This information may lead the users to reflect on and even change their current habits.

The reciprocal relationship between use practice and technology becomes intensively visible in this discussion: the 'configuration of the user' sets the boundaries for the users' use practices, as well as it is modified to enable more efficient, effective and satisfactory use practices, but the use practices may also be adapted, inspired or conceptualised anew (see [5]) when encountering the existing 'configuration of the user.

## 5  Concluding Discussion

This paper promised to examine the emergence of non-developer user innovations in an OSS usability discussion forum, at a detailed level in the natural setting without researcher intervention. Table 1 summarizes the key empirical findings of the examination.

**Table 1.** Characterizing User Innovation in OSS Development

| Aspect | Empirical Finding |
|---|---|
| **Encountering usability problems in everyday life** | - Emergence of need-based usability innovations: user recognizes a need<br>- Emergence of creativity or opportunity -based ('cool') usability innovations: user recognizes ways technology could be utilized in creative or opportunistic ways to improve usability |
| **Producing usability fixes** | - User invents usability fixes to the 'configuration of the user'<br>- Developer realizes the fixes in the OSS<br>- Producing fixes a collaborative process involving negotiation among users and between developers and users |
| **Reflecting on and adjusting** | - Users provide feedback to the refined 'configuration of the user'<br>- Developers adjust the 'configuration of the user' accordingly<br>- Users adjust their use practices accordingly |
| **Relationship between use practice and technology** | - Users identify usability fixes to the 'configuration of the user' after considering how technology could bring more efficiency, effectiveness or satisfaction to their goal achievement<br>- Users conceptualize anew (adding efficiency, effectiveness or satisfaction to) their use practices based on the encounter with the current 'configuration of the user' |

First of all, the results indicate that in the OSS development context end user usability innovation occurs, i.e. non-developer users innovate usability improvements to the OSS and those may also end up in the OSS. Next this process is characterized in more detail.

The process of user innovation is argued to include the following phases (modified from [5], [22]): a user recognizes a problem during her everyday life and experiences, she solves the problem by invention, she outlines a solution and finally she establishes its value in use, i.e. she reflects on the solution; how it fits the overall system and may also adjust the system as a result. All these phases could be identified in the usability discussion forum, but also clear differences and additions could the identified. Related to the user recognizing a usability problem during her everyday life and experiences, one clear distinction can be made between need-based and creativity or opportunity-based usability innovations (cf. [5]): some of the innovations are to meet the needs presented by the user, but other usability innovations are motivated as 'cool', even though still aiming at bringing efficiency, effectiveness or satisfaction to the users' use practices. Related to professional usability specialists working in the commercial context, one can argue that during user studies and such they are likely to encounter the usability problems and maybe also identify at least partly similar need-based usability innovations, but it might be very difficult for them to introduce the 'cool' factor into their innovations. Their 'cool' or 'eye-candy' may not equal the users' 'cool' or 'eye-candy'.

In the OSS development the users are speaking on behalf of themselves: they are reporting what they consider as needed, cool or eye-candy. The usability professionals, on the other hand, try to represent the users in the commercial software development context [13], i.e. they speak on behalf of these other people they are trying to learn to know. This 'speaking on behalf of other people' has been reported to exist also in the OSS development context, in which, however, it seems to be carried out by amateur intermediaries, i.e. by users who do so without professional background on the matter [13]. Despite that, the contribution from these users, whether speaking on behalf of themselves or other users, could be of interest also in the commercial software development context, in which there might be professional, hired usability professionals available, but in which the development could still benefit from this type of amateur usability contributions as well. It might be that there are totally different kinds of innovations created by people who encounter the software continuously in their everyday life than by people studying the software use in other people's everyday life or than by people developing the software as part of their everyday life, but not using it. However, one should very carefully consider the motives behind these amateur usability contributions to be able to judge the validity of the claims – it might be that in some cases they are not trying to improve the OSS usability, but instead to accomplish other goals (cf. [13]).

User innovation necessitates the user to solve the problem by invention. In the OSS development context it is evident that people not interested in or capable to code do invent usability fixes to the 'configuration of the user' without touching the source code. Their suggestions are connected to the introducing efficiency, effectiveness and satisfaction to the users' goal achievement. The users rely on their in-depth domain knowledge and reflection on their usage habits and practices in producing the suggestions (cf. [10], [21]), i.e. while they are introducing these new scripts to the OSS. The non-developer users however need the developers for realizing their usability innovations, since the non-developer users are by definition not capable to code. Therefore, usability innovations in this case are necessarily produced during a collaborative process between users and developers, collaboration among the users also emerging related to backing up each other and discussing and further refining the ideas.

After the implementation, the users can reflect on the innovations' value in use and adjust the system accordingly. Of course, this happens during the longer time span, which might also be observable in the OSS forum discussions. In this project it is evident that after the developers have changed the 'configuration of the user', the users eagerly experiment with it and provide feedback. User feedback gathering through OSS discussion forums has already been recommended in the literature [19], [23]. Also in this case the discussion forum has been built up for users to provide feedback and improvement ideas. Interestingly, even though most users are asking fixes to the 'configuration of the user' after considering how technology could be utilized to better support their use practices, in some messages also the question of how their use practices could be conceptualised anew with technology (cf. [5]) emerge: the users may conceptualise their use practices anew after experimenting with the existing 'configuration of the user'. These discussions, however, were clearly in minority in the discussion forum.

This study opens interesting avenues for user innovation researchers. This study provides a lens through which to examine user-developer interaction related to innovation in the context of technology development. The lens emphasizes that technological artefacts always include a 'configuration of the user' the users encounter and interpret. The users may wish to provide feedback and suggest fixes to the configuration, but they may also redefine their use practices anew after the encounter. The artefact already implies certain kind of future user with certain kind of future use practices, but the users can also interpret and modify those in new, creative, innovative ways. Other researchers are welcomed to adopt and refine this sensitising device.

Companies are showing increasing interest to enhance user innovation related to their products and services. Use of OSS communities for that has already been recommended (e.g. [3], [6]). The results of this study suggest that also end user innovation occurs in OSS development and online forums provide support for it. However, it might be a great challenge for companies to establish online forums and associated communities and to keep them vital. In traditional OSS development everything operates on a voluntary basis, and it might be a challenge to invite the non-developer users to participate in development and in the associated discussions. The forum examined in this paper has succeeded in inviting also the non-developer users to contribute to the development. However, this study does not provide tools for companies for online community building, but it informs the companies that end user innovation may emerge and evolve in online OSS forums. In addition, this study offers interesting insights for usability research, showing that 'amateur usability specialists' are contributing to OSS development. They are at least active and creative usability innovators, probably not replacing the professional ones but surely producing interesting ideas to be considered for the implementation.

This study is based on only one case, which is naturally particular in many ways. In the future more cases and more diversity (small vs. large projects, new-found vs. long-term projects, different kinds of application domains, projects with or without company involvement and so on) should be included in the future analysis in the OSS development context.

## Acknowledgements

This research has been partly funded by the Academy of Finland.

## References

[1] Andreasen, M., Nielsen, H., Schrøder, S., Stage, J.: Usability in Open Source Software Development: Opinions and Practice. Information Technology and Control 25(3A), 303–312 (2006)

[2] Benson, C., Müller-Prove, M., Mzourek, J.: Professional usability in open source projects: GNOME, OpenOffice.org, NetBeans. In: Extended Abstracts of CHI, pp. 1083–1084. ACM Press, New York (2004)

[3] Chesbrough, H.: The Era of Open Innovation. MIT Sloan Management Review 44(3), 35–41 (2003)

[4] Denzin, N., Lincoln, Y.: Introduction: The Discipline and Practice of Qualitative Research. In: Denzin, N., Lincoln, Y. (eds.) Handbook of Qualitative Research, 2nd edn., pp. 1–28. Sage Publications, Thousand Oaks (2000)

[5] Floyd, I., Jones, M., Rathi, D., Twidale, M.: Wab Mash-ups amd Patchwork Prototyping: User-driven technological innovation with Web 2.0 and Open Source Software. In: Proc. HICSS 2007. IEEE, Washington (2007)

[6] Franke, N., von Hippel, E.: Satisfying heterogeneous user needs via innovation toolkits: the case of Apache security software. Research Policy 32, 1199–1215 (2003)

[7] Franke, N., Shah, S.: How communities support innovative activities: an exploration of assistance and sharing among end-users. Research Policy 32, 157–178 (2003)

[8] Garcia, R., Calantone, R.: A critical look at technological innovation typology and innovativeness terminology: A literature review. Journal of Product Innovation Management 19(2), 110–132 (2002)

[9] Grint, K., Woolgar, S.: The Machine at Work. In: Technology, Work and Organization. Polity Press, Cambridge (1997)

[10] Hyysalo, S.: User innovation and everyday practices: Micro-innovation in sports industry development. R&D Management 39(3), 247–258 (2009)

[11] Iivari, N.: Discourses on 'culture' and 'usability work' in software product development. Acta Universitatis Ouluensis, Series A, Scientiae rerum naturalium 457 (2006)

[12] Iivari, N.: Usability in open source software development – an interpretive case study. In: Proc. ECIS, Galway, Ireland, June 9-11 (2008)

[13] Iivari, N.: "Constructing the Users" in Open Source Software Development – An Interpretive Case Study of User Participation. Information Technology & People 22(2), 132–156 (2009)

[14] ISO 13407. Human-centered design processes for interactive systems. International standard (1999)

[15] ISO 9241-11. Ergonomic requirements for office work with visual display terminals (VDT)s - Part 11 Guidance on usability. International standard (1998)

[16] Lettl, C., Herstatt, C., Gemuenden, H.: Users' Contributions to Radical Innovation: Evidence from Four Cases in the Field of Medical Equipment Technology. R&D Management 36(3), 251–272 (2006)

[17] Nichols, D., Twidale, M.: Usability Processes in Open Source Projects. Software Process Improvement and Practice 11, 149–162 (2006)

[18] Rogers, E.: Diffusion of Innovations, 5th edn. Free Press, New York (1995)

[19] Scacchi, W.: Understanding the requirements for developing open source software systems. IEE Proceedings – Software 149(1), 24–39 (2002)

[20] Twidale, M., Nichols, D.: Exploring Usability Discussions in Open Source Development. In: Proc. HICSS. IEEE, Washington (2005)

[21] van Oost, E., Verhaegh, S., Oudshoorn, N.: From Innovation Community to Community Innovation: User-initiated Innovation in Wireless Leden. Science, Technology & Human Values 34(2), 182–205 (2009)

[22] von Hippel, E.: The Sources of Innovation. Oxford University Press, New York (1988)

[23] Ye, Y., Kishida, K.: Toward an Understanding of the Motivation of Open Source Software Developers. In: Proc. ICSE, pp. 419–429. IEEE, Washington (2003)

[24] Zhao, L., Deek, F.: Improving Open Source Software Usability. In: Proc. AMCIS, Omaha, USA, August 11-14, pp. 923–928 (2005)

# Governance in Open Source Software Development Projects: A Comparative Multi-level Analysis

Chris Jensen and Walt Scacchi

Institute for Software Research
University of California, Irvine
Irvine, CA 92697-3455
{cjensen,wscacchi}@ics.uci.edu

**Abstract.** Open source software (OSS) development is a community-oriented, network-centric approach to building complex software systems. OSS projects are typically organized as edge organizations lacking an explicit management regime to control and coordinate decentralized project work. However, a growing number of OSS projects are developing, delivering, and supporting large-scale software systems, displacing proprietary software alternatives. Recent empirical studies of OSS projects reveal that OSS developers often self-organize into organizational forms we characterize as evolving socio-technical interaction networks (STINs). STINs emerge in ways that effectively control semi-autonomous OSS developers and coordinate project activities, producing reliable and adaptive software systems. In this paper, we examine how practices and processes enable and govern OSS projects when coalesced and configured as contingent, socio-technical interaction networks. We draw on data sources and results from two ongoing case studies of governance activities and elements in a large OSS project.

## 1 Introduction and Overview

In this paper, we contribute to this growing understanding for how to characterize the ways and means for affecting governance within and across OSS projects, as well as the participants and technologies that enable these projects and the larger communities of practice in which they operate and interact. Specifically, our contribution centers around providing an alternative perspective and analytical construct that offers multi-level analysis and explanation, as well as a framework for comparison and generalization based on empirical studies of OSS projects, work practices, development processes, and community dynamics [cf. 20]. The perspective draws from socio-technical interaction networks (STINs) [18] as a persistent organizational form for collective action with/through technical (computing) work systems, and also puts forward STINs as the analytical construct that serves as an organizing concept, configurational form [13], and adaptive process that both enacts and explains how governance in OSS projects is realized and directed.

Our belief is that the governance practices enacted through STINs found in OSS projects can be framed as possible options for understanding how these projects can

P. Ågerfalk et al. (Eds.): OSS 2010, IFIP AICT 319, pp. 130–142, 2010.

develop complex and reliable software without an explicit, centralized software project management regime. Further, these STINs act in a self-organizing manner to effectively realize a decentralized approach to organize, coordinate and control a dispersed, somewhat autonomous work force. This in turn can then be used to both understand the foundations for OSS organizational practices in the development, deployment, and support of complex software systems.

**Table 1.** OSS governance analytical levels and emergent themes

| Analyti-cal Level | Agents | Emergent Themes |
|---|---|---|
| Micro | Individual participants | Individual actions and resources, artifacts and resources as objects of interaction |
| Meso | Project teams | Collaboration, leadership, control, conflict resolution |
| Macro | Inter-project ecosystem | Coordination, leadership, control, conflict resolution |

## 2 Analytical Levels and Elements for Understanding Governance in OSS Projects

OSS work practices, engineering processes, and community dynamics can best be understood through observation and examination of their socio-technical elements from multiple levels of analysis [20]. In particular, OSS projects can be examined through a micro-level analysis of (a) the actions, beliefs, and motivations of individual OSS project participants, and (b) the social or technical resources that are mobilized and configured to support, subsidize, and sustain OSS work and outcomes [19]. Similarly, OSS projects can be examined through meso-level analysis of (c) patterns of cooperation, coordination, control, leadership, role migration, and conflict mitigation, and (d) project alliances and inter-project socio-technical networking [4]. Last, OSS projects can also be examined through macro-level analysis of (d) multi-project OSS ecosystems, and (e) OSS as a social movement and emerging global culture. As such, we will provide a multi-level analysis of the elements of OSS governance. Recent research on software development governance showed there are many issues critical to governing software development, including decision rights, responsibilities, roles, accountability, policies and guidelines, and processes [25]. The governance issues we have identified at these three levels in OSS bear similarities (see Tabel 1).

We engage in multi-level analysis of the elements of OSS governance using data sources and empirical results drawn from an ongoing, longitudinal case study of OSS projects. Our results have emerged from several years of research on how OSS practitioners organize themselves to get work done and what social and technical processes are employed in development, including recruitment and role migration or project participants, how software requirements are asserted, and how products are released. Our research is ethnographic, using a grounded theory approach to the analysis of project artifacts, including email discussions, chat transcripts, summary digests, (and others), as well as face-to-face interviews of project contributors.

The project of study is NetBeans, a sponsored OSS project focused on the development, support, and evolution of a Java-centered, Integrated Development Environment (IDE), which is a tool for developing Web-based enterprise software applications coded in the Java programming language that utilize other Java-based software products and services, such as those offered by Sun Microsystems Inc. [10]. NetBeans is a large OSS project with more than 400,000 active users, and tens of thousands of contributors.

Finally, it is our view that the elements of OSS governance span these multiple levels of analysis because they coalesce and are actively configured by OSS project participants into network forms for collective action—networks we designate as socio-technical interaction networks (STINs) [18]. Why? Our observation drawn from our own studies of OSS and those of others [4, 5, 13, 20] suggest to us that governance activities, efforts, and mechanisms are not disjoint or unrelated to one another, but instead are arrayed and configured by OSS project participants into networks for mobilizing socio-technical interactions, resources, rules, and organizational forms. Project participants are only accountable to each other, and not to corporate owners, senior executives, or stock investors. They can often suffice with lightweight governance forms that they configure and adapt to their needs and situations, rather than to budget, schedules, or profit growth. Accordingly, they choose organizational forms that are neither purely a decentralized market (a "bazaar") nor a centralized hierarchy (a "cathedral"), but instead choose a more agile network form that can be readily be adapted to local contingencies or emergent conditions that arise in the interactions among project participants, the technical computing systems/resources at hand, or the joint socio-technical system that is the OSS project. Thus, our multi-level analysis is one that is construed to draw attention to the persistent yet adaptive STINs that participants enact to span and govern OSS projects, practices, and processes that arise at different levels of socio-technical interaction.

## 3  Micro-level Analysis of OSS Governance Issues

Our analysis of OSS governance begins by examining what resources OSS project participants mobilize to help govern the overall activities of their project work and contributions. Much of the development work that occurs in an OSS project centers around resources that enable the creation, update, and other actions (e.g., copy, move, delete) applied to a variety of software development artifacts. These resources and artifacts serve as coordination mechanisms [16, 21, 22], in that they help participants communicate, document, maintain awareness, and otherwise make sense of how the software is structured/designed, what the emerging software system is suppose to do, how it should be or was accomplished, who did what, what went wrong before, and how to fix it. These artifacts help in coordinating local, project-specific development activities, whereas between multiple project communities, these artifacts emerge as boundary objects [10, 12] through which inter-project activities and relations are negotiated and revised. The artifacts may take the form of text messages posted to a project discussion list, webpages, source code directories and files, site maps, and

more, and they are employed as the primary media through which software requirements and design are expressed. These artifacts are software informalisms that are
collectively used to manage the consistency, completeness, and traceability of software functionality, development activities, and developer comprehension [17]. They
act as coordination resources in OSS projects since participants generally are not co-
located, do not meet face-to-face, often work asynchronously, and authority and expertise relationships among participants are up for grabs.

Accordingly, in order to explore where issues of collaboration, leadership, control
and conflict may arise within or across related OSS projects, then one place to look to
see such issues is in how project participants create, update, exchange, debate, and
make sense of the software informalisms that are employed to coordinate their development activities. This is the approach taken here in exploring the issues both within
the NetBeans project, as well as across the fragile software ecosystem of inter-related
OSS projects that situate NetBeans within a Web information infrastructure [10].

## 4   Meso-level Analysis of OSS Governance Issues

At the meso-level, have observed at least three kinds of governance elements that
arise within an OSS community like NetBeans. These are collaboration, leadership
and control, and conflict resolution.

### 4.1   Collaboration

According to the NetBeans website, individuals may participate by joining in discussions on  mailing lists, filing bug and enhancement reports, contributing Web content,
source code, newsletter articles, and language translations [11]. These activities can
be done in isolation, without coordinating with other community members, and then
offered up for consideration and inclusion.  Reducing the need for collaboration is a
common practice in the community that gives rise to positive and negative effects.
We discuss collaboration in terms of policies that support process structures that prevent conflict, looking at task completion guidelines and community architecture.

#### 4.1.1   Policies and Guidelines
The NetBeans community has detailed procedural guidelines for most common development tasks, from submitting bug fixes to user interface design and creating a
new release [24]. We can classify these guidelines as development task and design
style guidelines. Incidentally, the procedures for policy revision have not been explicitly specified, though social norms have developed to govern their revision.

Precedent states that policy and procedure revisions are brought up on the community or module discussion mailing lists, where they are debated and either ratified or
rejected by consensus. Consensus here means some support from at least one or two
other developers, along with the absence of strong conflicts or major disagreements
by other project contributors. Developers are expected to take notice of the decision
and act accordingly, while the requisite guideline documents are updated to reflect the
changes. In addition, as some communities resort to "public flogging" for failure to

follow stated procedures, requests for revision are rare and usually well known among concerned parties, so no such flogging is done within NetBeans.

Overall, these policies allow individual developers to work independently within a process structure that enables collaboration by encouraging or reinforcing developers to work in ways that are expected by their fellow community members, as well as congruent with the community process.

### 4.1.2 Separation of Concerns: An Architectural Strategy for Collaborative Success

Software products often employ a modular, plug-in application program interface (API) architectural style in order to facilitate development of add-on components that extend system functionality. This strategy has been essential in an open source arena that carries freedom of extensibility as a basic privilege or, in some cases, the right of free speech or freedom of expression through contributed source code. But this separation of concerns strategy for code management and software architecture also provides a degree of separation of concerns in developer management, and therefore, collaboration [cf. 2, 16, 9].

In concept, a module team can take the plug-in API specification and develop a modular extension for the system in complete isolation from the rest of the community. This flexibility is attractive to third-party contributors in the NetBeans community who may be uninterested in heavy involvement in the project, or who are unwilling or unable to contribute their source code back to the community. This separation of concerns in the NetBeans design architecture engenders separation of concerns in the development process [10]. Still, module dependencies limit development isolation.

Last, volunteer community members have observed difficulties collaborating with non-volunteer community members. At one point volunteer contributors experienced a lack of responsiveness of the (primarily Sun employed) user interface team[1]. This coordination breakdown led to the failure of usability efforts for a period when usability was arguably the most-cited reason users chose competing tools over NetBeans. Thus, a collaboration failure gave rise to product failure. After resolving collaboration issues NetBeans was able to deliver a satisfactory usability experience[2].

### 4.2  Leadership and Control

Ignoring internal Sun's organizational structure, there are five observable layers of the NetBeans community hierarchy. Members may take on multiple roles while migrating through different role sets [11]. Some of these roles span several layers of software functionality, development activity, commitment, and expertise. At the bottom layer are users, who can later migrate upward into roles as source contributors, module-level managers, project level release managers (i.e. IDE or development platform), and finally, community level managers at the top-most layer. Interestingly, the "management" positions are limited to coordinating roles; they carry no other technical or managerial authority. The release manager, for example, has no authority to determine what will be included in and excluded from the release[3] or the authority to assign people to

---

[1] http://www.netbeans.org/servlets/ReadMsg?msgId=531512&listName=nbdiscuss
[2] http://www.javalobby.org/thread.jspa?forumID=61&threadID=9550#top
[3] http://www.netbeans.org/community/guidelines/process.html

complete the tasks required to release the product. The same is true of module and community managers. Instead, their role is to announce the tasks that need to be done and wait for volunteers to accept responsibility. Overall, this practice at NetBeans resembles the adaptive hybrid mix of organizational governance mechanisms that O'Mahony and Ferraro [15] found in their study of the Debian project.

In NetBeans, we find that accountability and expectations of responsibility are based on precedent (prior practices) and volunteerism rather than explicit assignment. Such uncertainty has led to confusion regarding the role of parties contributing to development. Leadership is not asserted until a community member champions a cause and while volunteerism is expected, this expectation is not always obvious. The lack of a clear authority structure is both a cause of freedom and chaos in open source development. Though often seen as one of its strengths in comparison to closed source efforts, it can lead to process failure if no one steps forward to perform critical activities or if misidentified expectations cause dissent.

The coordination challenges across organizations occasionally brought up in the community mailing lists stem from the lack of a shared understanding leadership in the community. This manifests itself in two ways: a lack of transparency in the decision making process and decision making without community consent. While not new phenomenon, they are especially poignant in a movement whose basic tenets include freedom and knowledge sharing.

### 4.2.1  Transparency in the Decision Making Process

In communities with corporately backed development effort, there are often decisions made that create a community-wide impact that are made company meetings. However, these decisions may not be explicitly communicated to the rest of the project. Likewise private communication between parties may cause similar breakdowns. The lack of transparency in decision-making process prevented other community members from understanding and accepting the changes taking place. This effect surfaced in the NetBeans community recently following a discussion of modifying the release process[4]. Given the magnitude of contributions from the primary benefactor, other developers were unsure of the responsibility and authority Sun assumed within the development process. The omission of a stated policy outlining these bounds led to a flurry of excitement when Sun members announced major changes to the licensing scheme used by the community without any warning. It has also caused occasional collaboration breakdown throughout the community due to expectations of who would carry out which development tasks. The otherwise implicit nature of Sun's contributions in relation to other organizations and individuals has been revealed primarily through precedent rather than assertion.

### 4.2.2  Consent in the Decision Making Process

Without an explicit authority structure, OSS decisions in NetBeans are made through consensus, except among those over-arching or broad scope decisions that lack transparency. In the case of the licensing scheme change, some developers expressed their view that Sun was within its rights as the major contributor and the

---

[4]  http://www.netbeans.org/servlets/BrowseList?listName=nbdiscuss&by=thread&from=19116 &to=19116&first=1&count=41

most exposed to legal threat [5] while others saw it as an attack on the "democratic protection mechanisms" of the community that ensure fairness between participating parties[6]. A lack of consideration and transparency in the decision making process alienated those who are not consulted and eroded the sense of community.

### 4.3 Conflict Resolution

Conflicts in the NetBeans community are resolved via community discussion mailing lists. The process usually begins when one member announces dissatisfaction with an issue in development. Those who also feel concern with the particular issue then write responses to the charges raised. At some point, the conversation dissipates- usually when emotions are set aside and clarifications have been made that provide an understanding of the issue at hand.  If the problem persists, the community governance board is tasked with resolving the matter.

The governance board is composed of three individuals and has the role of ensuring the fairness throughout the community by solving persistent disputes. Two of the members are elected by the community, and one is appointed by Sun. The board's authority and scope are questionable and untested. While it has been suggested that the board intercede in the past, the disputes have dissolved before the board has acted.s

Board members are typically prominent members in the community. Their status carries somewhat more weight in community policy discussions, however, even when one member has suggested a decision, as no three board members have ever voted in resolution on any issue, and thus, it is unclear what effect would result. Their role, then, is more of a mediator: to drive community members to resolve the issue amongst themselves. To this end, they have been effective.

## 5  Macro-level Analysis of OSS Governance Issues

As noted earlier, the NetBeans project is not an isolated OSS project. Instead, the NetBeans IDE which is the focus of development activities in the NetBeans project is envisioned to support the interactive development of Web-compatible software applications or services that can be accessed, executed, or served through other OSS systems like the Mozilla Web browser and Apache Web server. Thus, it is reasonable to explore how the NetBeans project is situated within an ecosystem of inter-related OSS projects that facilitate or constrain the intended usage of the NetBeans IDE. Figure 1 provides a rendering of some of the more visible OSS projects that surround and embed the NetBeans within a Web information infrastructure [10]. This rendering also suggests that issues of like coordination  (integration of software products and development effort) and conflict can arise at the boundaries between projects, and thus these issues constitute  relations that can emerge between  projects in  a software ecosystem. With such a framing in mind, we look at coordination, leadership and control, and conflict resolution issues arising across projects that surround the NetBeans project.

---

[5] http://www.netbeans.org/servlets/ReadMsg?msgId=534707&listName=nbdiscuss
[6] http://www.netbeans.org/servlets/ReadMsg?msgId=534520&listName=nbdiscuss

## 5.1 Coordination

In addition to their IDE, NetBeans also releases a general application development platform on which the IDE is based. Other organizations, such as BioBeans and RefactorIT build tools on top of or extending the NetBeans platform or IDE. These organizations interact via bug reports, patches, and feature requests submitted to the NetBeans issue-tracking repository. Moreover, NetBeans (in part via its sponsoring organization) is a member of the Java.net and Java Tools communities, whose missions are to bring tool developers together to form standards for tool interoperability.

## 5.2 Leadership and Control

Leadership and control of the ecosystem is difficult to exert and more difficult to observe. However, at one point, NetBeans and its primary OSS competitor, the Eclipse Java IDE project (sponsored largely by IBM), considered merging as a single project. Ultimately, the union failed to emerge, largely due to (a) technical and organizational differences between Sun and IBM[7], including the inability or unwillingness to determine how to integrate the architectures and code bases for their respective user interface development frameworks (Swing for NetBeans and SWT for Eclipse), and (b) the potential for either company to be viewed as having lost in it's ability to assert technological superiority or design competence.

## 5.3 Conflict Resolution

Conflicts among communities in a software ecosystem can be especially complex considering differences in beliefs, values, and norms between organizations (both open and non-open source) in addition to technical hurdles.

NetBeans has a defined leadership and organizational structure, in part vis a vis its relationship with Sun Microsystems. Thus, Sun representatives play a significant role in macro-level conflict resolution involving the NetBeans community, as shown in the negotiations with Eclipse. Community member feedback extended beyond intra-community communication channels to include prominent technical forums (e.g. Slashdot and developer blogs). Unfortunately, many of these discussions occur after the collaborating developer has moved away from using NetBeans (often, in favor of Eclipse). Nevertheless, the feedback they provide gives both parties an opportunity to increase understanding and assists the NetBeans community by guiding their technical direction.

# 6 Discussion

The public communication channels we have seen used in OSS projects like NetBeans include mailing lists, defect repositories, requests for enhancement, Internet Relay Chat (IRCs), developer/stakeholder blogs and Web pages, trade forums, and developer conferences. Of these, mailing lists, defect repositories, and requests for

---

[7] http://www.adtmag.com/article.asp?id=8634, and
   http://www.eweek.com/article2/0,1759,1460110,00.asp

enhancement (RFEs) are intra-organizational--they exist within project community boundaries. IRC chats and developer conferences that facilitate communication may be intra or inter-organizational, in that they can be hosted by the community or by other organizations. On the other hand, stakeholder webpages and blogs and trade forums are purely inter-organizational. Communication channels provide means for enabling intrinsic governance in OSS projects through collaboration, leadership, control, and conflict negotiation processes. But they do not tell us much about how developers collaborate, lead, control, and resolve conflicts, nor what is collaborated on, led, controlled, and causing/resolving conflicts. We address these here.

In NetBeans, we have observed the following objects of interaction guiding OSS technical development and social integration processes: (a) project and software system architecture; (b) community vision/mission statement; (c) release plans and development roadmap; (d) community policies, task guidelines, and interaction guidelines; (e) defect reports and request for enhancements (RFEs); (f) mailing list discussions; and (g) private meetings (work done by organizations associated with the community). Arguing that project architecture is a primary coordination mechanism for software development, Ovaska and colleagues [16], and also Baldwin and Clark [2], collectively observed six coordination processes in multi-site software development like OSS projects. These include managing interfaces between system components, managing assembly order of system components, managing the interdependence of system components, communication, overall responsibility, and orientation (configuration) of the organization.

The link between organizational structure and system design has been known since Conway first published on the subject, however, in the NetBeans case, it is impossible to determine whether the system design evolved to reflect the desired organizational structure or vice versa. This observation also holds true for other large OSS projects. German [9] observes a similar coordination strategy in Gnome project: module interrelationships are kept to a minimum so each module can develop independently, thereby reducing the coordination burden across modules. Similar to NetBeans, Debian cross-module coordination is managed by a release team, whose role is to keep development on schedule. In contrast, system design can also restrict participation in OSS STINs. Core developers of the widely used Pidgin instant messaging client remain adamant that contributions to the project respect the strict isolation of user interface and communication protocol code even at the cost of added frequently requested functionality[8]. Of added note, the Gnome project does not have a single primary benefactor, like NetBeans, German reports similar governance and conflict resolution community structures.

Community interaction modes act as communication channels for governing, coordinating, and articulating of development tasks. Mission statements are important to the formation of the community social and technical infrastructure early in the community's lifespan when more concrete guidelines have not been explicitly stated (if established). They are the core instructions for the way individuals and organizations will interact with the community as a whole. But they are also a metric by which each release will be judged. Additional release planning activities in OSS typically consist of asserting the requirements for the release (what work will be done), the schedule of

---

[8] http://developer.pidgin.im/ticket/34

the release (when will the work be completed), and who will be responsible for what work (who will do what work) [17].

Defect/product recovery and redesign, as registered through submission of bug/defect reports is an integral coordination process. Like release planning, defect reports and RFCs (Request for Comments) tell developers both what work needs to be done as well as what has not been done yet, without an explicit owner or administrative supervisor to assign responsibility for doing it.

These observations suggest that governance processes are inherent in activities requiring coordination or leadership to determine which development tasks need to be done and when they need to be completed. This is analogous to what has previously been observed by management scholars (and also OSS developers) as adaptive "Internet Time" development practices [3] that enable a kind of project self-governance through adaptive synchronization and stabilization activities.

In some instances, leadership in coordinating development tasks is done in private meetings or communications between developers, for which little evidence is public or observable. However, we observed leadership and control of OSS project community through:

- Contribution of software informalisms (e.g., source, defect reports, requests for changes, news, internationalizations, etc. [17])
- Articulating and sharing technical expertise (e.g., on the mailing lists and defect repository reports, [7])
- Coordination of development and other tasks (e.g., through the role of the release manager, module maintainer, and source code contributors with "commit access" to shared source code repositories).

The NetBeans community is an unusual project: it receives the majority of its financial and developmental support from Sun Microsystems. Sun, as the primary benefactor and community founder, established the community vision, social and technical infrastructure, funds development by providing many core developers, and initiates most release plans, driving the development roadmap. Thus, Sun is most exposed to risks from community failure and external threats. As demonstrated by Sun's move to alter the project licensing scheme, exercising this authority unilaterally led to division within the community, risking breakdown of the project and development process. As such, social process conflict can give rise to conflict within the overall technical development process.

Drawing on this, sources of conflict that precipitate some form of active governance to deliberate and resolve may arise from: (a) community infrastructure, sociopolitical vision, and direction; (b) technical direction (what should be in the release, when should a release occur, which tools to use to develop software); (c) how developers can get involved in making decisions and what roles they play; and (d) relationships between and alignment of the diverse goals of many organized groups (e.g., corporations) and unaffiliated volunteers involved in the community. These conflicts are resolved through OSS governance activities in a variety of ways. When conflicts arise due to miscommunication or lack of communication between developers, or between developers and organized groups contributing to the community, resolution is reached by talking it out on community mailing lists. In more pronounced cases, it may take

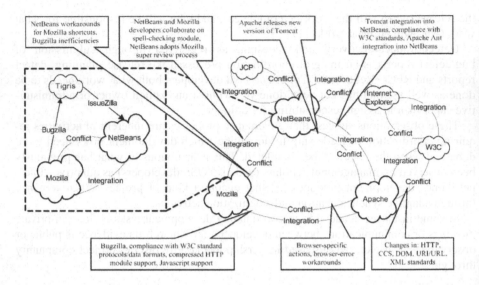

**Fig. 1.** An overview of integration and conflict relationships between NetBeans and other OSS projects that facilitate and constrain activities within NetBeans [10]

project veterans and highly influential community members to act as mediators. Failing this, in NetBeans, the project culture prescribes that developers shall bring the issue to the governance board for deliberation, who will issue a final decision on the matter. Board involvement is viewed as a last resort, and community members are encouraged to resolve their conflicts through other means.

We find social processes like collaboration, leadership and control, and conflict resolution are ways for governing OSS through articulating and reconfiguring the technical processes that are either unstated or understated. In a way, articulation is the background social process of making sure people understand the technical development process [18]. As such, when there is a breakdown, whose responsibility is it to address or resolve the breakdown? In the NetBeans project, accountability is only partially assigned but does exist in some fashions. No complete articulation of governance infrastructure exists in NetBeans. The emerging processes to do this are collaboration, leadership, control, and conflict negotiation, which are used to continually re-articulate the process and figure out what is going on at present. Based on our study, OSS is best understood neither as primarily a technical development or social process perspective, but instead as an inherent network of interacting socio-technical processes, where its technical and social processes are intertwined, co-dependent, co-evolving, and thus inseparable in performance.

# 7   Conclusions

The results and interpretations we present on  intrinsic governance forms, conditions, and activities as STINs are limited and therefore preliminary, though based on empirical case studies. They are limited in that our analysis focuses on two contrasting case studies, which differ in many ways, and thus represent merely an initial sample with little knowledge about whether what we have observed in representative of other types, sizes, or samples of OSS project communities. Additional studies may in turn lead us to revise our emerging model of how governance is realized in globally distributed OSS project communities. However, we believe that we have observed through empirical study of OSS (by us and others) the emergence of a comparatively small network of interacting socio-technical relationships that can serve as foundations that can account for how decentralized OSS projects can be self-governed. Such a result represents an alternative to the long dominant views that software development projects must be centrally controlled and explicitly managed, and must adhere to mature software development process capabilities, in order to produce complex yet reliable software systems.

## Acknowledgments

The research described in this report is supported by grants from the Center for Edge Power at the Naval Postgraduate School, and the National Science Foundation, #0534771 and #0808783. No endorsement implied.

## References

[1]  Augustin, L., Bressler, D., Smith, G.: Accelerating Software Development through Collaboration. In: Proc. 24th Intern. Conf. Software Engineering, pp. 559–563. IEEE Computer Society, Orlando (2002)

[2]  Baldwin, C.Y., Clark, K.B.: The architecture of participation: Does code architecture mitigate free riding in the open source development model? Management Science 52(7), 1116–1127 (2006)

[3]  Cusumano, M., Yoffe, D.: Software Development on Internet Time. Computer 32(10), 60–69 (1999)

[4]  de Laat, P.B.: Evolution of open source networks in industry. The Information Society 20(4), 291–299 (2004)

[5]  de Laast, P.B.: Governance of open source software: state of the art. J. Management and Governance 11(2), 165–177 (2007)

[6]  Elliott, M., Scacchi, W.: Free Software Development: Cooperation and Conflict in A Virtual Organizational Culture. In: Koch, S. (ed.) Free/Open Source Software Development, pp. 152–172. Idea Publishing, Pittsburgh (2005)

[7]  Elliott, M., Ackerman, M., Scacchi, W.: Knowledge Work Artifacts: Kernel Cousins for Free/Open Source Software Development. In: Proc. ACM Conf. Support Group Work (Group 2007), Sanibel Island, FL, pp. 177–186 (2007)

[8]  FOSSBazaar,org., https://fossbazaar.org (last accessed September 2, 2008)

[9]  Franck, E., Jungwirth, C.: Reconciling rent-seekers and donators –The governance structure of open source. J. Management and Governance 7(4), 401–421 (2003)

[10]  German, D.: The GNOME project: a case study of open source, global software development. Software Process–Improvement and Practice 8(4), 201–215 (2004)

[11]  Jensen, C., Scacchi, W.: Process Modeling of the Web Information Infrastructure. Software Process–Improvement and Practice 10(3), 255–272 (2005)

[12] Jensen, C., Scacchi, W.: Role Migration and Advancement Processes in OSSD Projects: A Comparative Case Study. In: Proc. 29th. Intern. Conf. Software Engineering, pp. 364–374. IEEE Computer Society, Minneapolis (2007)

[13] Lee, C.: Boundary Negotiating Artifacts: Unbinding the Routine of Boundary Objects and Embracing Chaos in Collaborative Work. Computer Supported Cooperative Work 16(3), 307–339 (2007)

[14] Markus, M.L.: The governance of free/open source software projects: monolithic, multi-dimensional, or configurational? J. Management. and Governance 11(2), 151–163 (2007)

[15] O'Mahony, S.: The governance of open source initiatives: what does it mean to be community managed? J. Management and Governance 11(2), 139–150 (2007)

[16] O' Mahony, S., Ferraro, F.: The Emergence of Governance in an Open Source Community. Academy of Management J. 50(5), 1079–1106 (2007)

[17] Ovaska, P., Rossi, M., Marttiin, P.: Architecture as a Coordination Tool in Multi-Site Software Development. Software Process–Improvement and Practice 8(4), 233–247 (2003)

[18] Scacchi, W.: Free/Open Source Software Development: Recent Research Results and Emerging Opportunities. In: Proc. European Software Engineering Conference and ACM SIGSOFT Symposium on the Foundations of Software Engineering, Dubrovnik, Croatia, September 2007, pp. 459–468 (2007b)

[19] Schmidt, K., Simone, C.: Coordination Mechanisms: Towards a Conceptual Foundation of CSCW System Design. Computer Supported Cooperative Work 5(2-3), 155–200 (1996)

[20] Simone, C., Mark, G.: Interoperability as a Means of Articulation Work. In: Proc. Intern. Joint Conf. on Work Activities Coordination and Collaboration, San Francisco, CA, pp. 39–48. ACM Press, New York (1999)

[21] Strauss, A.: The Articulation of Project Work: An Organizational Process. The Sociological Quarterly 29(2), 163–178 (1988)

[22] Shah, S.K.: Motivation, governance and the viability of hybrid forms in open source software development. Management Science 52(7), 1000–1014 (2006)

[23] NetBeans Issuezilla Issue Repository, http://www.netbeans.org/community/issues.html (last accessed November 28, 2009)

[24] NetBeans Community Guidelines, http://www.netbeans.org/community/guidelines (last accessed November 27, 2009)

[25] Workshop Summary: Software Development Governance 2008, http://www.cs.technion.ac.il/~yael/SDG2008/ (last accessed December 20, 2008)

# Evaluating the Readiness of Proprietary Software for Open Source Development

Terhi Kilamo, Timo Aaltonen, Imed Hammouda, Teemu J. Heinimäki, and Tommi Mikkonen

Department of Software Systems, Tampere University of Technology
Korkeakoulunkatu 1, FI-33720 Tampere, Finland
{firstname.lastname}@tut.fi

**Abstract.** As more and more companies are releasing their proprietary software as open source, the need for supporting guidelines and best practices is becoming evident. This paper presents a framework called R3 (Release Readiness Rating) to evaluate the readiness of proprietary software for open source development. The framework represents a checklist for the elements required to ensure a better open source experience. The framework has been applied to an industrial proprietary software planned to be released as open source. The evaluation has been carried out by both external and internal stakeholders. The early experiences of the case study suggest that the R3 framework can help in identifying possible bottlenecks before evangelizing the software to the open source community.

## 1 Introduction

Companies are getting more and more interested in releasing their closed source software products to open source communities. The two large scale examples of this are Sun Microsystems' opening of its Java platform during 2006 and 2007, and Nokia's actions to open the Symbian operating system during 2009-2010 [11]. As the trend is relatively recent, the phenomenon of opening industrial software is not well understood despite of the existence of general guidelines such as in [2,10]. In this paper we tackle the problematic of releasing industrial software.

Most often companies are not used to release the source code of their products. Their standard ways of behavior tend to be more biased to hiding than to releasing information. The processes, tools and infrastructure used by companies might turn out to be an obstacle for a successful release. The software itself might have been written so that open source developers run into troubles when trying to contribute. This suggests that there is a need for proper methodologies to evaluate the readiness of proprietary software for open source development. Such methodologies would help identifying possible bottlenecks before taking the software to the open. The bottlenecks are then resolved to in order to increase the success rate of community building around the software.

Given the above observations, the research questions we would like to explore include the following:

P. Ågerfalk et al. (Eds.): OSS 2010, IFIP AICT 319, pp. 143–155, 2010.

- What kind of evaluation criteria could be used to assess software readiness for open source development?
- How the evaluation should be planned and which stakeholders are involved?
- How to obtain data for the evaluation process?
- How to exploit the results of the evaluation process?

We argue that these issues have not been studied enough by the open source research community. The closest works to our study are the open source maturity models such as OSMM (Open Source Maturity Model$^{TM}$) [3], QSOS (Qualification and Selection of Open Source Software) [8], and BRR (Business Readiness Rating$^{TM}$) [1]. These models are typically used by companies that plan to use open source. The context of our research problem in this paper is just the opposite: taking software out from companies to open source communities.

The main contribution of the paper is two-fold. First, we discuss the specificities of the problem of opening proprietary software. Second, we present a framework called R3 (*Release Readiness Rating*) to evaluate the readiness of proprietary software for open source development. In order to demonstrate our approach, we have applied the framework to an industrial proprietary software planned to be released as open source.

The rest of this paper is structured as follows. In Section 2 we discuss related work and the challenges of opening proprietary software. The details of the R3 framework and the overall evaluation process are presented in Section 3. In Section 4, we evaluate the R3 framework in the context of two industrial case studies. Future work is discussed in Section 5 and finally, we conclude in Section 6.

## 2   Background

### 2.1   Open Source Maturity Models

Several methods have been developed assessing the maturity of open source software. For instance, Open Source Maturity Model$^{TM}$(OSMM) enables a quick assessment of the maturity level of an open source product. Products are ranked according to OSMM scores, which are evaluated in a three-phase process: 1) assess each product element's maturity and assign maturity score; 2) define weighting for each element based on the company's requirements and 3) calculate the score.

The Qualification and Selection of Open Source Software (QSOS) maturity model is a four-step iterative process: 1) define (and organize criteria), 2) assess (against the criteria), 3) qualify (define weighted scores, new and mandatory criteria) and 4) select (asses using the weights, and select). Another evaluation framework called Business Readiness Rating$^{TM}$(BRR) was proposed as a new standard model for rating open source software. The model consists of a four-phase process: 1) quick assessment filter (for quickly abandon bad candidates), 2) target usage assessment (for inputting the needs of the company), 3) data collection & processing (for collecting the actual information) and 4) data translation (which leads to one outcome: the rating).

Compared to the method we propose in this paper these maturity models take a totally different direction. Whereas our model attempts to study one software product which is going out from a company, these maturity models attempt to study a set of software products, one of which is selected to come in to the company. However, some ideas are still quite similar. For example, in both cases the architecture of the software plays an important role, and it can be evaluated similarly. On the other hand the infrastructure of an open source project might not be so important when evaluating open source software to be used, however it is a crucial element when building an open source community.

## 2.2  Opening Proprietary Software

Like any other online community [7], creating and maintaining a sustainable open source community for proprietary software can be considered as a multi-facet challenge. It is a complex process that is driven by various kinds of factors, which in turn can be grouped along six dimensions.

*Software.* Improved software quality may increase the success rate of community building. Quality can be enhanced by incorporating best practices, documentation, code cleanup, coding standards and convention. Furthermore, in order to support the community, source code may be accompanied with user manuals, API documentation, and architecture descriptions. It is vital to have the first experience with downloading, installing, deploying and using the software as easiest as possible.

*Infrastructure.* There are two key elements in any open source project: community and project repository. An open source engineering process should provide enabling tools and technologies to facilitate the planning, coordination, and communication between the community members. In addition, efficient mechanisms and tools are needed to facilitate the access and management of the project repository.

*Process.* Open source development can be regarded as an open maintenance process. A process needs to be established in order to handle decisions regarding the evolution of the software, maintenance actions, and release management. The process needs to balance between the practices of communities and the needs of the company.

*Legality.* The releasing company needs to select a license type (e.g. GPL versus LGPL) and a licensing scheme (e.g. single or multi licensing). In addition, source code should be legally cleared against IPR and copyright issues. Also, the availability of trademarks and names used in the software should be checked.

*Marketing.* Building an open source community can be regarded as a marketing challenge. Effective marketing strategies are needed to market the open source project to potential users and developers. Selecting an existing open source community as a target customer can be an important success factor.

*Community.* The releasing company should be ready to support the project community. For instance, community members should be provided with clear guidelines on how and what to contribute. Furthermore, company developers who are participating in the community should be trained for their new roles and should be given clear responsibilities. When the software is opened, it is vital that all information are made public and that private discussions are avoided. In addition, there should be trust among community members, zero tolerance of rudeness and no use of bad language both in the software artifacts and the communication among the members.

In the next section, we present a framework that addresses these questions by evaluating software in a pre-bazaar phase. The pre-bazaar phase helps in getting early feedback and experiences on using and evolving the software outside its original development environment. This may need the involvement of external stakeholders.

# 3    The Release Readiness Rating Framework

The Release Readiness Rating (R3) framework is a tool for planning the open sourcing of a software system. The goal of the framework is to help identifying possible bottlenecks and to eliminate them in so-called pre-bazaar phase, whose goal is to prepare the software to be released and the releasing company to the continuation of the life of the system as open source.

## 3.1    Framework Overview

The evaluation criteria for R3 consists of four different dimensions, including software itself, intended community and its roles, legality issues, and the releasing author. The output of the evaluation can be considered as a vector that determines the relative values of these different elements. The dimensions are further decomposed as indicated in Table 1. The table also lists the relative importance of the item.

The current weights are based on our experience on previous case studies. All dimensions are equal in weights when the individual items are associated with different weights.

## 3.2    Evaluation Criteria

In the following, we discuss the different views that should be considered when deciding the value set for the items representing different dimensions.

**Software.** The software itself forms an important aspect for any open source project for obvious reasons. Based on our experience, at least the following issues must be taken into account.

*Source code.* Fundamentally, any open source project deals with source code. When releasing a new software system to open source, there are numerous properties that the system itself, manifested in its code, should contain. These in particular include *quality of code, integrity, and coding conventions* that help

Table 1. R3 dimensions, items and relative weights

| Dimension | Item | Weight |
|---|---|---|
| Software | | *0.25* |
| | Source code | 0.5 |
| | Architecture | 0.4 |
| | Quality attributes | 0.1 |
| Community | | *0.25* |
| | Purpose and mission | 0.4 |
| | User community | 0.4 |
| | Partners | 0.2 |
| Legalitites | | *0.25* |
| | Copyright | 0.6 |
| | Licensing | 0.3 |
| | Branding | 0.1 |
| Releasing authority | | *0.25* |
| | Mindset, culture and motivation | 0.5 |
| | Process, organization and support | 0.3 |
| | Infrastructure | 0.2 |

other developers to participate in coding. The importance of coding conventions is highlighted, since introducing coding conventions as an afterthought can turn out to be impossible. Moreover, the source code should express *code of conduct*, which is a necessity for making the code public. Finally, *documentation* of the system is a practical necessity for attracting other developers.

*Architecture.* In order to make a software system easily approachable, it must be easy to understand by the developers. This in turn calls for an architecture that can be easily understood and communicated - and preferably documented. In addition to this, one should pay special attention to the design drivers of the architecture: Is the system designed as a monolithic system that solves a particular problem, or has the design taken into account extensibility, modifiability, and the use of the system as a subsystem in another system. Provided that the architecture has been designed with changes in mind, it is often easier for other developers to alter certain parts to create various types of new systems.

*Quality attributes.* In software, quality attributes are commonly associated with architectures. Indeed, many qualities, such as performance, scalability, and memory footprint, are often dictated by the architecture. However, when considering a completed software system, quality properties are often considered separately from the actual implementation, which makes the perceived quality of software being released as open source an important factor.

**Community.** In order to release a software system in open source, one generally has an idea on what kinds of developers should get involved. Careful planning of the intended community participants can have an impact on what to release and how.

*Purpose.* As already stated in [9], good work on software commonly starts by developers scratching their own itch. Therefore, we feel that in order to become attractive for developers, the released system should be of practical importance and relevant for the developers. This in turn enables one to solve their own problems, not further developing some random software for companies who seek profit in maintenance and creativity of others. Based on the above, the goal of the community is probably the most important single issue when releasing a piece of software as open source. Provided with a mission statement welcomed by developers, companies, and other organizations, a community can obtain support from numerous sources. The goal must be practical enough to be meaningful for the developers, as well as clear enough to manifest itself in the development. Unfortunately, estimating the attractiveness of a certain purpose is difficult, and therefore it is sometimes difficult to make assumptions in this respect. Moreover, since there commonly are numerous similar ongoing projects, the adequacy of the mission is only a prerequisite for a successful launch, not an automata for succeeding in community building.

*User community.* In addition to partners developing software, we feel that the potential for the user community is important. Based on recent findings, it seems that a community of 100 users can support one full-time developer. In contrast, provided with an active user community, development resources can be invested in actual development, and the user community can provide support for other activities, such as peer user guidance, documentation, and testing.

*Partners.* The definition of partners that join in the community can be straight-forward. For instance, if a releasing company has been subcontracting from another company, the latter may be automatically involved in the newly formed community. Moreover, the use of subcontracting may also imply that at least some documentation exists, which in turn simplifies introducing the system to other partners. In general, getting partners involved in a community guides the authoring company towards processes that liberate the development from company specific tools and practices. In contrast, if the company that is about to release a piece of software as open source has no partners that would share the interest in the development, there should be a clear plan to motivate others to join in the development effort.

**Legalities.** In the context of companies, one of the most commonly considered aspects of releasing software as open source is legalities. This is a wide topic to cover, and there can be several subtle differences in different contexts. Here, we assume a straightforward view where different concerns are discussed independently.

*Copyright and intellectual property rights (IPR).* Most commonly, companies release software whose copyright and IPR they own. However, things can be more complex, if subcontractors or open source communities have provided pieces of the system that is being released. If the company does not own the copyright, it can sometimes be obtained via different transactions.

*Licensing.* Provided with the copyright of the system to be released, the company is in principle somewhat free to determine its licensing scheme. However, the choice of license (or licenses if several alternatives are offered) also has an effect on how others perceive the community. This in turn potentially affects the willingness of developers to participate in the community effort. Therefore, since licenses that have strong copyleft, such as GPL [6], can be considered safe for community building, they bear some advantage over other licenses in this respect. However, since licenses with no copyleft, such as MIT [6] and BSD [6], are favored due to their liberal flavor in some other contexts, one should also take the mission of the community into account when defining the license. Moreover, license compatibility with related systems should also be considered. Furthermore, one can even compose a list of accepted open source licenses, which can be used during the development.

*Branding.* Availability of brand names is an issue for any project looking for a good name that can be used in public. While issues such as trademarks may bear little significance when developing an in-house product, once the product is released, branding becomes an issue. Moreover, since an open source project can be a long lasting one, selecting suitable brand names is important. The same applies to hosting the project, since in many cases it would be practical to reflect the name of the project also in the domain.

**Releasing Authority.** The final element we address in our framework is the releasing authority, most commonly a company in the scope of the framework, which is targeted to releasing in-house software as open source. However, also other parties can act as the releasing authority, including universities, non-profit organizations, and individuals.

*Mindset, culture, and motivation.* Sometimes the mindset of developers working in a company is somehow biased - either positively or negatively - towards open source development. This is particularly true when their pet project is about to be open sourced. In order to benefit from an open source community, the releasing authority should be mentally and culturally ready for dealing with developers outside the company under fair terms. The terms include equal access to code, similar guidelines and conventions, as well as mutual respect. We believe that the seeds of building such cooperative relation should somehow be sewn well before entering the pre-bazaar phase. Therefore, evaluating the readiness of the company to go for open source is fundamentally dependent on mindset, culture, and motivation.

*Process, organization, and support.* In order to gain benefits from open sourcing a system, the releasing authority should have a system in place that provides support for users and developers. This requires planning of a process that is to be followed, and putting the process in practice by the support organization. Establishing such support organization is a natural step to take towards the end of the pre-bazaar phase. However, it should be in place before the actual release, since support should be available from the very beginning.

*Infrastructure.* In order to establish an open source project, the releasing authority sometimes must be prepared to provide infrastructure. For instance, the company that releases a piece of software may provide web servers for hosting the system, as well as maintain a build system needed for compiling the code on top of certain reference hardware. While some systems do not need such support as such - it would be perfectly reasonable solution to release a vanilla Linux program in SourceForge - companies often wish to gain visibility through offering the download opportunity. Moreover, if a company is releasing a system targeted for the development of embedded systems, it is only reasonable to assume that also tools for composing builds are offered from the very beginning in open source.

The R3 evaluation model is organized into three main levels. For each dimension there are a number of categories. Each category is then associated with a number of measures (i.e. questions). This is illustrated in Figure 1 taking the software dimension as example.

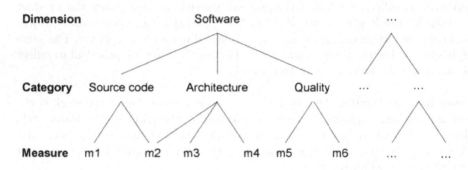

**Fig. 1.** The R3 framework model

### 3.3   Evaluation Process

The diversity of software products (and different goals of companies) makes it impossible to evaluate all software in similar fashion. The evaluation model itself must be tuned to take into account the characteristics of the product under release. Not all criteria make sense to all cases, and some crucial criteria might be missing. The proposed R3 model should be considered as a template which has to be instantiated to each case. Instantiation R3 means going through all aspects of the model and validating that they are appropriate to the case in hand.

At the concrete level the evaluation process starts with downloading R3 Spread Sheet Template from `http://tutopen.cs.tut.fi/R3/R3_Template.xls`. Instantiating the template requires removing and adding dimensions and criteria to the spread sheet. Also the evaluation weights require attention from the evaluator.

**On the Criteria.** The evaluation criteria form a continuum from a criterion that can stop the release process to others that can be easily fixed. Examples of the former are some legality issues, like possible copyright and IPR violations, and probably mindset of partners. Changing these is hard or even impossible. The latter group can be worked on during the releasing process: usability, and quality can be improved; infrastructure and process can be organized. An example continuum is depicted in Figure 2.

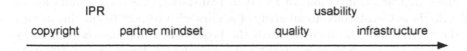

**Fig. 2.** The continuum of criteria

**Actual Process.** The evaluation process consists of three phases depicted in Figure 3.

Phase 1: Deciding the evaluation order. The evaluation is carried out according to continuum of criterion. This allows early no-go decisions. If, for example, the company does not own copyright of the product, it makes no sense to continue the process. However, there is no one unique and universal order of the criteria, but the order is fixed in the beginning of the process.

Phase 2: Data collection and processing. Most of the work is done in this phase. The criteria are evaluated one by one in the order fixed in the first phase. This activity is shaped by the framework model presented in Figure 1. First a dimension (e.g. software) is picked, then a specific category (e.g. source code) is selected, and finally concrete questions are answered. After each evaluation the decision of continuing the release process is made. Each measurement is filled to a standard spreadsheet with justification of the evaluation.

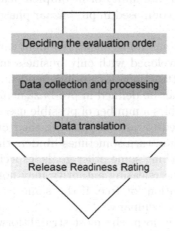

**Fig. 3.** Evaluation process

Measures require expertise from different fields: engineers, marketing people, legal experts and external open source experts. For example, the legal department of the company is often contacted in the beginning of the evaluation to verify the copyright and possible IPR issues of the release. Engineers take care of technically-oriented criteria. External open source experts have probably the best understanding of the whole release process, and they know how open source communities operate. The releasing process reminds much of marketing challenges, therefore, marketing people are valuable for the process.

Phase 3: Data translation. In the data translation phase the evaluation of the criteria is transformed to an array of scalars with respect to the dimensions of the framework. The final result of the process is a four-dimensional array of evaluations of each dimension: software, community, legalities and releasing authority.

## 3.4    Open Source Engineering

R3 assessment of proprietary software is not just a pass-or-fail process. The outcome of R3 evaluation is a set of recommendations based on which the software under evaluation and its development environment undergoes an open source engineering process. This process needs to be carried out before evangelizing the software to the open source community. The aim is to eliminate the problems and shortcomings identified during the assessment process. This will increase the success rate of community building and sustaining. The open source engineering process itself is driven by different kinds of influential factors that follow the same criteria as we used in the R3 framework.

In the case of the *software* itself, any considerable rework requires an extensive investment. Therefore there are numerous restrictions on what can be accomplished during the pre-bazaar phase. Still, it is possible to clean up the code, if there are some company specific remarks in comments. Since the code may already be in use in products, special attention must be paid to determine what to do with comments that indicate faulty or incomplete features. To some extend, documentation can also be composed in pre-bazaar phase, or simply included in the comments in the code.

Adding purpose to a *community* as an afterthought can be difficult. Assuming that a system has been developed with only business interests in mind, it can be difficult to introduce attractions for an independent developer. However, a mission for a community can be defined in pre-bazaar phase, provided that the software to be released enables a number of possible uses. Unfortunately companies can be somewhat biased towards supporting their own plans regarding the released system only, which in turn sometimes hinders the outside participation in the development for reaching some other goals, especially if the missions are conflicting. For instance, the releasing authority may not be willing to incorporate a community contribution for free, if the same feature can be sold for a commercial customer by the company.

*Legalities* most commonly form the most straightforward category of items. There is a lot of freedom to define the other legal aspects once the copyrights have been provided. However, copyrights can be difficult to obtain in pre-bazaar phase only.

The seeds of building cooperative relation between the *releasing authority* and the actual community should in our opinion be somehow sewn at the latest when entering the pre-bazaar phase. This can already be evidenced by existing ways of working and infrastructure, but they can also be introduced later on.

## 4   Case Study

In order to demonstrate our approach, we have applied the R3 framework to measure the open source readiness of an industrial software platforms: Wringer and Gurux. The Wringer software is a JavaScript binding platform for GNOME/GTK+ [4] using V8 [12] as JavaScript engine. It was originally developed by Sesca Mobile Oy. The Gurux software [5] is a platform for developing device communication systems.

### 4.1   The Wringer Case

The R3 evaluation of the Wringer platform has been carried out separately by the releasing authority as an internal stakeholder and by us as external open source experts. It was observed that we agree on most answers. However there are still a number of differences. For instance, in the software dimension, there were few differences with respect to rating the technology used, the use of well-known design principles, rating of bugs and warnings, and scalability level. We received more pessimistic answers from the releasing authority. A partial reason for this could be that the company is assessing Wringer, which is a prototype software, relative to other high quality products developed inside the company.

On the opposite, the answers with respect to the community perspective for instance have been mostly identical. This probably shows that both parties are aware and honest about the community-related properties of the software. Also this shows that both parties are fairly aware of related user and developer communities.

Taking a numerical perspective, we noticed that the software and legality dimensions received the best scores compared to the community and releasing authority dimensions. This confirms our hypothesis that companies are generally dealing with open source from legality point of view. We could also infer that software-related properties are considered important irrespective of whether the software is supposed to be used and developed as closed source or as open source.

The low rating of the community dimension suggests that the company have to work on making the software more attractive to open source communities, involve business partners if possible, and look for potential users of the software. As for the releasing authority dimension, the low rating suggests that the company is not fully ready for open source operations. Concrete remedial actions include training internal developers, setting clear open source related processes, and building an infrastructure for the project. As mentioned earlier, these remedial actions are to be carried out in the context of an R3 evaluation post activity called the open source engineering process.

## 4.2 The Gurux Case

Four people from the releasing authority carried out the R3 evaluation for the Gurux platform before it was released as open source in November 2009. The overall impression was positive and R3 was considered a valuable and useful tool in the pre-bazaar phase. The evaluation process acted as a good checklist for things that need to be considered when planning the release and showed well the items where most improvement is required. In addition, those items where improvement would be most beneficial were easily identified with R3, i.e. the releasing authority knew where to focus most of the effort.

Some items were not seen relevant in the case of the Gurux platform. However, this was not a significant problem for the process as irrelevant items were simply skipped by the people doing the evaluation.

Sometimes choosing the correct grading was found hard. Grades like "well" and "reasonably" may mean different things to different people as these types of grades depend on how things are seen by individuals doing the evaluation and what they value.

## 5    Future Work

We are in the process of improving the R3 framework based on our experiences with the case projects. This includes adjusting the weights, proposing new metrics, and covering other dimensions if found necessary. Currently the weights are chosen based on experience gained from earlier similar case studies. As the framework is applied to further cases, enough data will be gathered to enable us to better finetune the weights.

Furthermore, the case studies confirmed that it is difficult to come up with a one R3 framework template for all software projects. For instance, usability as a quality metric should be considered for end user software but was found less relevant in the case of software platforms like Wringer and Gurux. We plan to provide different templates for different kinds of software. Still, it is highly probable that the R3 framework template needs to be adapted to the needs of the subject software on a case by case basis.

## 6    Conclusions

Similarly to other major steps in software development, aiming at releasing a piece of in-house software as open source requires an engineering effort. Moreover, in order to estimate the outcome of the release, tools and techniques are needed for evaluating the potential of the emerging community as well as the attractiveness the system for external developers.

In this paper, we have introduced the concept of Release Readiness Rating Framework to determine how complete an in-house piece of software is for releasing in open source, and what its potential to attract external developers is - in essence evaluating how easily a cathedral could be transformed into a bazaar. We also discussed potential engineering actions that can be taken as a part of this transformation process, and provided a summary of two industrial cases.

# References

1. BRR. http://www.openbrr.org/ (Last visited March 2009)
2. Fogel, K.: How to Run a Successful Free Software Project. O'Reilly Media, Inc., Sebastopol (October 2005)
3. Golden, B.: Succeeding with Open Source. Addison-Wesley, Reading (2004)
4. GTK+, http://www.gtk.org/ (Last visited March 2009)
5. Gurux/open source, http://www.gurux.fi/index.php?q=OpenSource (Last visited December 2009)
6. Licences, http://www.opensource.org/licenses (Last visited February 2009)
7. Preece, J.: Online Communities: Designing Usability, Supporting Sociability. Wiley, Chichester (2000)
8. QSOS, http://www.qsos.org/ (Last visited March 2009)
9. Raymond, E.S.: The Cathedral and the Bazaar. O'Reilly Media, Sebastopol (1999)
10. Stürmer, M.: Open source community building. licentiate thesis (2005)
11. The Symbian Foundation, http://www.symbian.org/ (Last visited December 2009)
12. V8 JavaScript Engine, http://code.google.com/p/v8/ (Last visited February 2009)

# Where and When Can Open Source Thrive?
# Towards a Theory of Robust Performance

Sheen S. Levine[1] and Michael J. Prietula[2]

[1] Singapore Management University
50 Stamford Road
Singapore 178899
sslevine@sslevine.com
[2] Emory University
Atlanta, GA 30322
United States of America
prietula@bus.emory.edu

**Abstract.** While the economic impact of, and the interest in, open source innovation and production has increased dramatically in recent years, there is still no widely accepted theory explaining its performance. We combine original fieldwork with agent-based simulation to propose that the performance of open source is surprisingly robust, even as it happens in seemingly harsh environments with free rider, rival goods, and high demand. Open source can perform well even when cooperators constitute a minority, although their presence reduces variance. Under empirically realistic assumptions about the level of cooperative behavior, open source can survive even increased rivalry and performance can thrive if demand is managed. The plausibility of the propositions is demonstrated through qualitative data and simulation results.

**Keywords:** Innovation, Exchange, Performance, Agent-based Modeling.

## 1 Introduction

Interest in open source is booming. With its roots in freely shared software [1], the term has been expanding to include broader instances of product development [2], process innovation [3], and knowledge exchange among end-users [4].[1] All in myriad fields, including technology, science [7], medicine [10], and law [11], among others.

Open source software has become a viable alternative to commercial software. What has once been the domain of computer hobbyists (or "hackers"), has gained acceptance with major corporations and governments [12] and created hundreds of millions of dollars in value [13]. In software, where open source has frequently been studied, developers were found to contribute through collective action organized online and in absence of direct monetary compensation [1, 14, 15]. Not only that the

---

[1] We use "open source" as shorthand to refer not only to open source software, but also to what scholars have called "community-based innovation" [e.g., 5, 6], "commons-based peer production" [7, 8], "free software" and "software libre" [9]. We are not attempting to diminish the differences between them.

P. Ågerfalk et al. (Eds.): OSS 2010, IFIP AICT 319, pp. 156–176, 2010.

produced software is shared among contributors, but it is also freely available to non-contributors for personal and (often) commercial use.

The actions of such collectives, working in unison yet often without ever meeting each other to create products and services of economic value, have captured the attention of the media, general public, business practitioners and academics. Yet despite its growing economic impact, a theoretical explanation for its performance is still incomplete. While many have documented cases of open source and theorized about the motivation of participants, internal organization and market dynamics, there is little understanding of its performance. For instance, while open source has been discussed often in the case of software, recent accounts suggest that it may be much more widespread. In which industries, then, can we expect open source to compete with firms? Which goods can be successful produced by open source? Which environment does it require to succeed? Ultimately, what affects the performance of open source?

We build on previous explanations, mostly pertaining to open source software, and combine original fieldwork with agent-based simulation to generate propositions that offer analytical and predictive power as to the performance of open source. These can aid in pinpointing the elements that make such entities successful in achieving their collective goals. We proceed by reviewing the relevant literature and briefly presenting our fieldwork and the agent-based model and. We then present the computational experiments we conducted and the resulting propositions, the implications of which are later discussed.

## 2 Literature Review

The last decade saw booming interest in open source as a mechanism of production and innovation initially in software and more recently in realms beyond it [16, 17] A bibliometric query reveals a pattern of rapid growth in the use of "open source" and "open innovation" as terms in the academic literature. While in the three year period 1995-1998 "open source" was mentioned only 12 times, it appeared 32 times in 1999, 40 times in 2000, and more than double that in the following year. In 2009, the term appeared 687 times, representing an impressive seventeen-fold growth in a decade. [2]

### 2.1 Open Source Differs from Firm-Based Innovation

Scholars generally agree that open source represents a distinctive way of innovating as well as producing goods and services (hereafter: "goods"). Open source is "a fundamentally different organizational model for innovation and product development"

---

[2] The citation count was carried out in January 2010, using the Web of Science database, which provide access to current and retrospective bibliographic information, author abstracted, and cited references from over 10,000 leading journals of science, technology, social sciences, arts and humanities and over 100,000 book-based and journal conference proceedings. It provides access to seven databases: Science Citation Index (SCI), Social Sciences Citation Index (SSCI), Arts & Humanities Citation Index (A&HCI), Index Chemicus, Current Chemical Reactions, Conference Proceedings Citation Index: Science and Conference Proceedings Citation Index: Social Science and Humanities. See http://thomsonreuters.com/products_services/science/science_products/a-z/web_of_science

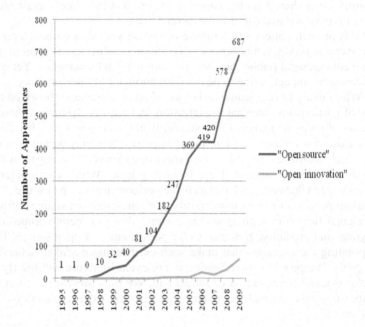

**Fig. 1.** Appearances of the terms "open source" and "open innovation" in the academic literature, 1995-2009

[18]. While open source may complement (or replace) firm-based innovation [5, 19], it is emphatically different in elements as diverse as the individual motivation to participate [20], production [21], governance [18, 22] and market behavior [23, 24].

The characteristics that distinguish open source from firm-based (and market-based) innovation are related to its unique benefits. First is the ability to build upon others' work in the most direct way because the architecture of the good is visible and openly available, such as the lines of code in software or the circuits of a mobile phone [25, 26]. Second, open source may enhance innovation and problem solving by "removing barriers to entry to non-obvious individuals" [27], allowing users to contribute as much or as little as they may wish (or able) to contribute, without requiring upfront commitment or pre-specified roles, unlike in most firm settings [7, 28]. For firms, open source can generate rents from innovations created not by its personnel, but by users or consumers [29-31]. Better performance can also come from engaging in open collaboration with other firms [32, 33]. At the societal level, open source can benefit public welfare [7, 34] as well as foster virtue [8].

## 2.2   A Lingering Puzzle of Performance

Yet those unique benefits of open source also pose several important puzzles [16]. It is intriguing that open source thrives on the internet [but also elsewhere, cf. 35], which is largely devoid of "the social signaling, cues, and relationships that tend toward moderation in the absence of law" [36]. These were thought to be necessary for such collaborative sharing behavior to occur [cf. 37]. The same conditions that give rise to open access and consumption with minimal intellectual property rights should have also caused it to quickly collapse under the weight of free riding, as theoretician have predicted [38, 39, cf. 40]. It is all the more puzzling because participants in open source effort could have easily regulated the sharing of costs and benefits by restricting access to those able and willing to contribute. Sharing does not *necessitate* open source. [cf. the notion of club goods, 41]. Nevertheless, open source seems resistant to free riding and highly skewed contributions [28, 42, 43].

While there is a large and growing body of knowledge on innovation in firms, the same has been developing for open source production and innovation. It is hard to imagine how the same conditions that promote innovation in firms would apply unchanged to open source, given the dramatic differences in the why innovations are produced [for a review see 5, 42]. Hence, we see a need to identify variables that exert important effect on the performance of open source organization, be it a coding community, file sharing or an advice forum [44]. Such theory would be essential in encouraging it, as some have urged to do [45, 46].

## 3 From Qualitative Data to a Model of Open Source

We build on qualitative data originating in a non-software environment, where a product of economic value is openly shared, to develop agent-based modeling. We use the two approaches jointly to investigate the phenomenon, utilizing the complementary nature of a grounded theory and formally expressed theory [47]. Qualitative accounts are rich in detail, evocative, describe processes lucidly, and possess high external validity [48, cf. 49]. However, qualitative data is hardly parsimonious, affords little in generalizability, and allows only limited field experimentation [cf. 50, 51]. Formal models have been employed in some of the pivotal studies in organizational theory [52, 53]. Agent-based models, of the kind we employ here, are increasingly seen as a promising way to study complex phenomena, whether social or organizational [54-56]. They have been used in the study of open source [57]. As recently demonstrated [58], agent-based models can capture features of the social order, such as embeddedness of actors and the emergence and dynamics of norms, that are difficult to represent by traditional analytic models.

### 3.1 Specifying the Model: Building Blocks of Open Source

In line with the boarder conceptualization of open source, which extends beyond software to other realms of innovation and production, we sought to indentify elements that were present in the specific field setting and also theoretically distinguishing between open source and non-open-source systems. While an all encompassing definition is beyond the scope of this paper, we present a working definition that marks the boundaries of our framework. It is applicable to all systems that feature the following elements.

**Table 1.** Modeled Elements of Open Source

| Elements | Description | Corresponding Model Manipulation | Theoretical and Empirical Referents |
| --- | --- | --- | --- |
| Open access to contribute and consume | Anyone can join the development process or partake in its outcome, regardless of their level of contribution | Any agent in the model can participate in development or consumption with no exclusions | [1, 17, 44, 59, 60] |
| Create products of economic value | Products have clear economic value. That is, organizations whose primary purpose is social interaction (e.g., primary social group) are excluded | The dependent variable is the overall value created by the efforts of the collective, reflecting the economic value created | [17, 19, 35, 61] |
| Interaction and exchange activities are central | Participants interact, exchange and reuse each other's work | Each agent can engage in its own work or exchange with others | [7, 25-28, 62] |
| Participants work purposefully yet loosely coordinated | Coordination, structure and hierarchy are emergent and less specified compared to a firm or market setting | Agents coordinate only when engaging in exchange events | [5, 18, 21, 22, 42, 62-64] |

The elements allow the model to capture multiple instances of what is generally regarded as open source, including beyond software [35], e.g. Wikipedia [28, 36, 65], user-run support forums [44, 66] and file sharing services.

### 3.2  Specifying the Model: Variables Related to Open Source Performance

While the variables presented above are thought to define open source as such, the manipulated variables are likely to affect the performance of the system.

**Degree of Rivalry.** A good is considered non-rival if for any level of production the cost of providing it to a marginal (additional) individual is zero [41]. Non-rivalry does not imply that the *total* production costs are low, but that the *marginal* production costs are low.[3] Thus, sharing a pure non-rival good does not decrease the utility of any individual from consumption (e.g., sharing a digital music file or software code). Few goods are perfectly rival or non-rival and one can imagine a continuum of rivalry [18, 71].

---

[3] For instance, a non-rival good such as national defense is extremely expensive to produce, but the cost is insensitive to the number of beneficiaries.

**Table 2.** Model variables Manipulated in the Studies

| Level of Analysis | Variable | Description | Corresponding Model Manipulation | Theoretical and Empirical Referents |
|---|---|---|---|---|
| Individuals | **Agent & Populations Characteristics** (Study 1) | Agent characteristic is defined in terms of its likelihood to contribute; population characteristic is defined in terms of the distribution of agent characteristics in a population. | Agent characteristics are drawn from three types: cooperator, reciprocator, and free rider; Population characteristics are systematically varied and their impact on performance is assessed. | [18, 38, 40, 42, 43, 67] |
| | **Demand** (Study 2 & 3) | Demand describes the uniqueness of a sought resource. | Levels of Demand were manipulated to discern effect on performance, and how Agent, Population & Rivalry characteristics interact with it. | [7, 27, 28, 68-70] |
| Goods | **Rivalry** (Study 2 & 3) | Goods differ on the cost associated with providing them to a marginal individual, i.e., the cost of contributing them to others. E.g., rival good cannot be used simultaneously by more than one agent. | Levels of Rivalry were manipulated to discern effect on performance, and how Agent, Population & Demand characteristics interact with it. | [19, 41] |

**Composition of Cooperative Types in the Population.** People's behavior with regards to cooperation is heterogeneous between- and stable within-individuals [67, 72-75] with remarkable stability across cultures [76]. An individual's type is so stable that "a group's cooperative outcomes can be remarkably well predicted if one knows its type composition" [67]. We reflect those recent findings by assigning agents to follow one of the three empirically observed types: 1) Cooperators – which contribute to others even at cost to self [77]; 2) Reciprocators – which contribute based on others' behavior [cf. 78], e.g., contribute if they observed others doing so; 3) Free Riders – which do not contribute, but still consume. Kurzban and Hauser's [67; hereafter: KH] empirical results (hereafter: KH ratio) suggest that 13% of individuals can classified as Cooperators,

53% as Reciprocators, and 20% as Free-riders. The behavior of the remaining 14% is not stable enough to be classified. In implementing the KH types in the model, a global parameter of group contribution behavior was defined in terms of the exchange ratio (exchanges/attempts), reflecting the population rate of contribution for a given period, εp.[4]

**Demand Homogeneity.** Of particular importance is the homogeneity of participants with regards to demand, which may be dependent on relevant skills [e.g., 7, 27, 28] and the motivation to contribute and consume [e.g., 4, 18, 69, 70]. Together, they both determine the extent to which participants are placing demand on the open source system.

## 4 Computational Experiments and Results

The model consists of a simple population of 100 agents ($a_1,....,a_{100}$), each of which follows the algorithm described in the figure.[5]

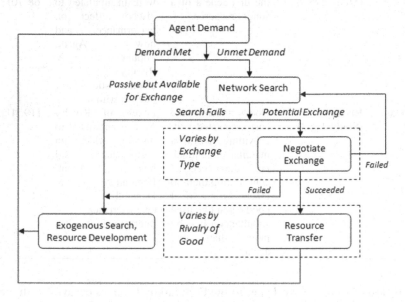

**Fig. 2.** Agent behavioral algorithm for the simulation

### 4.1 The Impact of Cooperative Types on Performance

We consider the impact of a range of cooperative type mixes, modeled after KH types, on the time (measured in steps) that it takes to complete the set of problems (goals) in an open source system. Hundred agents were randomly assigned 100 goals each to complete, with initial resources levels randomly drawn from a [0,100] flat distribution.

---

[4] Detailed description of implementation of KH types in the model is available from the authors.
[5] Complete description of the model is available from the authors.

All goals were unique (i.e., low demand for each resource) and all resources were non-rival. The values of each type were varied in steps of 10% ranging from 0.0% to 100%, with 100 runs made for each configuration.[6] Each run defines an instance of a problem faced by the population of the open source system. All problems are considered independent, such that there is no knowledge carryover from prior runs.

**Results.** The results are plotted the figure as overall performance means for each level of cooperator percentage in the population (solid line), with the vertical bars denoting 95% confidence intervals. In the figure, the x-axis depicts the percent of cooperators in the population (e.g., 5%). For each set of runs at a fixed cooperator percentage of the x-axis, the remaining percent (i.e., 95%) of the population contains reciprocators and free-riders whose mix is systematically varied from all reciprocators to all free-riders. An expected main effect of cooperators is apparent and confirms our intuition about cooperation in general – the more cooperators in the population, the better performance of the open source model.

*Proposition 1a: Cooperators Improve Performance.* Over a mix of reciprocator and free-rider levels, a higher ratio of cooperators leads to better performance.

Although the performance plot is increasing (see figure), that there is a distinct concavity in the graph. We continued by analyzing further the effect of cooperator ratio on performance to reveal a significant pattern of nonlinearity. The figure shows (in dotted line) the added benefit in performance due to higher percentages of cooperators. As is suggested in the figure, there is a decreasing marginal benefit to the addition of cooperators to the population.[7] The largest gain occurs when the percentage in the population jumps from 1% to 5%, but trends down as more cooperators enter the population.

*Proposition 1b: Decreasing Marginal Returns from Cooperation.* Over a mix of reciprocator and free-rider levels, increasing the ratio of cooperators has decreasing marginal positive effect on performance.

Finally, an examination of the confidence intervals in the figure indicates that the spread decreases (i.e., the intervals become "tighter") as the ratio of cooperators increases. That is, the variation in performance for a given ratio of cooperators decreases as there are more cooperators in the population.

*Proposition 1c: Cooperators Reduce Variance.* Over a range of reciprocator and free-rider levels, a higher ratio of cooperators reduces variation in performance. Stability, and thus predictability, in model performance is accommodated by increasing the ratio of cooperators in the population.

---

[6] For all three studies, cell sample sizes (replications) were planned in order to detect absolute effect sizes with $\alpha = 0.05$ for all main effects and interaction contrasts with likely power $\geq .80$ [79].

[7] An analysis of variance confirmed that differences in performance across percentage levels of cooperators were significant ($F(1,8588) = 4897.8$, $p < .001$). A post-hoc analysis (Games & Howell 1976) revealed that all means differed from each other significantly and the subtended line connecting the means had a best fit with a logarithmic model ($R^2 = .842$, SE = .075). All analyses conducted using SPSS 17.0 (www.spss.com) and Statistica 8 (www. statsoft.com).

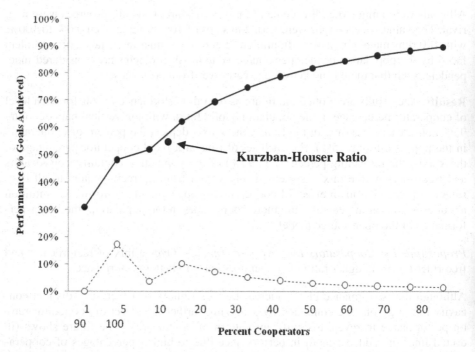

**Fig. 3.** Mean performance for cooperator ratio in the population (*solid line*), marginal improvement in performance (*dotted line*)

What accounts for the higher variation and unpredictability in performance with lower ratios of cooperators? To answer this question, we examined the sensitivity of performance to these specific mixes of reciprocators and free-riders. As can be seen in the figure, for each cooperator percentage, there is an embedded plot that shows performance decreasing from left to right, as less reciprocators (and more free-riders) enter the population. We interpret this as a distinct sensitivity to the percentage of free-riders. For example, as indicated in the figure, populations with 5% cooperators exhibit a wide variation in performance depending on the particular mix of reciprocators and free-riders (compare performance in point A in the figure with performance in point B). Thus, variance in performance is driven by the ratio of reciprocators to free-riders.

As described in Figure 4, performance drops can be mitigated: for example, a change of 95% reciprocators to 55% reciprocators results in only a 3% average drop in performance, after which a significant decline performance ensues (C in the figure). A pattern of compensatory substitution between cooperators and reciprocators is revealed in the data: high levels of reciprocators can compensate for low levels of cooperators, and vice versa. For example, in the figure, the Φ-line depicts a performance level of 42%, which can be achieved with a population with 1% cooperators, but also with 5% cooperators or 10% cooperators, all under various mixes of Reciprocators and Free-riders. However, this flexibility of these types of tradeoffs decreases as the percent of Cooperators increase in a population. In part, this is due to the decrease in sensitivity to the number of Free-riders in the population and, consequently, the decline in performance variance.

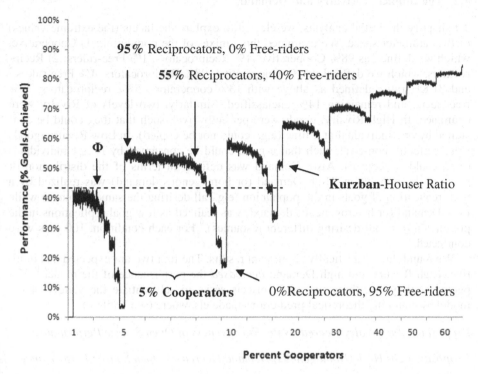

**Fig. 4.** Performance by varying types of agents. For each constant percent of cooperators (*x-axis*), the relative ratio of reciprocators to free-riders is varied (high to low reciprocators).

*Proposition 1d: Cooperator-Reciprocator Trade-off.* Reciprocators improve performance and higher-ratios provide performance improvements. But the performance improvements decline as more Cooperators are present.

Intuitively, the finding is driven by the varying presence of *reciprocators* in the population, which is reversely correlated with the presence of free riders. Empirically, reciprocators, who adjust their behavior according to that of those around them, are the largest section of human population [67, 76]. At low levels of cooperators (left side of the X axis) it means that a large chunk of agents are willing to "change colors" according to the situation. Performance is then driven by the group (cooperators or free riders) that serves as the "role model" for reciprocators. In each step in the analysis, we held the ratio of Cooperators constant while varying the ratio of Reciprocators to Free-riders. As the ratio of Free-riders grew, they exert growing influence on the behavior of reciprocators pushing them to behave as free-riders. The conversion processes accelerates, which leads to rapid decrease, indeed a collapse, in performance. As one travels to higher ratios of cooperators (by moving towards the right side of the X axis), there is more influence on reciprocators to mimic a cooperator-like behavior. Reciprocators are less likely to be influenced by free-riders, the variance in performance decreases and the substitution effect is lessened.

## 4.2 The Impact of Rivalry and Demand

To simplify the initial analysis, we elected to explore the facets (the extreme values) of the parameter space. We examined three mixes of the population: 1) Cooperative, which we defined as 98% Cooperative, 1% Reciprocators, 1% Free-riders; 2) Reciprocators, which we defined as 1% Cooperative, 98% Reciprocators, 1% Free-riders;[8] and 3) KH ratio, defined as above with 13% cooperators, 53% reciprocators, 20% free-riders, and remaining 14% unclassified. Similarly, two levels of Rivalry were examined. In High Rivalry, goods were perfectly rival, such that they could be consumed by one individual at a time (e.g., could not be copied). In Low Rivalry, goods were perfectly non-rival, such that a good could be consumed by N > 1 individuals (e.g., could be copied). Also, Demand was defined in terms of the distribution of goals in the population. High Demand (or homogenous demand) was realized by a high replication of goals in the population (e.g., all desiring the same resource) while Low Demand (or heterogeneous demand) was realized as few goal replications in the population (e.g., all desiring different resources).[9] For each condition, 100 runs were conducted.

We found three statistically significant results. The first two are expected and intuitive. High Rivalry and high Demand *decreased* the performance of the model.[10] Expected as they are, these two main effects serve to substantiate the validly of the model by matching theoretical predictions made elsewhere (see Table 2).

*Proposition 2a: Rivalry Decreases the Effectiveness of Open Source Performance.*

*Proposition 2b: High (Homogenous) Demand Decreases Open Source Performance.*

The third result is less expected. As visible in the figure, we found an interaction: under low Rivalry conditions performance is at its highest when Demand is high (50.9%); however, when Rivalry conditions are high, high Demand resources drive performance to its lowest level (21.1%). Recall that high demand conditions would be reflected in many agents seeking the same resource type and vice versa. The consequences for resource availability are specified by the Rivalry factor interacting with the Demand – high demand results high exchanges and resource redundancy when the Rivalry is low (e.g., exchanging music files). However, when both Demand and Rivalry is high, available resources existing in the network are quickly extracted and resistant to exchange. Therefore, available supply, and consequently performance, would vary across Demand conditions subject to resource availability within Rivalry constraints.

---

[8] One percent rather than zero was used in order to prevent conditions where no exchanges would occur in the case of cooperators, and therefore was incorporated in the other conditions for balance. Dominant free-riders (i.e., 98%) were eliminated from this analysis because of the extremely low performance.

[9] High Demand was determined by a random draw from a [90-100%] flat distribution of goal redundancy while Low Demand was determined by a random draw form a [0-10%] flat distribution.

[10] The overall main effects for Rivalry and Demand were $F(1, 1196) = 87.3$, $p < .001$ and $F(1, 1196) = 26.8$, $p < .001$ respectively. The interaction was $F(1, 1188) = 63824.0$, $p < .001$, and a post-hoc Tukey analysis indicated that all means differed significantly ($p < .001$).

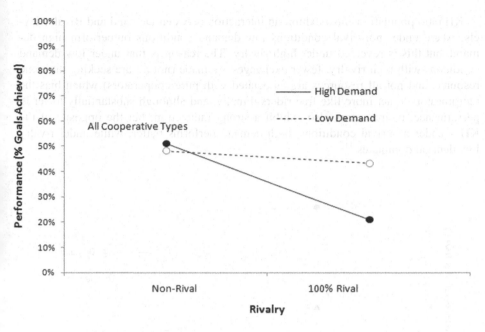

**Fig. 5.** Interaction between Rivalry levels and Demand levels, across the entire mix of cooperative ratios

But would such obvious constraints on exchange vary across agent types? To answer this, we   analyzed the interaction between these factors and the population types. In this analysis, we also included the Kurzban-Houser ratio, the most empirically valid ratio. The results are shown in Figure 6. Populations with high numbers of cooperators under low demand are insensitive to rivalry, but populations with KH ratio compositions are sensitive, resulting in significant drops in performance, but not near those populations dominated by reciprocators. The former populations are willing and able to share resources, but the latter populations are suffering from the impact of free riders. In KH populations, there seem to be sufficient cooperators to sway the behavior reciprocators repeatedly toward higher donation behaviors. On the other hand, with high demand goods, cooperators are more sensitive to rivalry than KH ratio populations. Under high rivalry conditions, cooperating population performance declines greater than the KH decline, and the two converge to virtually the same performance levels. Despite well-intended cooperators, exchanging high rivalry goods essentially does not alter the performance as supplies are fixed within the population. Variation in performance (i.e., low reciprocators) is accounted for by a substantial drop in cooperation where the base level of resources in the population (from initial conditions) is not extracted. Therefore, KH populations function to distribute resources as efficiently as high ratios of cooperators.

KH ratio populations also exhibit an interaction between Demand and Rivalry levels, where under non-rival conditions low demand conditions outperform high demand, but this is reversed under high rivalry. The reason is that under low demand conditions with high rivalry, fewer exchanges are made (not all are seeking the same resource, and not all resources are associated with pure cooperators) which bias the reciprocators to act more like free-riders. Finally, and although substantially lower in performance, reciprocators also exhibit a strong interaction, but the opposite of the KH – under non-rival conditions, high demand performs better, while under rivalry low demand dominates.[11]

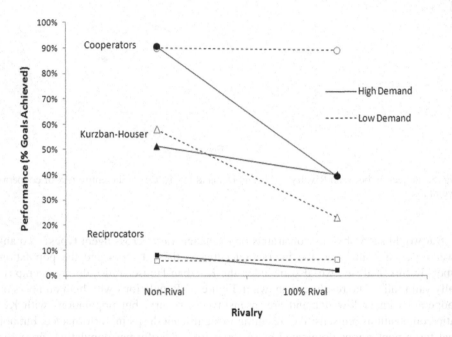

**Fig. 6.** Interaction between Rivalry levels, Demand levels, and three mixes of cooperator types

*Proposition 2c. Population Types Interact with Rivalry and Demand Levels Differently.* The impact of Rivalry and Demand on performance varies with the cooperative mix of a population.

Intuitively, lower rivalry alleges some of the damage done by free riding, because even massive free riding decrease only slightly the availability of goods to cooperators. In contrast, if the goods are highly rival, free riders will bear no cost but enjoy all benefits, while cooperators will still bear cost. Obviously, a system with more contributors will likely be better off, but the point here is that when goods are (perfectly) non-rival, the tragedy of the commons would be avoided even with free riding.

---

[11] Rivalry by Demand by Cooperator Type interaction: $(F(2,1188) = 21690.0, p < .001)$.

### 4.3   Typical Cooperation Type Ratios in the Population

With growing interest in the cooperative type distributions found in the human population [e.g., 67, 72-75], we elected to conduct further analysis on the KH ratio.[12] We examined multiple population compositions that meet KH ratio limits across three levels of Rivalry (0%, 50%, 100%) and three levels of Demand (1%: Low, 50%: Medium, 100%: High) with 100 runs per level, then plotted the surface mapped into performance levels. The result is shown in the figure.[13] Interestingly, the performance surface changes from a linear response (under High Rivalry, see α) to a non-linear response (under High Demand, see β). Thus,

*Proposition 3a. Rivalry has Non-Linear Effect on Performance.* Open source performance in Kurzban-Houser ratio responds linearly to Demand and non-linearly to Rivalry. The impact of Rivalry and Demand on performance varies with the levels of each.

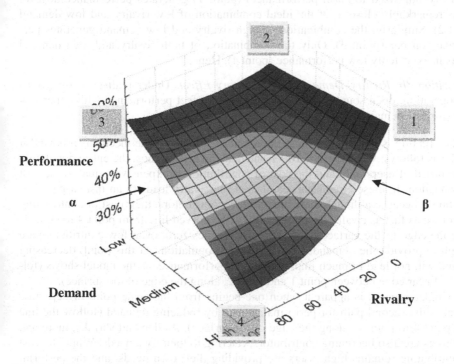

**Fig. 7.** Performance surface for Kurzban-Houser ratios, crossing Demand levels with Rivalry levels. *Colors* reflect bands of performance levels from maximum (*Dark Red*) to minimum (*Dark Green*).

---

[12] Where 13% were classified as cooperators, 53% as reciprocators, and 20% as free-riders, with the remaining 14% behaving inconsistently.

[13] 900 points were plotted done using Statistica 8.0 (www.statsoft.com) a using distance-weighted least squares fit model.

This finding begins to explain why open source does not collapse due to free riding. First, the impact of free riders is not large because, as KH showed, they are a minority to begin with. It is apparent in the figure that even with few cooperators and many reciprocators an increase in the free rider population affects performance appreciably only when they become a large part of the population. Second, the effect is further weakened thanks to the low rivalry (e.g., in point β), where the presence of free riders does not lead to decreased availability of goods. Because the goods are non-rival, the consumption of each marginal unit is zero, much like a downloader of music or video file does not directly reduce the availability of the same file for other users. The only effect of free-riders is in the opportunity cost of their time, which reduces the overall performance since they make no contributions. Overall then, the effect of free riders is muted.

The most interesting result comes from the interaction between the two variables. A combination of low rivalry (hypothesized to improve performance) and high demand (hypothesized to harm performance) (point 1) generates performance at level that is remarkably close to of the ideal combination of low rivalry and low demand (point 2). Similarly, the combination of high rivalry and low demand generates performance on par (point 3). Only the combination of high rivalry and high demand results in expectedly low performance (point 4). Hence,

*Proposition 3b. Rivalry–Demand Compensatory Effect.* Under realistic assumptions of cooperation, open source will exhibit close to perfect performance with either low rivalry *or* low demand.

The last finding is perhaps the most important of the three. The mechanism behind it may be as follows: when rivalry=100% and demand high, almost the entire population is composed of users only and the cost of contribution is expensive, equal to the cost of production. In these circumstances, decreasing rivalry leads to an instant performance boost, because although there are still few contributors, their cost of contributing goes down as fast as rivalry decreases (follow the curved line from point 4 to point 1, along the edge of the surface). When rivalry reaches zero, even few contributors are enough to provide the demands of a large user population. At this point, decreasing demand will not lead to much improvement in performance, as the model shows (follow the linear edge between point 1 and point 2, along the edge of the surface).

A similar process is apparent when one begins from the same point of departure and takes the second path for performance gain by reducing demand (follow the line from point 4 to point 3, along the edge of the surface). As demand shrinks, more and more users are also becoming contributors, catering to their own needs. While the cost of contributing remains high, users are providing their own needs, and the performance of the system improve. At the extreme (point 3*)*, all users are also self-sufficient as contributors. Then, decreasing rivalry leads to just small improvement in performance (follow the line connecting point 3 to point 2, along the edge of the surface).

## 5  Discussion and Conclusion

The results contribute to the emerging theory of open source. One, they validate the model and provide theoretical support of existing theory. Two, several propositions point out to novel effects, which we have not seen proposed hitherto. Importantly, we find decreasing marginal returns from cooperation (proposition 1b) and trade-off effect between cooperators and reciprocators in the population (proposition 1d). These findings may have significant implications for research and practice. For instance, they suggest that open source systems do *not* require a population of cooperators. Open source can thrive even when cooperators are just a small minority in the population, implying that it can be expected to appear in more places. Not only that, but also the performance of the system depends critically on just a small core group of contributors. Increasing that group leads to performance improvement but in a decreasing manner. Of particular interest is the finding that under the realistic cooperation assumptions of the Kurzban-Houser ratio, the most empirically valid ratio, performance of 42.5% can be achieved with only 13% cooperators, While researchers have observed that contributions in open source setting are highly skewed, our propositions aid in making sense of that. Having a core group of cooperators is not a deficiency in the system, but a rather expected feature. Cooperators, which may be difficult to find as they are a minority in human population can be replaced by reciprocators, which are much more common. For practitioners who are interested in facilitating open source, these findings can offer relief: creating a viable system may be easier than previously presumed.

While many cooperators are not necessary for open source systems to perform well, their presence has another impact: they reduce variance in performance (proposition 1c). This can lead a step forward in designing more fertile grounds for open source. Some systems, such as research and development teams, may be geared more towards high performance than reliability [cf. 80]. In such environments, a small percentage of cooperators with a majority of reciprocators may be sufficient to achieve the breakthrough sought. Additional effort to attract cooperators will better performance but can be inefficient in terms of cost versus benefit. This can explain, for instance, the focus on open sources in innovation systems, as opposed to production systems. However, even when reliable performance are sought, for instance in when providing an on-going service (e.g., *Wikipedia*), open source can still thrive. There, the effort to attract more cooperators may be efficient as it leads to more reliable performance over time.

Finally, the results suggest interactions between conditions of rivalry and demand, on the one hand, and population composition, on the other hand. We find that the impact of rivalry and demand on performance varies with the cooperative mix of a population (proposition 2d). For instance, increasing rivalry leads to a dramatic drop in performance, but mostly in populations that are made of cooperators. In the more empirically likely case of KH ratio population, increasing rivalry leads to a gentler drop in performance in the case of high demand, which is important in applications of open source. Once again, this is good news for observers of open source as well as practitioners who would like to benefit from it.

Finally, two propositions carry particular importance for the performance of open source in harsh environments, such as with rival goods or high (homogenous) demand. The findings suggest that Rivalry has non-linear effect on performance in the likely case of a KH ratio population (proposition 3a). The implication is that even a slight reduction in rivalry can bring about a boost in performance. For practitioners this can be an effective tool in improving the performance of open source systems.

Additionally, even systems with high rivalry can end up in close to perfect perform-ance if the demand is low (proposition 3b). Thus, even products that are close to per-fect rivalry (e.g., in producing food) can benefit from open source production, as long as demand is managed properly. This suggests that open source can be expected in greater variety of venues that we currently see.

Our analysis leaves much room for future work on the performance of open source. One future direction may include modeling of social institutions, such as those that regulate exchange. We know that individuals can exchange (or contribute) under multiple exchange configurations [81, 82]. In particular, two exchange configurations that were documented as legitimate in the institutional sense [83, 84]: *embedded exchange* [85, 86] and *generalized exchange* ("pay it forward", "gift economies"), where unacquainted participants help each other with the expectation that reciprocity will come from *any* other member, not necessarily the specific receiver [87, 88]. While such analysis is beyond the scope of this paper, we suspect that it may be fruit-ful in furthering our understanding of open source performance.

**Acknowledgements.** We are thankful for comments at the meetings of *The American Sociological Association* 2006 in Montréal, *The Academy of Management* 2008 in Anaheim, California, and the User and Open Innovation workshop 2009 at the Hamburg University of Technology. We thank Ian Zi Yang Lim for his research and editorial support.

# References

1. Raymond, E.S.: The Cathedral and the Bazaar. O'Reilly, Sebastopol (1999)
2. Henkel, J.: Champions of Revealing - The Role of Open Source Developers in Commercial Firms. Industrial and Corporate Change 18, 435–471 (2009)
3. von Hippel, E.: Open Source Software Projects as User Innovation Networks. In: Feller, J., Fitzgerald, B., Hissam, S.A., Lakhani, K.R. (eds.) Perspectives on Free and Open Source Software, pp. 267–278. The MIT Press, Cambridge (2005)
4. Franke, N., Shah, S.: How Communities Support Innovative Activities: An Exploration of Assistance and Sharing Among End-Users. Research Policy 32, 157–178 (2003)
5. Lee, G.K., Cole, R.: From a Firm-Based to a Community-Based Model of Knowledge Creation: The Case of the Linux Kernel Development. Organization Science 14, 633–649 (2003)
6. Shah, S.: Community-Based Innovation & Product Development: Findings From Open Source Software and Consumer Sporting Goods. Massachusetts Institute of Technology, Cambridge (2003)
7. Benkler, Y.: Commons-Based Strategies and the Problems of Patents. Sci. 305, 1110–1111 (2004)
8. Benkler, Y., Nissenbaum, H.: Commons-based Peer Production and Virtue. Journal of Political Philosophy 14, 394–419 (2006)
9. Berry, D.M.: The Contestation of Code: A Preliminary Investigation into the Discourse of the Free/Libre and Open Source Movements. Critical Discourse Studies 1, 65–89 (2004)
10. Maurer, S.M., Rai, A., Sali, A.: Finding Cures for Tropical Diseases: Is Open Source an Answer? PLOS Medicine 1, E56 (2004)

11. Bodie, M.T.: The Future of the Casebook: An Argument for an Open-Source Approach. Journal of Legal Studies 57, 10–35 (2007)
12. Guth, R.A.: Asia to Develop Software to Rely Less on Microsoft. The Wall Street Journal, B4 (2003)
13. Lohr, S.: Novell to Buy SuSE Linux for $210 Million, p. 6. The New York Times, New York (2003)
14. O'Mahony, S.: The Emergence of A New Commercial Actor: Community Managed Software Projects. Unpublished doctoral dissertation, Stanford University (2002)
15. Kogut, B., Metiu, A.: Open-source Software Development and Distributed Innovation. Oxford Review of Economic Policy 17, 248–264 (2001)
16. von Krogh, G., von Hippel, E.: The Promise of Research on Open Source Software. Management Science 52, 975–984 (2006)
17. von Krogh, G., von Hippel, E.: Special issue on open source software development. Research Policy 32, 1149–1157 (2003)
18. Shah, S.K.: Motivation, Governance and the Viability of Hybrid Forms in Open Source Software Development. Management Science 52, 1000–1014 (2006)
19. von Hippel, E., von Krogh, G.: Open source software and the private-collective innovation model: Issues for organization science. Organization Science 14, 209–223 (2003)
20. Roberts, J.A., Hann, I.-H., Slaughter, S.A.: Understanding the Motivations, Participation and Performance of Open Source Software Developers: A Longitudinal Study of the Apache Projects. Management Science 52, 984–999 (2006)
21. MacCormack, A., Rusnak, J., Baldwin, C.Y.: Exploring the Structure of Complex Software Designs: An Empirical Study of Open Source and Proprietary Code. Management Science 52, 1015–1030 (2006)
22. O'Mahony, S., Ferraro, F.: The Emergence of Governance in an Open Source Community. Academy of Management Journal 50, 1079–1106 (2007)
23. Casadesus-Masanell, R., Ghemawat, P.: Dynamic Mixed Duopoly: A Model Motivated by Linux vs. Windows. Management Science 52, 1072–1084 (2006)
24. Economides, N., Katsamakas, E.: Two-Sided Competition of Proprietary vs. Open Source Technology Platforms and the Implications for the Software Industry. Management Science 52, 1057–1071 (2006)
25. Häfliger, S., von Krogh, G., Späth, S.: Code Reuse in Open Source Software. Mangement Science 54, 180–193 (2008)
26. Majchrzak, A.: The Effect of Expertise Sharing and Integrating Behaviors in Wiki-based Organizational Intranets. Working Paper (2009)
27. Jeppesen, L.B., Lakhani, K.R.: Marginality and Problem Solving Effectiveness in Broadcast Search. Organization Science (forthcoming)
28. Anthony, D., Smith, S.W., Williamson, T.: Explaining Quality in Internet Collective Goods: Zealots and Good Samaritans in the Case of Wikipedia. Rationality and Society 21, 283–306 (2009)
29. Bonaccorsi, A., Giannangeli, S., Rossi, C.: Entry Strategies Under Competing Standards. Hybrid Business Models in the Open Source Software Industry Management Science 52, 1085–1098 (2006)
30. von Hippel, E.: Democratizing Innovation. MIT Press, Cambridge (2005)
31. West, J.: How open is open enough? Melding proprietary and open source platform strategies. Research Policy 32, 1259–1285 (2003)
32. Henkel, J.: Selective Revealing in Open Innovation Processes: The Case of Embedded Linux. Research Policy 35, 953–969 (2006)

33. Waguespack, D.M., Fleming, L.: Scanning the Commons? Evidence on the Benefits to Startups Participating in Open Standards Development. Management Science 55, 210–223 (2009)
34. Maurer, S.M., Scotchmer, S.: Open Source Software: The New Intellectual Property Paradigm. National Bureau for Economic Research Working Paper W12148 (2006)
35. Shah, S.K.: Open Beyond Software. In: Cooper, D., DiBona, C., Stone, M. (eds.) Open Sources 2.0: The Continuing Evolution, pp. 339–360. O'Reilly Media, Sebastopol (2005)
36. Zittrain, J.: The Future of the Internet – and How to Stop It. Yale University Press (2008)
37. Uehara, E.: Dual Exchange Theory, Social Networks, and Informal Social Support. American Journal of Sociology 96, 521–557 (1990)
38. Hardin, G.: The Tragedy of the Commons. Sci., 1243–1248 (1968)
39. Olson, M.: The Logic of Collective Action: Public Goods and the Theory of Groups. Harvard University Press, Cambridge (1965)
40. Milinski, M., Semmann, D., Krambeck, H.-J.: Reputation helps solve the 'tragedy of the commons'. Nature 415, 424–426 (2002)
41. Cornes, R., Sandler, T.: The theory of externalities, public goods, and club goods. Cambridge University Press, Cambridge (1986)
42. Mockus, A., Fielding, R.T., Herbsleb, J.D.: Two Case Studies of Open Source Software Development: Apache and Mozilla. In: Feller, J., Fitzgerald, B., Hissam, S.A., Lakhani, K.R. (eds.) Perspectives on Free and Open Source Software, pp. 163–209. The MIT Press, Cambridge (2005)
43. Lerner, J., Tirole, J.: Economic Perspectives on Open Source. In: Feller, J., Fitzgerald, B., Hissam, S.A., Lakhani, K.R. (eds.) Perspectives on Free and Open Source Software, pp. 47–78. The MIT Press, Cambridge (2005)
44. Lakhani, K.R., von Hippel, E.: How Open Source Software Works: Free User to User Assistance. Research Policy 32, 923–943 (2003)
45. Lessig, L.: Open Code and Open Societies. In: Feller, J., Fitzgerald, B., Hissam, S.A., Lakhani, K.R. (eds.) Perspectives on Free and Open Source Software, pp. 349–360. The MIT Press, Cambridge (2005)
46. Free Software Foundation, http://www.gnu.org/philosophy
47. Strauss, A., Corbin, J.: Basics of qualitative research. Sage Publications, Newbury Park (1990)
48. Glaser, B., Strauss, A.: The Discovery of Grounded Theory: Strategies for Qualitative Research. Aldine De Gruyter, New York (1967)
49. Maanen, J.V.: Different Strokes - Qualitative Research in the Administrative Science Quarterly from 1956-1996. In: Maanen, J.V. (ed.) Qualitative Studies of Organizations, vol. 1. Sage Publications, Thousand Oaks (1998)
50. Perlow, L.A.: The Time Famine: Toward a Sociology of Work Time. Administrative Science Quarterly 44, 57 (1999)
51. Lewin, K.: Action Research and Minority Problems. Journal of Social Issues 2, 34–46 (1946)
52. Nelson, R.R., Winter, S.G.: An Evolutionary Theory of Economic Change. Harvard University Press, Cambridge (1982)
53. Cyert, R., March, J.G.: A Behavioral Theory of the Firm. Prentice Hall, Englewood Cliffs (1963)
54. Black, L.J., Carlile, P.R., Repenning, N.P.: A Dynamic Theory of Expertise and Occupational Boundaries in New Technology Implementation: Building on Barley's Study of CT Scanning. Administrative Science Quarterly 49, 572–607 (2004)

55. Rudolph, J.W., Repenning, N.P.: Disaster Dynamics: Understanding the Role of Quantity in Organizational Collapse. Administrative Science Quarterly 47, 1–31 (2002)
56. Repenning, N.R., Sterman, J.D.: Capability Traps and Self-Confirming Attribution Errors in the Dynamics of Process Improvement. Administrative Science Quarterly 47, 265 (2002)
57. Bonaccorsi, A., Rossi, C.: Why Open Source software can succeed. Research Policy 32, 1243–1258 (2003)
58. Moss, S., Edmonds, B.: Sociology and Simulation: Statistical and Qualitative Cross-Validation. American Journal of Sociology 110, 1095–1131 (2005)
59. Zeitlyn, D.: Gift economies in the development of open source software: anthropological reflections. Research Policy 32, 1287–1291 (2003)
60. Kollock, P.: The Economics of online cooperation: Gifts and public goods in cyberspace. In: Smith, M.A., Kollock, P. (eds.) Communities in Cyberspace, pp. 220–239. Routledge, New York (1999)
61. Benkler, Y.: The Wealth of Networks: How Social Production Transforms Markets and Freedom. Yale University Press, New Haven (2006)
62. von Krogh, G., Späth, S., Lakhani, K.R.: Community, joining, and specialization in open source software innovation: a case study. Research Policy 32, 1217–1241 (2003)
63. Koch, S., Schneider, G.: Effort, Cooperation and Coordination in an Open Source Software Project: GNOME. Information Systems Journal 12, 27–42 (2002)
64. Kuk, G.: Strategic Interaction and Knowledge Sharing in the KDE Developer Mailing List. Management Science 52, 1031–1042 (2006)
65. Giles, J.: Internet encyclopaedias go head to head. Nature 438, 900–901 (2005)
66. Jeppesen, L.B., Laursen, K.: The role of lead users in knowledge sharing. Research Policy 38, 1582–1589 (2009)
67. Kurzban, R.O., Houser, D.: An experimental investigation of cooperative types in human groups: A complement to evolutionary theory and simulations. Proceedings of the National Academy of Sciences 102, 1803–1807 (2005)
68. Thomas-Hunt, M.C., Ogden, T.Y., Neale, M.A.: Who's Really Sharing? Effects of Social and Expert Status on Knowledge Exchange Within Groups. Management Science 49 (2003)
69. Dahlander, L., Mckelvey, M.: Who is not developing open source software? non-users, users, and developers Economics of Innovation and New Technology 14, 617–635 (2005)
70. Alexy, O., Henkel, J.: Promoting the Penguin: Who is Advocating Open Source Software in Commercial Settings? Working Paper (2007)
71. Leach, J.: A course in public economics. Cambridge University Press, Cambridge (2004)
72. Fischbacher, U., Gächter, S.: Heterogeneous Social Preferences And The Dynamics Of Free Riding In Public Good Experiments. Working Paper The Centre for Decision Research and Experimental Economics, School of Economics, University of Nottingham (2008)
73. Kim, J., Lee, S.M., Olson, D.L.: Knowledge Sharing: Effects of Cooperative Type and Reciprocity Level. International Journal of Knowledge Management 2, 1–16 (2006)
74. Simpson, B., Willer, R.: Altruism and indirect reciprocity: The interaction of person and situation in prosocial behavior. Social Psychology Quarterly 71, 37–52 (2008)
75. Fehr, E., Fischbacher, U., Gächter, S.: Strong Reciprocity, Human Cooperation and the Enforcement of Social Norms. Human Nature 13, 1–25 (2002)
76. Ishii, K., Kurzban, R.: Public Goods Games in Japan: Cultural and Individual Differences in Reciprocity. Human Nature 19, 138–156 (2008)
77. Fehr, E., Fischbacher, U.: The Nature of Human Altruism. Nature 425, 785–791 (2003)

78. Fowler, J.H., Christakis, N.A.: Cooperative Behaviour Cascades in Human Social Networks. Working Paper (2009)
79. Lenth, R.V.: Some Practical Guidelines for Effective Sample-Size Determination. The American Statistician 55, 187–193 (2001)
80. Hannan, M.T., Freeman, J.: Structural Inertia and Organizational Change. American Sociological Review 49, 149–164 (1984)
81. Biggart, N.W., Delbridge, R.: Systems of Exchange. Academy of Management Review 29, 28–49 (2004)
82. Levine, S.S., Prietula, M.J.: Towards a Contingency Theory of Knowledge Exchange in Organizations. In: Weaver, K.M. (ed.) Best Paper Proceedings, Academy of Management, Atlanta (2006)
83. Thomas, G.M., Walker, H.A., Morris, Z.J.: Legitimacy and Collective Action. Social Forces 65, 378–404 (1986)
84. Meyer, J.W., Rowan, B.: Institutionalized Organizations: Formal Structure as Myth and Ceremony. American Journal of Sociology 83, 340–363 (1977)
85. Granovetter, M.: Economic Action and Social Structure: The Problem of Embeddedness. American Journal of Sociology 91, 481–510 (1985)
86. Uzzi, B.: Social Structure and Competition in Interfirm Networks: The Paradox of Embeddedness. Administrative Science Quarterly 42, 35–67 (1997)
87. Ekeh, P.P.: Social Exchange Theory: The Two Traditions. Harvard University Press, Cambridge (1974)
88. Baker, W.E., Levine, S.S.: Mechanisms of Generalized Exchange. Working Paper (2009), http://ssrn.com/abstract=1352101

# How Open Are Local Government Documents in Sweden? A Case for Open Standards

Björn Lundell and Brian Lings

University of Skövde, Sweden
{bjorn.lundell,brian.lings}@his.se
http://www.his.se

**Abstract.** There is in Europe an increasing recognition of the need for governmental organisations to support and promote the effective curation of electronic
data, including public documents, for easy public access and reuse. Such a vision can stand in stark contrast with reality. In this paper we address the question: to what extent are local government documents preserved electronically
for discovery and re-use? Our goal is to establish the level to which calls for the
greater use of open document standards is being heeded, and to understand the
potential consequences of not heeding the advice. We find that availability of
electronic copies of documents is very variable, and accessibility is poor. In
particular, there is little evidence of policy to maintain electronic copies of
documents, and little awareness of open standards and their importance in data
curation. This is in stark contrast to stated central Government policy. The
study highlights a lack of strategy in organisations regarding the effective curation of electronic data.

## 1 Introduction

On 26th March 2009 the Swedish Government took a decision to set up a delegation
for e-Governance (Regeringen 2009). The delegation was mandated to first draft a
strategy for e-Government. In doing so, a number of principles were to be observed.
Notable amongst these were principles relating to:

- accessibility and usability in ICT;
- the use of open standards and "software based on open source software";
- solutions that gradually liberate the administration from dependence on specific
  platforms and solutions;
- long-term digital preservation.

The proposed strategy was published 19th October 2009 as SOU 2009:86 (SOU
2009) and contains strategies for considering open source and open standards in public sector procurement. The reference to open standards is important; standardisation
is not in itself considered to be sufficient. The definition of an open standard in SOU
2009:86 (SOU 2009) is identical to that included in the European Interoperability
Framework version 1.0 (EU 2004), namely:

P. Ågerfalk et al. (Eds.): OSS 2010, IFIP AICT 319, pp. 177–187, 2010.
© IFIP International Federation for Information Processing 2010

1. The standard is adopted and will be maintained by a not-for-profit organisation, and its ongoing development occurs on the basis of an open decision-making procedure available to all interested parties (consensus or majority decision etc.);
2. The standard has been published and the standard specification document is available either freely or at a nominal charge. It must be permissible to all to copy, distribute and use it for no fee or at a nominal fee;
3. Intellectual property – i.e. patents possibly present – of (parts of) the standard is irrevocably made available on a royalty-free basis;
4. There are no constraints on the re-use of the standard."

The principles and strategy are in line with current best practice on the transmission and archiving of electronic data (see, for example, DCC (2005)). The development of effective strategies is seen as essential for the preservation of access to data long-term, something which is of particular importance in the public sector.

Sweden is not alone in setting such objectives. Belgium, the Netherlands, Denmark and Norway already have guidelines on the use of an ISO approved open format for documents used in public administration. PDF and ODF have been specifically identified amongst the few applicable formats. According to Morten Andreas Meyer, Norwegian Minister of Modernisation, in a press announcement on 2nd July 2009:

"When exchanging documents attached to emails between the Government and users, it is from 1 January 2011 mandatory to use the document formats PDF or ODF." (Regjeringen 2009)

Many articles have been written about the problem of legacy data, i.e. data for which the originating software or hardware is no longer available. Such data is at best difficult and costly to recover, and at worst no longer accessible. In practice, many organisations recognise the potential problems this may cause. In the words of Gordon Frazer, managing director of Microsoft UK:

"Unless more work is done to ensure legacy file formats can be read and edited in the future, we face a digital dark hole." (BBC 2007)

The need is pressing in the case of open standards for document formats, not just for public access to current data but also for maintaining the archives of data received and generated by governmental organisations.

In this paper, we consider the extent to which, in the absence of a clear policy, electronic access to local government data is being supported. We do this through a quantitative and qualitative study of the availability and accessibility of electronic copies of executive board meetings of Swedish municipalities.

## 2 Open Standards for Document Formats

The concept of a standard in ICT is well understood. According to Berkman (2005) they are the "mortar holding interoperable ICT systems together." Standards enable interoperability between diverse systems. Add to this the concept of openness, and the concept is less well understood. Krechner (2005) suggests ten important rights that enable open standards, covering everything from IPR to how the standard is developed. However defined, open standards are increasingly recognised as central to interoperability – and have been credited with making the internet revolution possible.

Two of Krechner's 'ten rights', access to documentation and free usage of open standards, are often considered as the most important features. Perhaps the most important effect of openness is that it encourages free competition and thereby diversity, which in turn protects against reliance on one product or platform. In other words, open standards lower risk. As put by Bird (1998): "(an) open standard is one which is used as a basis for producing interoperating products from a large number of providers – who can compete on any of a multitude of competitive advantages to the market buying their product."

The primary purpose of open standards for document formats is to make documents independent of the systems which generated them. This is of paramount importance for any organisation wishing to promote long term accessibility, including interoperability. Otherwise, in the worst case, a specific tool must be purchased and maintained in order to access an organisation's data; and this tool may well not be available on all ICT platforms. A separate advantage of open standards for document formats is that they act as enablers of fair competition in the marketplace, encouraging the development of tools which can compete because of their ability to interchange documents. The two most cited standards in government contexts are PDF/A and ODF.

PDF/A is an ISO standard for using PDF format (see www.pdfa.org). It is designed for the long-term archiving of electronic documents, and is currently based on PDF Reference Version 1.4. It restricts PDF in order to better guarantee prospects for archiving; in particular, it ensures visual reproduction of the document (PDF/A-1b compliance) and document structure, to allow searching and reuse (PDF/A-1a compliance).

Open Document Format (ODF, docs.oasis-open.org/office/v1.2/) is a standard developed by the Organization for the Advancement of Structured Information Standards (OASIS), and also published as an ISO/IEC standard. It is an XML-based format specification for office applications, including text and spreadsheets amongst others. It has gained traction as an open standard adopted within the open source OpenOffice.org application suite.

## 3 Research Approach

The research question addressed through this study is the following. To what extent are local government documents preserved electronically for discovery and re-use? In particular, we address three related sub-questions. To what extent are official records available digitally? And, of those available, to what extent are they accessible, that is can be opened with the latest versions of current applications? Importantly, of those which are available and accessible, to what extent are they searchable? In essence, a searchable document is available for re-use. This clearly rules out scanned PDF documents as reusable assets.

The question is made easier to answer in Sweden, which has a very strict policy on governmental responses to questions: all questions must be responded to.

We emailed a questionnaire to each municipality (290 in all) requesting minutes of selected executive board meetings. These minutes represent the decisions of the most senior board in each municipality, and hence need to be preserved. In fact, these minutes are archived and kept indefinitely (over hundreds of years) – usually in paper copy. Although rules on the long term electronic preservation of official documents are under consideration by the National Archives, there is currently no official policy.

Three requests were made. The first was for the minutes of the last meeting in 2008. The second was for the minutes of the first meeting in 1999; this was the oldest which could be expected in some cases as municipalities are allowed to selectively but systematically purge certain less important records after 10 years. Some municipalities interpret this as including electronic versions of minutes. The third was for the oldest minutes available electronically. The requests were sent in late January 2009 to the official email address of the registrar for each. If the request was not answered then a final reminder was sent in mid June. Responses to each email were recorded, together with all attached minutes.

The request was specifically for documents as currently stored electronically. It was made clear that the documents sent should not be specially created from physical archives or transformed from their stored format.

The study resulted in both quantitative and qualitative data. Quantitative data was analysed to answer the three specific questions on the extent to which minutes are available, accessible and searchable. The text of email responses was analysed qualitatively, to give some indication of factors affecting availability.

## 4  Quantitative Analysis

Of the 290 municipalities contacted, 267 (92%) have responded to date. Of these, 264 were able to provide documents. Of the three that did not, one does not save any electronic copies; another will start in 2009; the third only stores scanned documents and understood that these were not of primary interest to us. Of the respondents, only 88% responded promptly. This may be significant, as the request for 1999 was at the edge of a 10 year window and a significantly delayed response could affect availability due to deletion policies. The oldest available document gives some insight into the possible significance of this. In the worst case, 4 authorities may have removed the requested document during the period of data collection, sending a later 1999 document as the oldest. However, all of these responded early so this is a low probability.

### About the Documents Received

Table 1 details the documents and formats sent by municipalities. In that table, "Other" includes files with the following extensions: .htm(6), .pro(4), .wpd(3), .tif(2), .dot(1), .027(1), .rft(1) and no extension (4). In all cases except .htm, .tif and .rft it was possible to open the document as an .rtf file. RFT is an (outdated) IBM binary format used on their mainframes, requiring special software to open it. In several cases, the filename extensions did not correspond to the actual format of the file which caused some confusion (e.g. some files were sent without extension, and some with a .doc extension were actually formatted as RTF).

Table 2 details the applications used by the municipalities for generating the documents. In analysing each file it was possible to identify which application had been used to generate it with a high degree of certainty. For almost all files sent to us in PDF we were able to identify the application used. Interestingly, no municipality that used OpenOffice.org sent us their minutes in ODF format.

We also looked at the application which generated each PDF file. Table 3 shows the results.

Clearly, if a document format was unavailable at the time that the minutes of a meeting were created, then reformatting has subsequently taken place. This is very evident with PDF files for 1999. Clearly, at least 66% of the PDF files sent for the year 1999 are not contemporary.

**Table 1.** Responses by document format

| Format of response | Minutes from 2008 | | Minutes from 1999 | | Earliest Minutes available | |
|---|---|---|---|---|---|---|
| DOC | 57% | (154) | 74% | (120) | 66% | (175) |
| PDF | 38% | (104) | 18% | (29) | 24% | (64) |
| RTF | 3% | (9) | 6% | (9) | 4% | (10) |
| DOCX[1] | 1% | (2) | 0% | (0) | 0% | (1) |
| Other | 1% | (2) | 2% | (4) | 5% | (14) |
| TOTAL[2] | 100% | (271) | 100% | (162) | 100% | (264) |
| Paper only | | (3) | | (106) | | (3) |
| Did not reply | | (23) | | (23) | | (23) |

**Table 2.** Application used to generate documents

| Application generating the document | Minutes from 2008 | | Minutes from 1999 | |
|---|---|---|---|---|
| Word | 89% | (242) | 96% | (153) |
| WordPerfect | 0% | (0) | 1% | (2) |
| IBM DisplayWriter | 0% | (0) | 0% | (0) |
| OpenOffice.org | 1% | (3) | 0% | (0) |
| Scanner | 10% | (26) | 3% | (5) |
| TOTAL | 100% | (271) | 100% | (160) |

**Table 3.** PDF versions, 2008 and 1999

| Adobe PDF version number | Year | Minutes from 2008 | | Minutes from 1999 | |
|---|---|---|---|---|---|
| 1.2 | 1996 | 3% | (3) | 14% | (4) |
| 1.3 | 1999 | 20% | (21) | 21% | (6) |
| 1.4 | 2001 | 61% | (63) | 38% | (11) |
| 1.5 | 2003 | 10% | (10) | 14% | (4) |
| 1.6 | 2005 | 7% | (7) | 14% | (4) |
| TOTAL | | 100% | (104) | 100% | (29) |

---

[1] One of the documents from 1998 was sent in .docx format, which was not available at the time of the meeting (see next section for related discussion).
[2] Up to 3% of municipalities provided their minutes for a given year in both DOC and PDF formats.

We further analysed documents for signs of post-facto curation. In particular, we excluded PDF documents created by scanning paper copies from the archive. We also noted documents created significantly after the date for the meeting, implying that the document was re-saved after further processing (for example, generated as PDF for a more recent initiative to place documents on the web).

Maintaining electronic access to documents is of significant and ongoing concern to archivists. Two primary methods are used when document formats are no longer supported: reformatting and emulation. In the former, documents are transformed to a current format and re-archived. In the latter, access to old formats is maintained by emulating the applications which originally produced them. Looking at the DOC formatted documents, there is clear evidence of reformatting – the most extreme case being the 10 year old minutes in .docx format.

## Availability

To consider availability, we analysed the oldest document provided by each municipality. First we analysed each of these documents irrespective of its format and how it was created. When viewed cumulatively (Figure 1), it is possible to gain a sense of the level of electronic availability of documents by year over all municipalities. The data is not fully accurate. As our request only concerned one specific meeting (in 1999) it is evident from the comments in the responses that several municipalities also have other gaps, for various reasons, in what has been kept in electronic form (see further next section). We are in fact aware of some drop-outs – cases in which an older document than for 1999 has been provided but not one for 1999 itself.

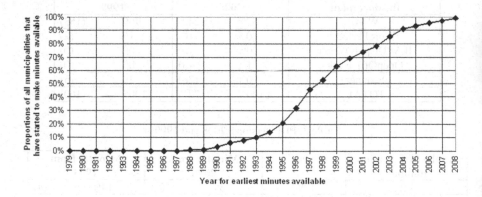

**Fig. 1.** Earliest available document by year

## Accessibility

For accessibility we looked at both the 1999 and 2008 documents, but were particularly interested in 2008. This can give an indication of the extent to which e-government initiatives have penetrated local government, specifically in areas which are not directly controlled through legislation.

We classify a word processing document as accessible if it can be opened in the latest, currently available version of an appropriate word processing application in a manner which preserves its content and layout (to the extent that it can be read as intended, so for this analysis we ignored detailed formatting issues). We consider both proprietary and Open Source options. The only clear proprietary candidate was Microsoft Word (hereafter referred to as Word), as all word processing documents from 2008 were produced in Microsoft formats.

Table 4 shows the Word versions used in 2008 and for 1999 (for comparison). Of the total of 156 files received for 2008 two were docx-files and the other 154 were documents in "Word.Document.8" format. For the minutes received from 1999, 64 documents were supplied in "Word.Document.8" format and 56 were supplied in "Word.Document.6" format.

**Table 4.** Word versions used for 2008 (and 1999 for comparison)

| MS Word version | Minutes from 2008 | | Minutes from 1999 | |
|---|---|---|---|---|
| MS Office Word for Windows 95 | 0% | (0) | 28% | (34) |
| MS Word 6.0 | 0% | (0) | 24% | (29) |
| MS Word 8.0 | 3% | (4) | 32% | (39) |
| MS Word 9.0 | 18% | (28) | 0% | (0) |
| MS Word 10.0 | 17% | (27) | 6% | (7) |
| MS Office Word | 63% | (95) | 9% | (11) |
| MS Office Word (for docx) | 1% | (2) | 0% | (0) |
| TOTAL | 100% | (156) | 100% | (120) |

We classify any other document as accessible if it can be opened in the latest, currently available version of an appropriate application. Again we considered both proprietary and Open Source options. As all but two of these documents are in PDF, we conducted the tests with Adobe Reader 9.1 and Sumatra (Beta) v0.9.4. The two remaining documents were in .htm and .tif formats.

In testing the documents there were three primary concerns:

1. can the document be opened with the appropriate proprietary software application?
2. can the document be opened with the appropriate Open Source software application?
3. is the document searchable in each application?

The first step was to attempt to open all of the set of 'oldest' documents, in appropriate current applications. This is a test of their current accessibility. Clearly reformatting may have improved or compromised accessibility, and may have compromised reusability (for example if a scanned paper copy was maintained).

We initially considered the level of success in opening word processor documents using Word 2007 and OpenOffice.org 3.1. We considered the oldest documents first. It is documented on the vendor website that early versions of the .doc format are not supported in recent versions of their software (Word 2003 and Word 2007). This was confirmed when we tried to open documents stored in Word for Windows 1.x (1 document) and 2.0 formats (6 documents in all). Interestingly, all documents, including these, could be opened in OpenOffice.org 3.1. However, the Windows 1.x document was treated as a text file. The text of the minutes could be discerned, but most of the formatting was lost and the binary encodings were presented as unicode text. Both applications were able to open all other documents with file formats .doc or .rtf. Both were also able to open a number of documents with other file formats, namely Word-Perfect 5.x and 6.x and the single occurrence of a docx file (clearly curated).

All documents from the 2008 set could be opened in both Microsoft Word 2007 and OpenOffice.org 3.1, although a number of problems were still encountered. These included problems related to: formatting that breaks; macros and the use of different variables and templates that are not kept embedded in the document; locked and password protected documents; documents that try to access embedded SQL-queries (with dependencies on other applications); and Word comments that cause problems (the document opens with comments that are generated in a previous version of Word).

All PDF documents could be opened with Adobe Reader 9.1 and with Sumatra (Beta) v0.9.4. Although we were able to open all files there was for one PDF-file a font problem identified when using Adobe (resulting in an error message which warned that the file may not display and print correctly), whereas such an error was not indicated when we opened the same file using Sumatra. We therefore concentrated on the question of whether the document was searchable – this being an indicator of whether a document was reusable.

**Reusability**

For studying reusability, we are interested primarily in the information within a document, not its styling. We consider the data sets from early 1999 and late 2008, giving us a view of two time points roughly 10 years apart.

Looking at the minutes submitted of the first meeting of each municipality held in 1999 (29 documents in PDF format), we found that 5 (17%) were produced by some form of scanning and were not searchable. All of the other documents were searchable.

In looking at the details of each non-scanned document, it could be seen that the PDF had been generated using a variety of versions of Word. Considering these together with the .doc documents from the same period, we obtain a snapshot of the latest versions of Word used to save each document (see Table 5). This does not necessarily reflect the Word version used to create the file, since PDF generation may have been done at a later point in time, and the document may have been opened and saved in a later version of Word at a later date – for example to update its format.

Finally, we considered the documents supplied from 2008. The majority of these were .doc format (see Table 1), but the proportion in PDF format had significantly increased (from 18% to 38%). This is unsurprising, as many municipalities now routinely publish meeting minutes on the web, but few extend back as far as 1999. Some municipalities even supplied URLs in their response rather than attaching documents.

The increase in use of PDF for maintaining electronic copy of documents raises further questions. In particular, are the documents in a form of PDF which is reusable, accessible and open? For this research we checked whether each PDF file could be searched (a necessary prerequisite for reusability) and whether it was open (in a format compatible with PDF/A, and in particular the weaker requirement PDF/A-1b). PDF/A is a subset of PDF designed to be more suitable for the long-term archiving of documents. Disappointingly, only one document from 2008 was found to be compliant with PDF/A-1b.

Of the 104 PDF-documents from 2008, the percentage non-searchable was found to be 23%, whereas for the 29 documents from 1999 the percentage non-searchable was 17%. It should also be noted that not all non-scanned documents were searchable. Interestingly, one of the documents was generated by OCR using the Adobe Acrobat paper capture plug-in.

# 5 Qualitative Analysis

Analysing the document attachments gives insight into the extent of availability and accessibility of files. However, any explanation of why certain documents are unavailable or inaccessible requires a qualitative study. For this, we analysed the content of the emails to which documents were attached. In most cases an explanation was given if there was any problem in meeting the request. We present these for insights into availability (and hence archiving policies), and also, where relevant, for accessibility.

In 40% of cases municipalities could not provide the requested minutes from 1999. Reasons varied. At one extreme, there was a practice of keeping no electronic archives. In some cases, there was a policy of periodically pruning back. For example:

"Electronically stored copies are selectively deleted and this has been the case with the minutes from 1999"

Such policies are not always strictly systematic. The minutes of these meetings – although related to the top level committee of the municipalities – are not protected by law except in their paper form. Hence this next response:

"The minutes written before 2000 have probably been deleted at the time the minutes were signed and archived. There is no record kept about how this was handled exactly in the year 1999."

This lack of obligation to protect electronic copy of minutes has led to a lack of policy, allowing this informal approach. This potentially impacts both on availability and accessibility. Hence, even where minutes are made available this seems to be as a result of other informal processes. For example,

"(The municipality) does not have any electronic storage of documents ... . However, the minutes are temporarily available as PDF and stored in an ordinary folder on disc (and) used for presentation on our home page."

Most had a less systematic reason for unavailability. Most pertinent to our enquiry is loss of unavailability due to legacy systems and formats:

"I can unfortunately not find anything from 1999. A different system was in use then, which we do not have access to today."

"Unfortunately we do not have the minutes from 1999 for technical reasons."

It is clear that even after only ten years there are problems related to accessing or interpreting files which are known to exist. In a number of instances this was because the file uses a proprietary format which is only interpretable by the legacy tool which created it. This may imply extra cost and delay in meeting a request:

"The oldest minutes are not available and the minutes from 1999 should be available but the tool ... has been phased out from the organisation. Your request has triggered our IT department to resurrect the software."

In fact, it may be significant enough for the organisation to seek a way of not incurring the expense:

"From 1994-08-10 we used Ergo-ord for electronic documents. We have not been able to recreate the first document from 1993-01-13, a specific environment is necessary for this which we have not set up. You will have to be satisfied with a document from 1994-08-10, which is the first Word version. Please come back to me if you insist on the minutes from 1993."

In other cases, it was not possible to completely reconstruct the document even though the organisation was willing to make the effort. In one case this resulted in significant data loss:

"For the documents from 1990 I have been forced to ask the IT section to convert so that the documents can be read. The meeting minutes from 1990-01-09 were inserted into a pre-printed template, hence this is missing in the document. The minutes from 1999-01-12 contain only 2.5 paragraphs and 4 pages. According to the original minutes in paper form there should be 35 pages."

In a different case, layout information was lost even for recent minutes, and the older minutes also suffered data loss (in the response a paragraph refers to a distinct item on the agenda):

"The minutes from 2008 are digitally preserved in Word but in that the paragraphs are not in the correct order. The 1999 minutes are only partially preserved: the first page and a few of the paragraphs are missing" ... "We had the principle of one Word file for each paragraph at the time, so it is messy to organise these."

Such access problems were not limited to documents generated by tools which are no longer used. In a number of instances old files, in a proprietary format, could not be opened natively in the latest version of the tool which created them. Interestingly these same files could be opened natively in an open source tool, OpenOffice.org.

## 6  Discussion and Conclusions

There is no evidence from our study that municipalities have a data curation policy with respect to executive minutes. In the absence of a direct duty to preserve electronic copy, curation is left to the work practices of individuals.

Where electronic copy is kept, proprietary and closed formats are overwhelmingly used for public documents, even though there is experience of losing access to, or increased cost of access to documents because of formats which are no longer supported. Further, and perhaps more significantly, we find no evidence that this situation is changing.

Our general finding is that availability of electronic copies of executive board minutes for municipalities in Sweden is very variable, and accessibility is poor. In particular, there is little evidence of the existence of policies to maintain electronic copies of documents, and little awareness shown of open standards and their importance in data curation. It is striking that no municipality provided a document in a reusable, open standard document format. This stands in stark contrast with stated central Government policies. The study also highlights a consequent lack of strategies in organisations regarding effective communication and archiving of electronic data.

As a result, there are already many gaps in the electronic data record even for the most recent 10 year period. In this very real sense, the mooted digital dark hole in public records is fast becoming a reality.

# References

BBC, Warning of data ticking time bomb, BBC News (July 3, 2007), http://news.bbc.co.uk/2/hi/technology/6265976.stm (accessed 2009-12-23)

Berkman, Roadmap for Open ICT Ecosystems, Berkman Centre for Internet & Society at Harvard Law School (2005)

Bird, G.B.: The Business Benefit of Standards. StandardView 6(2), 76–80 (1998)

DCC Digital Curation Manual: Instalment on Open Source for Digital Curation (August 1, 2005), http://www.dcc.ac.uk/resource/curation-manual/chapters/open-source/

EU, European Interoperability Framework for pan-European eGovernment Services, European Commission, Version 1.0 (2004), http://ec.europa.eu/idabc/servlets/Doc?id=19529

Krechmer, K.: The Meaning of Open Standards. In: Proceedings of the 38th Hawaii International Conference on System Sciences – 2005. IEEE Computer Society, Los Alamitos (2005)

Regeringen, Kommittédirektiv: Delegation för e-förvaltning, Dir., March 19–26 (2009) (in Swedish), http://www.sweden.gov.se/content/1/c6/12/40/02/ec50b88b.pdf

Regjeringen, Nye obligatoriske IT-standarder for staten vedtatt, Fornyings- og Administrasjonsdepartementet, Pressrelease (July 2, 2009) (in Norwegian), http://www.regjeringen.no/nb/dep/fad/pressesenter/pressemeldinger/2009/nye-obligatoriske-it-standarder-for-stat.html?id=570650

SOU, Strategi för myndigheternas arbete med e-förvaltning, Statens Offentliga Utredningar: SOU 2009:86, e-Delegationen, Finansdepartementet, Regeringskansliet, Stockholm (October 19 , 2009) (in Swedish), http://www.sweden.gov.se/content/1/c6/13/38/13/1dc00905.pdf

# Bug Localization Using Revision Log Analysis and Open Bug Repository Text Categorization

Amir H. Moin and Mohammad Khansari

Department of IT Engineering, School of Science & Engineering,
Sharif University of Technology, International Campus, Kish Island, Iran
`moin@kishlug.ir, khansari@sharif.edu`

**Abstract.** In this paper, we present a new approach to localize a bug in the software source file hierarchy. The proposed approach uses log files of the revision control system and bug reports information in open bug repository of open source projects to train a Support Vector Machine (SVM) classifier. Our approach employs textual information in summary and description of bugs reported to the bug repository, in order to form machine learning features. The class labels are revision paths of fixed issues, as recorded in the log file of the revision control system. Given an unseen bug instance, the trained classifier can predict which part of the software source file hierarchy (revision path) is more likely to be related to this issue. Experimental results on more than 2000 bug reports of 'UI'component of the Eclipse JDT project from the initiation date of the project until November 24, 2009 (about 8 years) using this approach, show weighted precision and recall values of about 98% on average.

## 1 Introduction

Both the total number of open source software projects and the total amount of open source code in the world, are growing at an exponential rate [1]. In addition, the number of developers interested in working in this field, are increasing tremendously fast. For example, the number of developers involved in the Linux kernel development project has doubled over the past three years [2]. Hence, one should expect a very high rate of bug reporting to the issue tracking system of large open source projects. As an example, consider the case of the Eclipse open bug repository, with an average bug reporting rate of above 50 issues per day from January 1 until November 24, 2009 [3]. Suppose that each issue takes an average of ten minutes from a developer in order to be localized in the software source file hierarchy. This simply means, at least 8 professional person-hours per day is required merely for searching where the buggy piece of the code is located, which is indeed an invaluable and rare resource for most open source projects.

In this paper, we present a new approach for automating bug localization, i.e. finding the most relevant part of the software source file hierarchy to a bug reported to an open bug repository. Firstly, we analyze the history of source revisions, available in the log file of the version control system, in order to find the bug IDs and their corresponding revision path (path of the revised file during

P. Ågerfalk et al. (Eds.): OSS 2010, IFIP AICT 319, pp. 188–199, 2010.

a successful bug fix). Secondly, we send a query to the open bug repository of the project in order to obtain summary and description of the extracted bug IDs. Then, we prepare our dataset in the proper and acceptable format for training the classifier. Afterward, we perform the classification via training a Support Vector Machine (SVM) classifier. Finally, given a new bug, we can localize the bug in the software source file hierarchy using the trained classifier.

Our approach is novel in that we use the large amount of valuable information in the open bug repositories of open source projects rather than performing analysis on the software source repositories to find latent software defects [14] [15] [16] [17] [18] [19] [20] [21]. Moreover, rather than looking for an exact bug-related source file and its line number, we localize the bug in one higher level of the software source file hierarchy (file path). One of the possibly useful applications of this approach could be in bug triage, i.e. deciding each reported issue should be assigned to which developer in order to be fixed [4] in a time and cost effective manner. In that problem, the triager could assign bugs with respect to the field of expertise [5] and level of interest [6] of developers in that particular part of the software source file hierarchy.

The paper is organized as follows. Section 2 provides some background about revision control systems, open bug repositories and machine learning. In section 3, we present the proposed approach for bug localization. Section 4 provides validation and experimental results, and Section 5 reviews related work. Finally, we draw conclusion and suggest future work in section 6.

## 2   Background

To understand the proposed approach one should be familiar with various areas including version control systems, open bug repositories and machine learning. We review the related concepts of these topics by giving examples from Eclipse projects.

### 2.1   Version Control Systems

A Version Control System (or more accurately, revision control system) is a combination of technologies and practices for tracking and controlling changes to a project's files, in particular to source code, documentation, and web pages. The main role of such a system is change management via identifying each change to the project, annotating it with relevant metadata such as the date, author, and possibly the reason of that change, and finally replaying these facts to whoever asks, in the desired format. In other words, it is an inter-developer communication mechanism where a change is the basic unit of information. The most widely used revision control system in the Free[1]/Open Source Software (FOSS) world is Concurrent Versions System (CVS). Although it has become the default choice along the time and most experienced developers are already

---

[1] Here, Free is a matter of liberty, not a matter of price. For more information please visit http://www.gnu.org/philosophy/free-sw.html

RCS file:
/cvsroot/eclipse/org.eclipse.jdt.core/compiler/org/eclipse/jdt/inte
rnal/compiler/parser/RecoveredField.java,v
...
revision 1.39.2.1
date: 2009-06-19 15:38:30 +0430; author: daudel; state: Exp;
lines: +20 -0; commitid: fb0c4a3b71ac4567;
R3_5_maintenance - Bug 277204

**Fig. 1.** A small part of a sample CVS log, the Eclipse JDT Project. ('...' represents the omitted lines.)

familiar with CVS, it has few disadvantages which consequently has led to the emergence of a number of alternatives such as Subversion (SVN), Git, Bazaar, and Mecurial [7].

Fortunately, the log files of the version control system for different components of a software project could be queried and saved in separate files. Figure 1 shows a very small part of the CVS log file for 'Core'component of the Eclipse JDT project.

## 2.2   Open Bug Repositories

Providing a bug tracking system (or more accurately, issue tracking system) is one of the necessary tools of open source software development [7]. A bug tracking system usually consists of a database known as bug repository which contains information about the bug reports. Almost any open source project is supported by an open bug repository in which anyone could have a username and password and either report an issue or put a comment on an existing report.

There are various bug tracking software such as Bugzilla and JIRA. Furthermore, some projects like Debian GNU/Linux have their own bug tracking system [8].

### Structure of Bug Reports

One of the better known bug tracking systems is Bugzilla. A typical bug report in Bugzilla consists of various parts including the predefined fields, free-form text, attachments and dependencies [4].

Figure 2 depicts the predefined fields in a sample bug report of the Eclipse bug repository. This bug report corresponds to the CVS log shown in figure 1. Some predefined fields such as the bug ID or reporter are specified when the report is created and fixed over the life cycle of the bug report (this life-cycle is covered in the next subsection). Other fields, either change successively while the bug report is tossed among the developers, i.e. forwarded from the developer to whom it is initially assigned to another one [9], like the Assigned To field, or change occasionally such as the Importance or the CC list[2] [10].

---

[2] The CC list is the list of the email addresses of people who are interested to be kept up-to-date about the status of the issue.

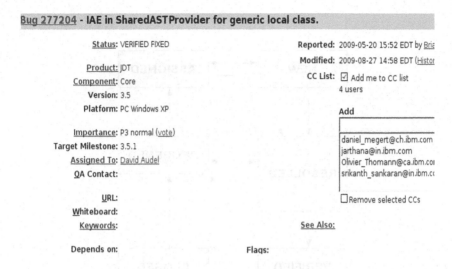

**Fig. 2.** Predefined fields in a sample Bugzilla bug report

The free-form text includes a one line summary of the issue, also known as its title, a detailed description of the report which should help a developer reproduce the bug and finally a number of comments on this issue which might refer to other similar bugs [4].

Other parts of bug reports include attachments and dependencies. Attachments are usually non-textual information such as screen-shots. Moreover, the bug tracking system tracks bugs which their resolution depend on fixing a specific bug report [10].

## Life-cycle of Bug Reports

Initially, when a new issue comes to the open bug repository of the Eclipse projects, its status field is set to NEW. Then either it is assigned to a developer by the triager or a volunteer developer accepts its responsibility. Consequently, it is tagged with ASSIGNED.[3] At the end, when there is no remaining task due to the resolution of the bug report, it is marked as RESOLVED. If the triager finds that this issue is already reported, it is marked as RESOLVED DUPLICATE. If the report is not indeed a bug report, for example it states a natural feature of the software which is mistakenly thought to be a bug, the report is tagged with RESOLVED INVALID. When the erroneous behavior is not repeatable, perhaps because of poor description of the problem, the developer sets the status to RESOLVED WORKSFORME. Otherwise, the resolution might need applying changes in the source code which causes the issue to be marked as RESOLVED FIXED. If a bug is believed to be unsolvable for any reason, it will be tagged with RESOLVED WONTFIX [11].

---

[3] There exist few cases in which bug reports are not assigned to developers and resolved immediately by the triager.

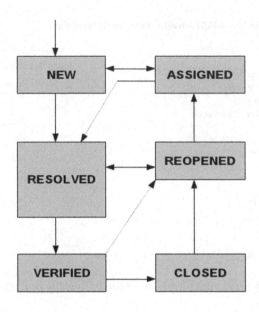

Fig. 3. Life-cycle of bug reports in Eclipse projects

The resolution status of the RESOLVED reports may later change to VER-IFIED and then CLOSED. One is allowed to reopen a previously RESOLVED, VERIFIED or even CLOSED issue at any time. Figure 3 shows the typical life-cycle of the Eclipse bug reports [11].

In this paper, we only care about the bug reports which are either RESOLVED FIXED, VERIFIED FIXED or CLOSED FIXED.

## 2.3 Machine Learning

Machine learning is a discipline concerned with design and development of algorithms in order to allow computers to learn how to recognize complex patterns in data, to be able to make smart decisions. In the context of machine learning, the training data consist of a number of examples which are called instances. Each instance bears a number of input objects known as attributes or features which are usually encapsulated in a vector. In supervised machine learning an output value is assigned to each instance of the training data in advance and the problem is to deduce a function in order to predict the output value of any similar valid input vector. If the output value is a continuous value, the problem is called regression; otherwise, the output value is called the class label, the function is named as classifier and the problem is called classification. One of the many applications of this kind of classification is in text categorization, where the classifier is expected to assign a relevant category to an arbitrary text document based on a number of previously seen examples [12] [13].

# 3   The Proposed Approach

Given a new bug report from the open bug repository of an open source software project, our approach uses a Support Vector Machine (SVM) classifier to suggest the part of the source file hierarchy which is more likely to be related to this issue. The suggestion is made based on a number of previously seen examples, i.e. fixed bug reports in the past. Various components engaged in the proposed approach are presented in figure 4.

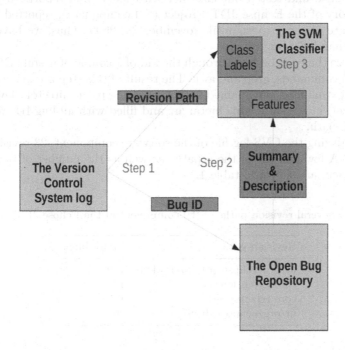

**Fig. 4.** Various components of the proposed approach

Our approach has three steps:

1. Analyzing the Revision Logs

   When a bug is fixed by a developer, the revision path, i.e. the path of the file which is revised due to this bug resolution, is not mentioned anywhere in the open bug repository. Consequently, one should analyze the entire change history of a specific software component in the log file of the version control system for that component, in order to find patterns such as 'fix for bug no ...'or similar among the comments of developers. The extracted bug IDs are used in step two, and the revision paths are used as class labels of the classifier in step three.

2. Querying the Bug Repository

   For each extracted bug ID in the previous step, we send a query to the bug tracking system and ask for the summary and description of that bug ID.

3. Training the Classifier

   After conducting previous steps, we have the revision path in software file hierarchy as well as the summary and description for each resolved bug ID. We use this information to train a Support Vector Machine (SVM) classifier.

## 4    Validation and Experimental Results

We have trained and tested our classifier with fixed[4] bug reports in the open bug repository of the Eclipse JDT Project ('UI'component), reported from the initiation date of the project until November 24, 2009. Thus, we have worked with more than 2000 bug reports.

We analyze the revision logs through the aid of a couple of useful GNU/Linux (and UNIX) commands, grep and awk. The result of this step is expected to be a number of textual files, each named with an existing path (directory level rather than file level) in the source file hierarchy and filled with all bug IDs related to that specific path.

After analyzing the CVS log file of the software component, 23 revision paths were found. A few number of these paths as well as the number of bug reports related to each, are shown in table 1.

**Table 1.** Several revision paths of 'UI'component of the Eclipse JDT project

| Revision paths | No. of bugs |
| --- | --- |
| ui/org/eclipse/jdt/internal/ui/ | 1392 |
| core extension/ | 133 |
| core refactoring/ | 137 |
| ui/org/eclipse/jdt/ui/ | 203 |

We have developed a Java application in order to connect to the Bugzilla open bug repository of the Eclipse JDT project using XML Remote Procedure Call (XML-RPC), a well known protocol for performing remote procedure calls over HTTP. For each of the bug IDs gathered in the previous step, we send a query to the bug tracking system and ask for the summary and description of that bug report. Eventually, we save the collected information about each bug in a separate textual file, named the same as the bug ID. One should keep every file related to a specific revision path in a distinct directory which is named after the revision path, in order to create a dataset to be used in the following steps.

In order to implement our approach, we use the Free/Open Source Software (FOSS) suite for machine learning written in Java, called WEKA. WEKA requires both the training and testing datasets to be in a standard format called ARFF. Fortunately, there is a converter, named TextDirectoryLoader in WEKA. This converter, receives a number of directories which contain a set of text files,

---

[4] Trivially, the revision path for unfixed bugs is meaningless.

and then treats the directory names as class labels, the text files as instances of each class and the information within each text file as features of that instance. The output of this converter is an ARFF file as desired.

Since the classifier which we use in the next step cannot handle String attributes, we must apply an appropriate filter to the dataset, i.e. the ARFF file, in order to perform TF-IDF (Term Frequency-Inverse Document Frequency) transformation. This transformation is often used in information retrieval and text mining problems in order to give a weight to each term, based on the number of occurrence of the term. The basic assumption is that the more times a specific term appears in a text document, the more important it is to that document [25]. There is a filter in WEKA, called StringToWordVector which does the needed transformation easily. The output is still an ARFF file.

The classifier also cannot handle numeric attributes. However, our ARFF file contains a number of such attributes. The solution is applying another filter available in WEKA, named NumericToNominal. Now, the resulted ARFF file is ready to be used for training the SVM classifier.

After gathering and preparation of the dataset, the next step is to train the classifier and validate the learned model.

We use an improved Support Vector Machines (SVMs) algorithm, called Sequential Minimal Optimization (SMO) with linear kernel. SMO is much faster and more memory-efficient than the initial SVM algorithm [26] [27]. We use binary SMO implementation with linear kernel which is available in WEKA as BinarySMO. This implementation replaces all missing values and transforms nominal attributes into binary ones. It also normalizes all attributes by default. The multi-class problem is solved by using pairwise classification [28]. Table 2 shows several normalized attribute weights of our dataset.

**Table 2.** Several normalized attribute weights

| Attribute(term) | Weight |
|---|---|
| JavaCore | 0.0847 |
| Synchronizer | -0.0328 |
| WM_CHAR | 0.0173 |
| container | 0 |

As in any other machine learning problem, we should somehow evaluate the performance of our approach. We use ten fold cross validation for training and validation of the linear SVM classifier. The detailed evaluation results are provided in table 3.

The True Positive (TP) rate is equivalent to Recall. It measures how much part of the class is captured. In other words, the TP rate (Recall) is the proportion of the instances which are classified as class A, among all instances which indeed have class A.

**Table 3.** Detailed evaluation results of the binary SMO classifier

| TP Rate | FP Rate | Precision | Recall | F-Measure | Class |
|---------|---------|-----------|--------|-----------|-------|
| 0.992 | 0.062 | 0.99 | 0.992 | 0.991 | 0 |
| 0.938 | 0.008 | 0.951 | 0.938 | 0.944 | 1 |
| 0.985 | 0.055 | 0.985 | 0.985 | 0.965 | Weighted Avg. |

The False Positive (FP) rate is the proportion of the instances which are classified as class A, but belong to a different class, among all instances which are not of class A.

The Precision is the proportion of the instances which indeed have class A, among all those instances which are classified as class A.

Since, often there is a trade-off between precision and recall, it is common to measure the classification performance via a mixture of both, called F-Measure [29].

$$F - Measure = \frac{2 * Precision * Recall}{Precision + Recall}$$

Finally, Accuracy is the proportion of the total number of correctly classified instances among all instances. Our accuracy through the classification has been 98.5137%.

## 5  Related Work

We are aware of a number of valuable efforts in the field of bug localization automation. One possible approach is trying to find bugs through checking either a well-specified program model [14] or real code directly [15] [16] within the software source code. This approach is called static analysis [17].

Gyimothy et al. [18] use two groups of machine learning algorithms, decision trees and neural networks to predict buggy classes with a static code analysis approach.

The second approach is called dynamic analysis which is concerned with the comparison of the run-time behavior of correct and incorrect executions in order to localize suspicious segments of the source code [19] [20]. This approach only labels program executions as correct or incorrect and needs no prior knowledge of the semantics of the software project [17].

Brun and Ernst [21] use Ernst's Daikon dynamic invariant extractor [22] to capture invariant features from the software source code with known errors and with errors removed. Then two groups of machine learning algorithms, Support Vector Machines (SVMs) and decision trees are employed to classify invariants as either fault-invariant or non-fault-invariant.

Most recently, Kim et al. [23] [24] has proposed a new technique for predicting latent software bugs, called change classification. They use Support Vector Machines (SVMs) to predict whether a specific change to the software source is more likely to be buggy or clean, based on the previous change history.

Kim et al.'s approach is similar to ours in a couple of aspects. Firstly, they analyze log files of the version control system of software projects to find related bug fixes in order to label that change in the source code as buggy. Similarly, we analyze those files in order to find bug-fixing revisions. However, we have nothing to do with the source code. Instead, we use the bug ID which is mentioned in the revision log to query the corresponding bug report from the open bug repository of the software project. Secondly, both works use machine learning algorithms for classification, in particular Support Vector Machines (SVMs), While the features (in the machine learning sense), class labels and also the aim of the two approaches are completely different. Our goal is to predict the most related part of the software source file hierarchy to a newly reported bug. In contrast, they try to predict whether a particular change made by a developer to the source code is more likely to be buggy or clean. Further, we use textual information of bug reports in open bug repositories to form our features. However, they use properties of the change made to the software. An example of such property has been mentioned as the frequency of words that are present in the source code, before and after performing the change. Finally, our class labels are various revision paths in the software source file hierarchy, while their class labels are clean and buggy.

## 6    Conclusion and Future Work

In this paper, we have presented a new approach to localize bugs in the source file hierarchy of open source software projects. We have used Support Vector Machines (SVMs) for predicting the file path which is more likely to be related to a given software bug report, using its summary and description. The classifier has been trained using the information of fixed bugs in the past.

We have evaluated our approach on 'UI'component of the Eclipse Java Development Tool (JDT) project. Both precision and recall values are about 98%. Applying this approach on other FOSS projects remains as future work.

Removing stop-words and performing stemming are two common data preparation tasks in text categorization problems. Here, since the experimental results are satisfying even without such preparations, we decided not to get involved with them through this work. However, it is a worthy effort to examine the effects of those techniques on other FOSS projects in future work.

One part of our future work involves applying other machine learning algorithms to the same dataset and comparing the results. We are also interested in using our approach, in the field of automated bug triage, as discussed in Section 1.

Finally, one could extend the proposed approach in order to localize the bug, either in file level or on its exact line of code, instead of our hierarchical directory level bug localization effort. Moreover, using our approach one could find the more buggy parts of the code in order to prioritize development tasks.

## References

1. Deshpande, A., Riehle, D.: The Total Growth of Open Source. In: The 4th International Conference on Open Source Systems, OSS 2008 (2008),
    http://homepages.uc.edu/%7Edeshpaaa/oss-2008-total-growth-final.pdf
    (Retrieved on November 27, 2009)

2. Kroah-Hartman, G., Corbet, J., McPherson, A.: Linux Kernel Development, How Fast it is Going. The Linux Foundation Publications (2008), https://www.linuxfoundation.org/publications/linuxkerneldevelopment.php (Retrieved on November 27, 2009)
3. Eclipse Bug Repository, https://bugs.eclipse.org/bugs (Verified on November 24, 2009)
4. Anvik, J., Hiew, L., Morphy, G.C.: Who Should Fix This Bug? In: Proc. 28th International Conference on Software Engineering, ICSE 2006 (2006)
5. Anvik, J., Morphy, G.C.: Determining Implementation Expertise from Bug Reports. In: 4th IEEE International Workshop on Mining Software Repositories, MSR 2007 (2007)
6. Baysal, O., Godfrey, M.W., Cohen, R.: A Bug You Like: A Framework for Automated Assignment of Bugs. In: 17th IEEE International Conference on Program Comprehension, ICPC 2009 (2009)
7. Fogel, K.: Producing open source software, 1st edn., pp. 60–79. O'Reilly, Sebastopol (2005)
8. Debian Bug Tracking System, http://www.debian.org/Bugs/ (Verified on December 9, 2009)
9. Jeong, G., Kim, S., Zimmermann, T.: Improving Bug Triage with Bug Tossing Graphs. In: The 7th joint meeting of the European Software Engineering Conference (ESEC) and the ACM SIGSOFT Symposium on the Foundations of Software Engineering, FSE (2009)
10. Anatomy of Eclipse Bugs, Retrieved from http://www.bugzilla.org/docs/2.18/html/bug_page.html (December 19, 2009)
11. Life-cycle of Eclipse Bugs, Retrieved from http://www.bugzilla.org/docs/2.18/html/lifecycle.html (December 19, 2009)
12. Witten, I.H., Frank, E.: Data Mining, Practical Machine Learning Tools & Techniques, 2nd edn. Elsevier, Amsterdam (2005)
13. Bishop, C.M.: Pattern Recognition and Machine Learning. Springer, Heidelberg (2006)
14. Clarke, E., Grumberg, O., Peled, D.: Model Checking. MIT Press, Cambridge (1999)
15. Visser, W., Havelund, K., Brat, G., Park, S.: Model checking programs. In: Proceeding of the 15th IEEE International Conference on Automated Software Engineering, ASE 2000 (2000)
16. Musuvathi, M., Park, D., Chou, A., Engler, D., Cmc, D.D.: A pragmatic approach to model checking real code. In: Proceeding of the 5th Symposium on Operating System Design and Implementation, OSDI 2002 (2002)
17. Liu, C., Yan, X., Fei, L., Han, J., Midkiff, S.P.: SOBER: Statistical Model-Based Bug Localization. In: The 3rd joint meeting of the European Software Engineering Conference (ESEC) and the ACM SIGSOFT Symposium on the Foundations of Software Engineering, FSE (2005)
18. Gyimothy, T., Ferenc, R., Siket, I.: Empirical Validation of Object-Oriented Metrics on Open Source Software for Fault Prediction. IEEE Trans. on Software Eng. 31(10), 897–910 (2005)
19. Cleve, H., Zeller, A.: Locating causes of program failures. In: Inverardi, P., Jazayeri, M. (eds.) ICSE 2005. LNCS, vol. 4309. Springer, Heidelberg (2006)
20. Liblit, B., Naik, M., Zheng, A., Aiken, A., Jordan, M.: Scalable statistical bug isolation. In: Proc. of ACM SIGPLAN 2005 International Conference on Programming Language Design and Implementation, PLDI 2005 (2005)

21. Brun, Y., Ernst, M.D.: Finding Latent Code Errors via Machine Learning over Program Executions. In: Proc. of 26th International Conference on Software Engineering (ICSE 2004) (2004)
22. Ernst, M.D., Perkins, J.H., Guo, P.J., McCamant, S., Pacheco, C., Tschantz, M.S., Xiao, C.: The Daikon System for Dynamic Detection of Likely Invariants. Science of Computer Programming (2006)
23. Kim, S., Whitehead Jr., E.J., Zhang, Y.: Classifying Software Changes: Clean or Buggy? IEEE Trans. on Software Eng. 34(2), 181–196 (2008)
24. Shivaji, S., Whitehead Jr., E.J., Akella, R., Kim, S.: Reducing Features to Imrove Bug Prediction. In: Proceeding of the 15th IEEE International Conference on Automated Software Engineering, ASE 2009 (2009)
25. Salton, G., Buckley, C.: Term-weighting approaches in automatic text retrieval. Information Processing & Management 24(5), 513–523 (1988)
26. Plat, J.C.: Technical Report, MSR-TR-98-14, Microsoft Research (April 21, 1998)
27. Plat, J.C.: Advances in Kernel Methods - Support Vector Learning, pp. 41–65. MIT Press, Cambridge (1998)
28. WEKA 3-7-0 source comments, weka.classifiers.functions.SMO
29. The official WEKA manual, Retrieved from http://www.cs.waikato.ac.nz/ml/weka/ (December 25, 2009)

# T-DOC: A Tool for the Automatic Generation of Testing Documentation for OSS Products

Sandro Morasca, Davide Taibi, and Davide Tosi

Università degli Studi dell'Insubria,
Dipartimento di Informatica e Comunicazione, Via Mazzini, 21100 Varese, Italy
{sandro.morasca,davide.taibi,davide.tosi}@uninsubria.it

**Abstract.** In the context of Open Source Software (OSS), the lack of project documentation is one of the most challenging problems that slows down the widespread diffusion of OSS products. The difficulty of providing up-to-date and reasonable documentation for OSS products relates to two main reasons. First, documenting development activities and technological issues is viewed as a tedious and unrewarding task. Second, data and information about an OSS project (such as source code, project plans, testing requirements, etc.) are scattered and shared via unstructured channels such as unofficial forums and mailing lists.

In this paper, we focus on technical documentation related to testing activities. In this context, the lack of documentation is exacerbated due to the use of the available testing methods that drastically increase code fragmentation. We propose T-doc, a tool that simplifies the generation of testing documentation. In particular, T-doc supports (1) the automatic generation of test cases documentation, (2) the generation of reports about test case results, and (3) the archiving of testing documents in central repositories. The automatic generation of documentation is facilitated by the adoption of built-in testing methods that simplify the aggregation of testing data.

We apply the tool to the OSS RealEstate Java application to show the applicability and the real benefits of our solution.

**Keywords:** Open Source Software testing, testing documentation, testing tools.

## 1 Introduction

Open Source Software (OSS) is experiencing an increasing diffusion and popularity in industrial sectors. However, this spreading is slowed down by the frustration a lot of potential users have when they start evaluating an OSS product that they would like to adopt. This is primarily due to the lack of reasonable and up-to-date user documentation that deeply describes the intent and the technical aspects of the project.

Most of the available OSS projects are currently released without up-to-date user manuals and technical documents. The lack of documentation in OSS is even more serious in the context of testing activities. It is very rare to find

P. Ågerfalk et al. (Eds.): OSS 2010, IFIP AICT 319, pp. 200–213, 2010.

well-structured documents, manuals, and reports about all the testing phases performed during the development of OSS products. Documenting OSS projects is a tedious and unrewarding task that is made more complicated by the scattering of data and information typical of OSS projects.

In this paper, we focus on the problem of documenting testing activities and we propose a tool (we called T-doc) that supports the automatic generation of unit, integration, regression testing documentation, the report of test results, and the aggregation of these data in dedicated central repositories we called "testing tracker systems." The automatic generation is simplified by the use of built-in testing methodologies that put together the code of methods and test cases in a single component to avoid the fragmentation of source code and to simplify the aggregation of the testing data [3]. T-doc provides a three-layered support:

- automatic generation of test cases documentation (in a java-doc like style);
- automatic generation of suggestions about integration and regression testing activities that should be performed by each developer and for each component of the project;
- automatic generation of reports about the results of test suites execution.

All the documents and testing data are then collected and archived in the testing tracker system of the project to favor data discovery and data sharing. This paper is a step towards our final goal, which is the development of a standard framework that OSS developers can use whenever they start testing their OSS products. In this paper, we apply an initial implementation of T-doc to the RealEstate Java application [2] to show the simplicity, the real benefits, and the level of automation provided by our solution.

The paper is structured as follows: Section 2 reports the analysis we conducted to confirm the low availability of testing documentation, and discusses the limits of a set of existing testing tools; Section 3 introduces the motivations that are at the basis for adopting built-in testing in the context of OSS products; Section 4 separately discusses the three layers of the T-doc tool, and shows how T-doc comes into play when applied to the RealEstate Java application; and finally we conclude in Section 5.

## 2    The Lack of OSS Documentation

The perception we normally have surfing the web portal of OSS products, observing OSS forums/blogs/discussions, and using OSS products in our every-day work is that most of the available OSS projects are released without user manuals and technical documents.

To have an empirical evidence of this perception, we conducted a two-fold analysis: first, we interviewed 151 OSS users (end users, developers, managers, OSS experts) and then, we analyzed the web portal of 32 well-known OSS projects[1]. The first analysis aimed to identify the importance the factor

---

[1] An extensive report of these experiences can be found in
www.qualipso.eu/node/45 and /node/84

"availability of technical documentation / user manual" has for OSS users. We discovered that in a scale from 1 (negligible importance) to 8 (fundamental importance), the factor "availability of technical documentation / user manual" took a very high score equal to 6,5. The second analysis aimed to check the actual availability of technical documentations and user manuals related to the 32 analyzed projects. We discovered that: 69% of the projects have up-to-date user manuals while the remaining 31% have not updated or available user manuals; 49% of the projects have an up-to-date technical documentation, while the remaining 51% have not an updated or available technical documentation.

This deficiency is exacerbated when we look at testing documentation: in our analysis, only 1 product (out of 32) provides a complete documentation about its internal testing activities. Only JBoss [www.jboss.org] exposes a detailed and up-to-date documentation about testing plans, testing methodologies, test cases description, and test suite results. We believe that this is primary due to three main reasons: first, the use of classical testing methodologies that are based on external testing (i.e., test cases are independent components that are separated from the applicative code) drastically augment the fragmentation of data, thus further complicating the process of documenting testing activities; second, the lack of well-agreed best practices on how to test OSS products increases the effort required for testing applications, thus stealing effort in documenting testing activities. Finally, the lack of tools, which support and automate the documentation of testing activities, leaves too much effort to the side of developers. The results obtained by our second exploration are in contrast with the requirements OSS users have. This analysis confirms our intuition and demonstrates the need for a tool that supports the automatic generation of testing documentation.

Currently, open source tools or frameworks that support the whole documentation of testing activities are not yet fully available. The famous portal [www.opensourcetesting.org] gathers a lot of testing tools that support a specific aspect of the test life cycle, but none of them are able to manage and create the documentation, the results report and the collection of these information. For example, Testopia [www.mozilla.org/projects/testopia] is a test case management extension for Bugzilla that tracks test cases and allows for testing organizations to integrate bug reporting with their test case run results. However, Testopia covers only a part of the functionalities provided by T-doc. Fitness [http://fitnesse.org] is a software development collaboration tool, which simplifies the management of testing documentation, test reports and the collaborative definition of acceptance tests. T-doc, is able to automatically generate testing documentation and it is not limited to acceptance tests. Moreover, T-doc is able to automatically suggest the integration and regression testing activities that should be performed. Other tools, such as TPTP [www.eclipse.org/tptp/] or Salome-TMF [https://wiki.objectweb.org/salome-tmf/], are complex frameworks that cover the entire test life cycle but are not able to automatically create testing documentation.

The next section discusses why a built-in testing method is preferred to classic testing solutions.

# 3   Built-in Test in OSS

Built-in self-test (BIST) and Built-in test (BIT) approaches for software systems originated in the context of component-based systems to simplify the integration of third-party black-box components and enhance software maintainability [9]. A BIT component (or BIT class) is a traditional component that puts together applicative code with testing code [3]. A BIT component can operate in a normal mode (i.e., testing capabilities are switched off to the user) or in maintenance mode (i.e., the user can test the component in his environment by exploiting the built-in testing capabilities) by interacting with the application or the testing interface, respectively. Listing 1 shows a code excerpt for a typical component with built-in testing abilities, where test cases are declared and implemented directly into the applicative class.

In the context of OSS, the heterogeneity of the developers/contributors increases the fragmentation of the source code and makes unfeasible the adoption of available testing methods, programming rules, and testing tools that could favor the whole comprehension of fragmented testing activities. Simple programming rules (as shown in Listing 1) may help standardize a common programming style that can improve the testing activity, decrease the testing effort, and simplify the generation of testing documentation. Whenever a developer/contributor of an OSS product introduces or modifies a functionality of a component, he or she designs and codes unit tests, integration tests and optionally non-functional tests into the component to provide BIT abilities. Modified components are then uploaded into the repository that stores the project and are integrated to generate the OSS product with comprehensive BIT abilities (as shown in Figure 1).

Putting together application code and testing code into single classes has several advantages: (1) it improves the visibility and inheritance of test cases. Test cases are coded as classic methods thus, when a class extends another class, the former inherits not only the application methods but also the test case methods. This simplifies the reuse of available test cases; (2) it favors the standardization

```
1   Class class_name {
2      // application interface
3      Data declaration;
4      Constructor declaration;
5      Destructor declaration;
6      Methods declaration;
7
8      // testing interface
9      TestCases declaration;
10
11     // application code
12     Constructor;
13     Destructor;
14     Methods;
15
16     // testing code
17     TestCases;
18  }
```

**Listing 1.** Code excerpt of a BIT component

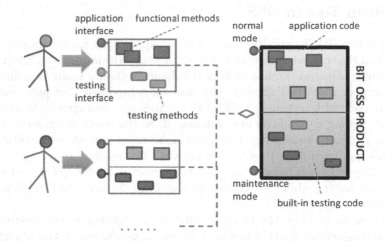

**Fig. 1.** Aggregating components into an OSS product with BIT abilities

of testing interfaces. Test cases are developed following the coding rules of the target programming language in use for the application, thus limiting the creativity of the developers. This improves the readability of the testing code; (3) it increases the aggregation of data. Test cases are grouped into single classes instead of into different packages, components, or libraries. This simplifies the discovery of testing data and their correlation with coding elements; (4) moreover, the documentation of test activities and the report of test case results is made easier, thus simplifying regression testing activities. Regression testing is made upon the availability of test cases and test results. The more test cases and test results are not available or they are disaggregated, the more the regression testing activity is tricky; (5) it favors run-time testing [8]: the system can be executed at run-time in maintenance modality [7], thus simplifying the detection of bugs that are undetectable in a controlled testing environment. In OSS, often components are separately tested at development time by each contributor that develops a small unit and tests its behavior in isolation. This leaves undetected a lot of integration failures. Moreover with BIT, the test suite can be executed over different hw/sw platform configurations, thus simplifying system, configuration and performance testing. Every time a user installs the application on his environment, he/she becomes a new tester of the application and he/she uses his/her hw/sw configuration as a new scaffolding of the testing activity. Hence, the "eye bird" ability, which is typical of OSS products (i.e., the capacity to evaluate a product by the large glance of the OSS community), can be fully exploited and can be complemented by testing activities.

However, BIT also introduces risks and limitations that need to be faced when designing the T-doc tool: run-time testing can move the system in an inconsistent state that may compromise the stability of the system. To mitigate this risk, the test suite must be executed in background only once, during the OSS product installation (or during critical updates). Moreover, BIT is an intrusive mechanism that can lead to security and privacy-related problems. To mitigate this risk, final users must be aware that the OSS product is under BIT, so they can block the BIT abilities if they so wish, and user-related data must not be collected by the framework. Finally, if built-in tests are executed without a control, system performance can degrade. The execution of the built-in test suite in background, during the OSS product installation, alleviates this problem.

To the best of our knowledge, we believe that the use of BIT abilities, instead of classic testing mechanisms, is a valid way to support and simplify the generation and the gathering of testing documentation in the domain of OSS.

## 4  The T-Doc Tool

Here, we present the architecture of the T-doc tool and we detail its threefold support by separately discussing: the automatic generation of test cases documentation, the automatic generation of suggestions about integration and regression testing activities, and finally the generation of reports about the results of the test suite execution. Figure 2 shows a high level architecture of T-doc.

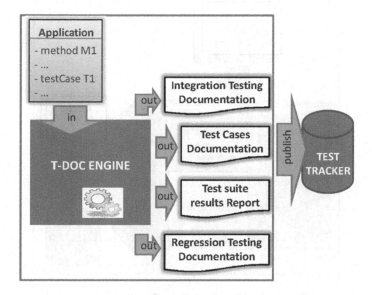

**Fig. 2.** High level architecture of the T-doc framework

### 4.1  Test Case Documentation

This first layer of support aims at simplifying and automating test case and test suite documentation generation. The generated documentation should increase the readability of the technical aspects of each test case, and should favor an overall comprehension of the testing activity. To allow for the automation of this process, built-in test cases must be surrounded by *doc* comments (i.e. short sentences that describe the test case, its purpose, and its behavior) and keywords in a way similar to the way comments and block taglets surround methods and functionalities in Java source code. Testing doc comments (T-doc comments) and block taglets are then parsed and processed by the T-doc engine to generate the test case documentation much the same way as the Javadoc tool operates. Javadoc is a tool from Sun Microsystems for generating API documentation out of declarations and documentation comments in Java source code. Javadoc produces HTML documentation describing the packages, classes, interfaces, methods, etc. of a software system. The output format of the Javadoc can be customized by means of doclets. Javadoc parses special tags embedded within a Java doc comment. These doc tags are used to automatically generate a complete, well formatted API from the source code. All tags start with a (@), e.g., @author. The tags are used to add specific information like a method's parameters (@param), return type (@return), and exceptions (@exception).

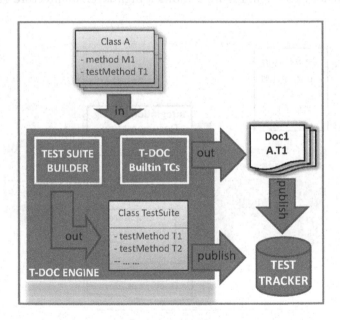

**Fig. 3.** Architecture of the first T-doc layer

To minimize the effort of developers and contributors in writing testing documentation, favor standardization, and avoid subjective interpretations of data, we clearly define a set of new conventions and a set of ad-hoc tags that developers and contributors should follow whenever they add a T-doc comment. An example of a real T-doc comment can be found in Figure 4.

In compliance with Javadoc, the conventions we defined are:

1) the first line contains the begin-comment delimiter (/**)
2) write the first sentence as a short summary of the test, as T-doc
   engine automatically places it in the summary table of the test
3) insert a blank comment line between description and the list of tags
4) the first line that begins with an ''@" character ends the description
5) there is only one description block per T-doc comment
6) the last line contains the end-comment delimiter (*/)

The tags, useful for commenting a test case, are listed below:

```
@param    (name of the parameter, followed by its description)
@return   (omit @return for tests that return void; required otherwise)
@succeedIf  (summarize the conditions under which the test case succeeds)
@failIf  (summarize the conditions under which the test case fails)
@qualityAttribute (specify the quality aspect addressed: performance, etc.)
@scope   (specify the test case purpose: unit, integration, etc.)
@author    (author name/surname)
@version   (version number + checkout date)
@see   package.Class#method(Type,...) (ref to the function under test)
```

Figure 3 shows a subset of the functionalities provided by the T-doc Engine. The T-doc Engine takes in input the set of classes that are added/modified by the developer. Each class is analyzed separately to discover and isolate the built-in test cases and their T-doc comments. The **Test Suite Builder** component aggregates all the built-in test cases into a single test suite, and the **T-doc TCs** component parses all the t-doc comments to generate the complete documentation of the test suite. Finally, the engine publishes the documentation to the central repository (Test Tracker) of the project to avoid fragmentation and versioning problems of the documentation. Versioning problems are also avoided by means of the introduction of the new tag @version.

To favor the comprehension of this layer, we exemplify the writing of a T-doc comment for a built-in test case we derived for the RealEstate OSS Java application [2]. The RealEstate is a Java application created at North Carolina State University that reproduces the Monopoly game. The RealEstate application will be used throughout the whole paper as proof-of-concept of our work. Figure 4 shows the source code of the built-in test case surrounded by a T-doc comment and T-doc tags. The purpose of this Figure is not to present the internal code of the test, but to highlight the structure of a T-doc comment.

The documentation automatically generated by the T-doc engine for this test case looks like as follows:

```
ID001:: UNIT Test: testGainMoneyCardAction
V1.0.2 06-02-09

Tests the behavior of the applyAction() functionality.
Checks whether the account of the current player's
CCard is properly updated when a gain of money is performed.

Succeeds if: getMoney() returns a value=1550$
Fails if: getMoney() returns a value!=1550$
See: edu.ncsu.realestate.MoneyCard()
```

```
public void applyAction() {
    currentPlayer.setMoney(currentPlayer.getMoney()+amount);
}
/**
 *  Tests the behavior of the applyAction() functionality. Checks whether the account of
 *  the current player's CCard is properly updated when a gain of money is performed.
 *
 *  @succeedIf      getMoney() returns a value = 1550 $
 *  @failIf         getMoney() returns a value != 1550 $
 *  @scope          unit testing
 *  @author         Davide Tosi
 *  @version        1.0.2   06/02/09
 *  @see            edu.ncsu.realestate.MoneyCard()
 */
public void testGainMoneyCardAction() {
    Card gainMoney = new MoneyCard("50$", 50, Card.CARD_TYPE_CHANCE);
    GameMaster.instance().getGameBoard().addCard(gainMoney);
    card.applyAction();
    TestCase.assertEquals(origMoney+50, GameMaster.instance().getCurrentPlayer().getMoney());
}
```

**Fig. 4.** A built-in test case with T-doc comments for the RealEstate application

The T-doc engine generates a documentation that is compliant with the visual representation of Javadoc comments, with small differences (such as the use of a label for each test ID00X), in order to maximize both the compatibility and also the readability of the documentation. Currently, this T-doc module has been fully implemented and its is fully compatible with the Eclipse IDE.

### 4.2   Regression and Integration Testing Documentation

This second layer of support aims at suggesting and documenting the integration and regression test cases that OSS contributors should develop during the update/maintenance of their OSS products. The generated documentation should simplify the contributors' task of writing these test cases. To this end, the dependencies among methods and components must be detected by the T-doc engine and visually reported to the developer. The T-doc engine uses the idea of *change points* and *call graphs* [4] [5] to automatically detect the source code location in which a code change has been performed, and to automatically create the graph of calls related to the method in which the change has been detected. These graphs are used by the T-doc engine as the starting point to create the suggestions for integration and regression testing activities.

Figure 5 shows the three main modules of this layer: the T-doc Integration module, the T-doc Regression module and the Call Graph tool.

The T-doc Integration module is responsible for suggesting integration testing scenarios that should be implemented by the OSS contributors whenever a new method is added or whenever an existing method is modified (i.e., the @version tag of the associated test case is updated). Integration testing checks dependencies among objects of different classes. Class A uses class B if objects of class A make method calls on objects of class B, or if objects of A contain references to objects of B [6]. The T-doc Integration takes as input the documentation generated by the T-doc TCs module (Doc1 A.T1) and automatically

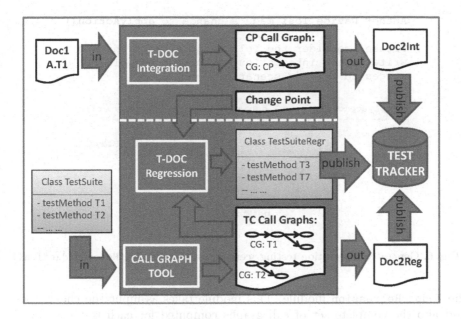

**Fig. 5.** Architecture of the second T-doc layer

generates the call graph for the change point (CP) that is related to the documented test case. To avoid graph size explosion, we chose to limit the computation to the third level of method's dependencies. We are conducting additional experiments to understand the code coverage we obtain with this limit.

Referring to our RealEstate example of Figure 4, the OSS contributor is working on the MoneyCard class, by modifying the applyAction() method and writing the built-in test case testGainMoneyCardAction(). First of all, the T-doc Integration module computes the call graph for the change point applyAction(), then it produces the integration testing scenario for this change. Figure 6 shows the result of this computation (Doc2Int). The root of the graph is the CP applyAction(), while leaves are the methods that directly or indirectly interact with the applyAction() method. The T-doc Integration module integrates the functionalities provided by CallGraph [www.certiv.net/projects/] to automatically create call graphs starting from a change point.

The T-doc Regression module is responsible for automatically detecting the subset of relevant test cases for regression activities whenever a change into the code is performed. Without this support, OSS contributors are forced to manually rerun all the test cases in the test suite for regression purposes. This task is very expensive for contributors that are not interested in testing. For instance, rerunning the complete test suite for the OSS WEKA application [sourceforge.net] require 45 mins in a fully dedicated machine. Moreover, other problems are: who runs the test suite? Where to store and collect the test cases that should be re-executed? When must the test cases be rerun? Where are the results of the test suite execution reported? All these problems are addressed by

Author **Davide Tosi** made a change to **applyAction()**

Please, consider the following interactions and write ad-hoc integration tests that exploit the suggested testing scenario:

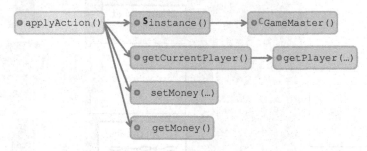

**Fig. 6.** Generated integration testing scenario for the `testGainMoneyCardAction()`

the `T-doc Regression` module. This module takes as input the change point and also the complete set of call graphs computed for each test case by the `Call Graph Tool` module. Then, the `T-doc Regression` module scans all the call graphs to detect the subset of graphs that are affected by the change point (i.e., the change point is present into the graph). The subset of relevant call graphs indicates the meaningful test cases that should be re-executed with respect to the change that has been performed. Here, we show the algorithm that the `T-doc Regression` module uses to detect the subset of meaningful test cases:

```
Input: test cases, CP
Output: documentation of the subset of meaningful regression tests

1. derive the call graph for each test case ending at the 3rd level
   of dependencies;
2. select a graph as starting entry;
3. scan the graph to detect whether the change point is present;
4  if the change point is present:
   select the test case for regression;
   else: jump to step 2.
5. when all the graphs have been evaluated, generate the regression
   documentation as the list of test cases wrt the CP
```

For the RealEstate application the `T-doc Regression` module takes as input, from the `Call Graph Tool`, 30 graphs and generated the following documentation (Doc2Reg). For space reason, we do not show the complete set of graphs computed by the T-doc engine.

```
Doc2Reg:
 This is the set of regression test cases
 for the applyAction() change point:
 01) testGainMoneyCardAction()
 02) testMovePlayerCardAction()
 03) testLoseMoneyCardAction()
 04) testJailCardAction()
 05) testJailCardUI()
 06) testLoseMoneyCardUI()
 07) testMovePlayerCardUI()
```

All the data provided by this second layer (Doc2Int, Doc2Reg and the regression test suite) are published into the central Test Tracker system.

## 4.3  Test Case Execution Report

This third layer of support aims at homogenizing and collecting all the outputs coming from the T-doc tool and the results obtained by the execution of the test cases. In this section, we only introduce the design of this layer since its implementation is not yet available. This layer is composed of two main entities: the Test Tracker system and the part of the T-doc engine that is responsible for collecting and manipulating the test case results.

The Test Tracker system is responsible for managing: (1) the class containing all the built-in test cases that are incrementally added (or modified) to the test suite (Class TestSuite); (2) the class of integration test cases (if available); (3) the class containing the regression test cases derived by the T-doc Regression module. The Test Tracker system stores the documentation of each test case (Doc1 A.T1, Doc1 A.T2, Doc1 A.Tn) and aggregates this documentation in a single document that describes the complete behavior of the test suite. Moreover, the Test Tracker system stores the documentation related to integration and regression test cases (Doc2Int and Doc2Reg), and it aggregates this documentation in a single file. Finally, the Test Tracker system provides search abilities among all the T-doc documents that are published by the T-doc engine. As in Bug tracker systems (such as Bugzilla [www.bugzilla.org]), T-doc documents can be searched and filtered by means of ad-hoc keywords. These keywords are identical to the tags we defined in Section 4.1. For example, you can filter your search by @author (T-doc documents are grouped regarding to the owner of the test cases) or by @scope (T-doc documents are grouped according to the purpose of test cases).

As mentioned in Section 3, built-in test cases favor the execution of run-time testing [7]. The T-doc engine exploits this feature and it is able to collect the results of the run-time execution of the test suite. Figure 7 shows the modules involved in this task. The two T-Report modules collect the results of the test cases execution. Hence, the two modules correlate these results with the run-time HW/SW configuration of the execution environment in which test cases have been executed. The output of these correlations are two reports (Report a and Report b) that document the results of the run-time testing activity. Currently,

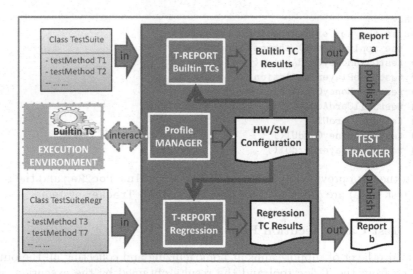

**Fig. 7.** Architecture of the third T-doc layer

we are working on the identification of the profile information that should be collected by the **Profile Manager** module (such as log files, active processes, hw/sw capabilities, etc.), and we are implementing this third T-doc layer to support the testing documentation of Java OSS projects.

### 4.4   Validation Remarks

Though the RealEstate demo application has shown the feasibility and the benefits of the T-doc tool, we are extending the validation of the tool with additional case studies. For example, in our labs, we are implementing a complex OSS project to validate the approach and to understand its potentialities. The project (we called MacXim) is a static-analysis tool (15000 LOC in 118 classes) that exploits the solution presented in this paper [1]. The MacXim test suite (composed of acceptance, unit, integration and regression tests) has been designed with in mind the guidelines proposed in this paper, and each test case has been documented with a T-doc comment that describes the purpose of the test.

This controlled project will provide important feedbacks about the potentialities and the weaknesses of the T-doc tool, and will be the basis for developing a stable tool that will be fully exploited in real-life OSS projects and in uncontrolled development environments.

## 5   Conclusions and Future Work

In this paper, we proposed T-doc, a tool that simplifies the generation of testing documentation in the context of OSS projects. We showed how T-doc supports the automatic generation of test cases documentation, the generation of reports about test case results, and the archiving of testing documents in central repositories. The automatic generation of documentation is facilitated by the adoption of built-in testing methodologies that simplify the aggregation of testing data. To understand the T-doc working in practice, we applied the tool to the OSS RealEstate Java application.

Currently, we are integrating all the modules of the T-doc tool and we are validating T-doc with the real-life Java application MacXim.

## Acknowledgments

The research presented in this paper has been partially funded by the IST project QualiPSo (http://www.qualipso.eu/), sponsored by the EU in the 6th FP (IST-034763); the FIRB project ARTDECO, sponsored by the Italian Ministry of Education and University; and the projects "Elementi metodologici per la descrizione e lo sviluppo di sistemi software basati su modelli" and "La qualità nello sviluppo software," funded by the Università degli Studi dellInsubria.

## References

1. MacXim: a static code analysis tool. Web published:
   http://qualipso.dscpi.uninsubria.it/macxim/ (Accessed: December 2009)
2. The RealEstate demo application. Web published,
   http://agile.csc.ncsu.edu/SEMaterials/realestate/ (Accessed: December 2009)
3. Beydeda, S.: Research in testing COTS components - built-in testing approaches. In: Proceedings of the ACM/IEEE International Conference on Computer Systems and Applications (AICCSA), pp. 101–104 (2005)
4. Mao, C., Lu, Y., Zhang, J.: Regression testing for component-based software via built-in test design. In: Proceedings of the ACM Symposium on Applied Computing (SAC), pp. 1416–1421 (2007)
5. Orso, A., Harrold, M.J., Rosenblum, D.S., Rothermel, G., Soffa, M.L., Do., H.: Using component metacontent to support the regression testing of component-based software. In: Proceedings of the IEEE International Conference on Software Maintenance (ICSM), pp. 716–725 (2001)
6. Pezzè, M., Young, M.: Software Testing And Analysis. Process, Principles, and Techniques. Wiley, Chichester (2007)
7. Suliman, D., Paech, B., Borner, L., Atkinson, C., Brenner, D., Merdes, M., Malaka, R.: The MORABIT approach to runtime component testing. In: Proceedings of the International Computer Software and Applications Conference (COMPSAC), pp. 171–176 (2006)
8. Vincent, J., King, G., Lay, P., Kinghorn, J.: Principles of built-in-test for run-time-testability in component-based software systems. Software Quality Control 10(2), 115–133 (2002)
9. Wang, Y., King, G., Wickburg, H.: A method for built-in tests in component-based software maintenance. In: Proceedings of the IEEE European Conference on Software Maintenance and Reengineering (CSMR), pp. 186–192 (1999)

# Open Source Introducing Policy and Promotion of Regional Industries in Japan

Tetsuo Noda and Terutaka Tansho

Shimane University
1060 Nishikawatsu-cho, Matsue-City, Shimane Pref. 690-8504, Japan
nodat@soc.shimane-u.ac.jp, tansho@riko.shimane-u.ac.jp

**Abstract.** The development style of open source has a possibility to create new business markets for Regional IT industries. Some local governments are trying to promote their regional IT industries by adopting an open source in their electronic government systems. In this paper, we analyze the data of open source application policy of the Japanese government and case studies of promotion policy of local industries by local governments; for example, Nagasaki Prefecture and Matsue City. And it aims to extract the issues in the open sources application policy of local governments and the promotion policy of regional industries in Japan.

## 1 Introduction

In Japan, it is public organizations, especially central government offices that have guided adoption of open source. The Ministry of Internal Affairs and Communications (MIAC) that aims at the spread of the e-municipality system, started to put Linux in the list as one of the choices of OS when the electronic government system was introduced in 2004. Before this, the Ministry of Economy Trade and Industry (METI) also reported "Survey on Usage of Open Software; A Guideline for its Introduction" in 2003. In this report, such matters were examined as using OSS including Linux positively as a choice, a guideline for examining the arrangement of its introduction, and a legal problem, etc. Thus, open source software promotion organization "OSS Center", as an affiliated association of METI, was established in 2006. The OSS center has been collecting and offering technological information in cooperation with domestic major IT vender enterprises.

As a result of the open source promotion plan that centers on these central government offices, in the IT-Solution market in Japan, the market of public organizations occupies an especially large percentage. The IT solution market using open source was 917 billion yen in fiscal year 2006 (8.8% of the entire IT solution market). It expanded to 1.05 trillion yen in 2007; an increase of 10.5 points compared to the previous year (10.8% of the entire IT solution market), and is expected to expand to 1.16 trillion yen, an increase of 14.5 points compared to the previous year in 2008. And, the ratio of the usage by public organizations to the entire open source market was 254 billion yen (24.2% in composition ratio) in fiscal year 2006, and 227 billion yen (24.8%) in fiscal year 2007. In Japan the public organizations occupy a big specific gravity in open source market, and they have still been guiding the growth of the entire open source market.

P. Ågerfalk et al. (Eds.): OSS 2010, IFIP AICT 319, pp. 214–223, 2010.

In addition, after 2006, the open source introduction policy in Japanese local governments has been advanced by "Open source software use infrastructure agenda" of the OSS Center. Though this is a policy of doing financial support by the OSS Center when regional municipalities introduce open source in their local electronic government systems, the support load and other expenses are not considered for later years. Therefore, some local governments have difficulties in continuing the maintenance of their systems when the financial supports break off. Moreover, the number of enterprises with technologies that can continuously support open source in provinces is small, and this is also a factor that the introduction policy doesn't continue in local governments. Consequentially, the market is only created for the major IT vender enterprises that were involved in the initial introductions to the local governments.

Thus the introduction and the spread of open source policy have been advanced around the central government offices and the major IT vender enterprises, as the policy has been concentrated on the adoption and introduction of open source. As the result, it led to the expansion of the IT-Solution market among the major IT vender enterprises. It somehow succeeded in protecting the IT-Solution market of domestic IT vender enterprises from the foreign IT enterprises (Table 1 and Table 2)[1]. But development method = Cathedral type of the top down has not been able to be changed.

**Table 1.** Major IT enterprises' share in IT solution market using open source (*domestic IT vender enterprises)

|          | 2006 | 2007 | 2008 |
|----------|------|------|------|
| Fujitsu* | 7.3% | 8.6% | 9.3% |
| IBM      | 7.2% | 7.1% | 6.9% |
| NEC*     | 6.2% | 6.7% | 6.9% |
| HITACHI* | 5.0% | 5.3% | 5.9% |
| HP       | 1.4% | 1.4% | 1.5% |
| Sun      | 0.6% | 0.6% | 0.6% |

**Table 2.** Major IT enterprises' share in IT solution market in the public organizations using open sources (*domestic IT vender enterprises)

|           | 2006  | 2007  | 2008  |
|-----------|-------|-------|-------|
| Fujitsu*  | 10.1% | 10.3% | 10.7% |
| NTT-Data* | 9.3%  | 9.4%  | 9.5%  |
| NEC*      | 8.3%  | 9.3%  | 9.8%  |
| HITACHI*  | 7.9%  | 8.5%  | 9.1%  |
| IBM       | 4.4%  | 4.3%  | 4.2%  |

On the other hand, the development style of open source is extending beyond the boundary of organizations, so it has the possibility to create new business markets for regional IT industries. So some local governments are trying to promote their regional IT industries by adopting open source in their electronic government systems. In the later sections, we analyze the data of open source application policy of Japanese government

---

[1] MIC Economic Research Institute (2008).

and case studies of promotion policies of local industries by local governments, for example, Nagasaki Prefecture and Matsue City of Shimane Prefecture, located in a typical local country area in Japan.

## 2  Introduction of OSS and Promotion of Regional Industries

### 2.1  The Method of Divided Orders by Using OSS in Nagasaki Prefecture

Actually, in introducing OSS at the local government level, in addition to Introduction/Management-cost reduction of the IT system of the municipality, local industry promotion has been advanced in Japan. A typical case is the construction of "Electronic Prefectural Government System" in Nagasaki Prefecture.

Nagasaki Prefecture invited CIO from a private organization in 2001, and adopted open source system for the basic technology of the electronic Prefectural government systems which consist of three functions (document management, application and tenders). Then, the orders, that used to be placed as a whole with major IT vender enterprises in Tokyo, are now divided to small systems before ordering.

Why did it become possible? Because, (1) The staff in the prefecture made specifications of the systems, (2) When the external program development was consigned, the divided orders decreased the budget for a matter, (3) The technology was decided before system was constructed. These processes enabled local small and medium-sized IT enterprises in Nagasaki to participate in the development of these systems. By this method, expenditure related to the computer system of the prefecture was greatly reduced with 694 million yen in fiscal year 2003 from 991 million yen in fiscal year 2002. This cost continues for five years for the leasing contract of the server system. But, when the server systems consisting of a mainframe, are completely replaced by the Linux server in five years, Nagasaki Prefecture estimates that the entire cost will be compressed into 30% of that of 2002 (Figure 1)[2].

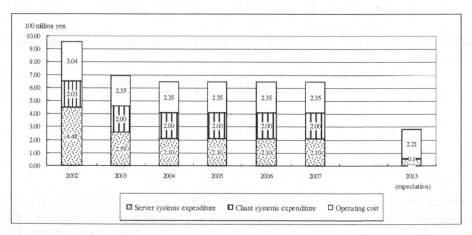

**Fig. 1.** Introduction/Management cost reduction of the IT system by using OSS in Nagasaki Prefecture

---

[2] Nagasaki Prefecture (2008).

Moreover, it becomes possible also for the local IT enterprises to participate in the system development by the divided order method. As a result, they received 48/100 orders directly in 2002- 2003 fiscal year (15.1% in terms of monetary amounts), and 73/96 orders in 2004 fiscal year (32.7% in terms of monetary amounts) (Table 3).

Table 3. Ratio of order receipt for local IT enterprises in the Nagasaki e-municipality system

|  | Number of Cases | Amount of Money |
|---|---|---|
| Before 2001 | No Result | No Result |
| 2002 | 47.9% | 15.2% |
| 2003 | 48.1% | 15.1% |
| 2004 | 76.0% | 32.7% |
| 2005 | 75.4% | 46.3% |
| 2006 | 82.1% | 69.3% |
| 2007 | 88.1% | 62.8% |

Adopting the open source policy and divided orders made it possible for them to expand the market for the local IT enterprises. But, this would not last longer than several years because the entire cost of the electronic government systems in Nagasaki was compressed in 2003. However they have expanded the ratio of the market, although it is obvious that the total pie has been cut back. They have to develop new market of open source. Nagasaki Prefecture's regional-industry promotion plan needs to move on to the next stage.

## 2.2 Opening of Source Codes and Regional-Industry Promotion

In Nagasaki, by this method of divided orders, local IT enterprises were encouraged to participate in the processes of making specifications and entry into the processes of the project and decision making for specifications that had been major IT vender enterprises' "Role", so that they could not only increase their orders of "Electronic Prefectural Government System" but also improve their abilities in project management.

In 2004, Nagasaki Prefecture announced the plan to shift legacy system from mainframe to Linux within eight years. Moreover, the source codes of their three systems, "Vacation system", "System of WEB list of government officials" and "Document keeping system" were opened to the public in 2005. These systems had already been decided to be introduced also to other prefectures such as Tokushima Prefecture and Wakayama Prefecture. This shows the possibility of expanding the market for the IT enterprises in Nagasaki Prefecture (The orders from outside of Nagasaki Prefecture have also increased).

Thus, there has already been a success case of promotion of the local IT enterprises, by introducing OSS, and to expand Market in Japan. In the system construction of a local government, it not only adopts OSS which local IT industries have developed, as a user = a purchaser, but also, if products are open to the public on Internet as OSS, it can promote many chances that the local IT industries can expand their own market on a nationwide scale. "Cathedral type development method" or "Vender lock-in" by major

IT vender enterprises can be released, so the local IT industries can participate in the process of software development and receive orders of the systems that have been divided in small sizes.

However, to make this process possible, as was shown in the case in Nagasaki Prefecture, the staff of the administration sector must participate in the upper processes of the software development, such as processes of "Requirement definition" or "Design". The reason why major IT vender enterprises secured orders of their own was that they had participated in these processes. Nagasaki Prefecture's case was, so to speak, an indirect development aid by the administration whose staff in the prefecture were related to these processes of "Requirement definition" and "Design", so it enabled them to achieve regional-industrial promotion plan.

Moreover, if they only aim to lock up the market of any local electronic government systems, they will come up against the cutting back of these total markets, and the scramble for the market would result. As is shown in Table 3, the local IT enterprises' order ratio in amount of money had increased till 2006. But it decreased by 6.5 points from 2006 to 2007. This is because that the other local IT enterprises out of Nagasaki invaded the market of the electronic Prefectural government systems in Nagasaki. The scramble for the open source market had already started. This may be the fact, but there will be another method for the regional-industry promotion plan.

## 3  OSS Development Style and Regional-Industry Promotion

### 3.1  Situation of Regional IT Industries in Shimane Prefecture

As we have mentioned above, in introducing OSS to local governments, their staff need to participate in the upper processes of the software development. And, the staff in Nagasaki Prefecture exactly did it. The reason why Nagasaki Prefecture decided to start this regional-industry promotion project is that most of the regional IT industries in Japan depend on their regional public organizations' orders for their market. And it is obvious that the market will reduce in later years because of the critical financial situation in Japan.

The situation is also the same for Shimane Prefecture. As is shown in Figure 2, their main market for IT enterprises is "Information Service Industries" themselves, amounting to 37%[3]. It is so high a ratio compared to that of the whole of the country (28.3%)[4]. It shows the multiple chains of commission and entrust among the same trade relations in Japanese IT industry, which is a typical water fall model of development. The local IT enterprises depend on the major IT vender enterprises in Tokyo for their receiving of orders.

And "Administration Sector" (22%) and "Construction Industry" (9%) follow in the commercial report of Shimane. In the national report those ratios are 8.6% and 1%, respectively. On the other hand, "Service Industry" (16%) and "Finance and Insurance" (1%) are extremely low compared to the whole in the country. It is shown that the orders are given by major enterprises outside Prefecture.

---

[3] Shimane Prefecture (2005).
[4] The Ministry of Economy Trade and Industry of Japan (2005-2008).

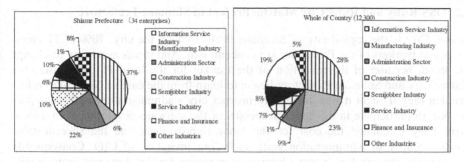

**Fig. 2.** The receipt of orders for IT industries from different industry categories in Shimane and Japan (Whole of country)

The orders from "Administration Sector" and "Construction Industry" had already been decreasing in nationwide scale. If they had depended mainly on the orders in these fields indefinitely, the sales would have clearly declined fast. Shimane Prefecture faced the need to make a choice whether waiting for the reduction of the market or starting regional industry promotion policy.

Additionally, when we asked what skills were mostly insufficient in these enterprises, the first listed answer was not the ability of "System Development"(26%) but the ability of "Project Management"(65%) which is dominated by major IT vender enterprises in Tokyo, and the local IT enterprises in Nagasaki acquired it by participating in the electronic Prefectural government systems (Figure 3).

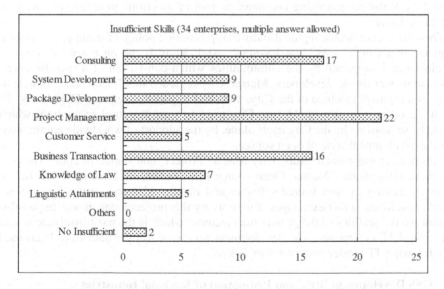

**Fig. 3.** Insufficient Skills of the IT enterprises in Shimane Prefecture

## 3.2  OSS Ruby and Ruby City Matsue Project in Shimane Prefecture

Matsue City is the capital city of Shimane Prefecture. In the city, 70% of IT enter-prises and workers of the Prefecture are concentrated. And it was necessary to change this trend, because of the reduction of the market for the local IT enterprises. So, Matsue City and Shimane Prefecture began to learn from the case of Nagasaki Prefec-ture, but they thought it was difficult to instruct city's administration staff to acquire the skill to participate in the upper processes of the software development in a short term. They calculated the cost for the labor improvement and the specification-decision ability of administration staff, including invitation of CIO. Consequently they came to the conclusion that adopting open source for the electronic municipality systems could not necessarily cut the budget of the administration totally.

At the same time, Matsue City, having their own regional open source resource "Ruby", intended to make open source development style in the industrial promotion plan. The City adopted another method, a little different one from Nagasaki Prefec-ture's. It is not the style that the administration adopt open source for its electronic municipality systems but the style of promoting open source innovation.

Ruby is the Object-oriented Script Language invented by Mr. Yukihiro Matsumoto called "Matz" in the community, and opened to the public in 1993. Matz lives in Mat-sue City and has been developing Ruby with many developers all over the world through the Internet. At first, Ruby did not spread in business uses except among some fanatic engineers. But in 2004, David Heinemeier Hansson, a programmer in Denmark, released "Ruby on Rails" which is a web application framework. Hence, Ruby came to attract attention and to be used also in enterprise areas. Then Matsue City started the project, which attempts to create a new regional city based brand around the Ruby programming language, as part of its efforts to regenerate the City and its environs.

There has never been a regional-industrial promotion policy of using programming language that can be said "Statelessness". And, Ruby is the open source software developed and supported by the communities which consist not of a specific enter-prise but of worldwide developers. Matsue City regarded this "Open" and "Stateless-ness" as a regional resource of the City, and tried to promote regional IT industry. To put it the other way around, Matsue City tried to start industrial promotion, which would be impossible by the City itself alone, by the support of worldwide community, like the development style of open source.

This project was named "Ruby City Matsue Project" and started in 2006. The pro-ject aims, through the "Matsue Open source Lab" facility, at creating a hub for all activities relating to open source software and making Matsue a centre for OSS re-search, development and exchanges. But, only by this project alone, it was impossible to improve the abilities of the project management which is the most insufficient skill in the local IT enterprises. The development power of Ruby would have been used only for major IT vender enterprises in Tokyo.

## 3.3  OSS Development Style and Promotion of Regional Industries

To support and advance this project, the organization of open source, "Open source Society Shimane" was also established in 2006, by enterprises, technicians, researchers,

students, and users who specialize in open source software for them to be able to exchange knowledge and information, and to improve their technological development
and project management abilities through utilizing "Matsue Open source Lab". Through
such activities Matsue City will be able to become a national centre for open source
software development, duly leading to new technological innovation, increased competitiveness in the field of OSS, and above all else the development of a modern workforce well versed in the intricacies of the IT world. Ultimately, it is expected to make
Matsue City a world-renowned Mecca for Ruby and OSS.

In order to materialize its mission of promoting the local IT enterprises through
open source initiatives, Open source Society Shimane conducts many varieties of
activities in both fields of policy and technology. One of the Society's main activities
is to hold "Open source Salon" in Matsue Open source Lab. This "Salon" is a series
of study meeting or seminar in highly casual forms. The theme of the salon also varies; with government officials explaining their policy initiatives, university professors
presenting the recent OSS or IT-related research results, and the Ruby or OSS engineers illustrating the cutting-edge OSS technologies, know-how and applications.

The salon has been held almost once a month for four years. With the number of
audiences amounting to approximately 30 to 40, this salon becomes the "hub" for
connecting people and information. With regard to the business area, the Society
conducts "Business Exchange Meeting" when receiving the visitors from ICT-related
enterprises outside of Shimane. This business meeting is exclusively for the members.
Hence, the Society plays the hub role connecting people, information, technology, and
businesses.

These are, so to speak, the Superficial Result, but most important purpose is industrial promotion by creating abilities in project management in the local IT enterprises.
In Nagasaki Prefecture, the administration sector of the prefecture did it. On the other
hand, in Matsue City "Open source Society Shimane", the organization of industry-
government-academia complex in the region has been doing this role. By improving
the abilities in project management, the local IT enterprises have been increasing the
chances to participate in the upper processes of the software development. For example, "Medical and Nursery Care System of Matsue City" was constructed by "Techno
Project" using Ruby in 2007, "Matsue SNS -Collaboration Effect of Regional SNS
Connection" was opened by "Wacom IT" using Ruby on Rails in 2008, and "Knowledge Management System of Matsue City" by "NaCl" was constructed also using
RoR in 2008. These are all local IT enterprises in Matsue City. Moreover, Shimane
Prefecture developed "Shimane Prefecture CMS" using RoR to construct its own Web
Site, and opened the source of "Shimane Prefecture CMS" to the public in 2008.
Then, many other local governments come to get interested in this CMS and prepare
to adopt it for their Web site. "Shimane Prefecture CMS" is, of course, open source,
but to construct and maintain Web sites, the roles of private businesses are needed,
and the orders mostly come to the local IT enterprises in Matsue City. As a result,
according to the investigation report book by Shimane Prefecture Information Industry Association[5], sales and the number of starting works of the IT enterprises in Shimane Prefecture show the expansion more than the whole country in fiscal year 2008
since fiscal year 2006.

---

[5] Shimane Prefecture Information Industry Association (2009).

**Table 4.** Transition of Amount of Sales in IT industries (million yen)

|  | Whole of Country | rate of increase | Shimane Prefecture | rate of increase |
|---|---|---|---|---|
| 2006 | 13,751,730 |  | 10,452 |  |
| 2007 | 13,409,700 | -2.5% | 12,060 | 15.4% |
| 2008 | 14,817,900 | 10.5% | 13,241 | 9.8% |

**Table 5.** Transition of Number of Persons Engaged in IT industries (man)

|  | Whole of Country | rate of increase | Shimane Prefecture | rate of increase |
|---|---|---|---|---|
| 2006 | 567,498 |  | 1,022 |  |
| 2007 | 501,807 | -11.6% | 1,389 | 35.9% |
| 2008 | 557,263 | 11.1% | 1,537 | 10.7% |

One of the key success factors of the project of Matsue City and Shimane Prefecture is the development power jointly collaborated with the nationwide open source community. By touching upon the nation wide expertise of the open source, Matsue City and the project-related people have recognized the crucial importance to cooperate with the open source of the whole country which connects further to the worldwide open source community. In addition, a personnel training and industrial promotion are advanced by the cooperation of the industry-government-academia complex in the region. In Nagasaki Prefecture, the prefecture itself participated in the upper processes of the software development and divided their own orders to local IT enterprises. On the other hand, in Matsue City, local IT enterprises in the city gather and collaborate through "Open source Society Shimane", the industry-government-academia complex as a catalyst, and improve their abilities of project management to acquire the orders and expand Markets. In both cases, adoption of the development style of open source can make the possibilities to expand more Markets outside of the regions 3 OSS Development style and Regional-Industry Promotion.

## 4  Conclusions

In the development style of open source, so to speak "Bazaar Style Development", a lot of researchers, developers, and also enterprises voluntarily participate in the organization, extending the boundary of the organizations. So it has the possibility to create new business markets to regional IT industries. But, the open source application policy or the adoption assistance to local governments requires the capabilities in technique, development, and project management, both for adoption sides and provider sides. As the result, it tends to lead to the expansion of the IT-Solution market among the major IT vendor enterprises.

However, as was shown in the case in Nagasaki Prefecture, once the staff of the administration sector (the adoption side) participates in the upper processes of the software development, they can get the ability in project management and divide the orders of their own to be placed. This process enables them to expand the market for the local IT enterprises. And, acquiring the orders by participating in developing processes, they can improve the abilities in technologies corresponding to the development of open source including the project management and decision making for specifications. Moreover, if the deliverables are opened, the market will expand much further for the IT enterprises in Nagasaki Prefecture. This is the indirect development aid by the administration to promote the regional IT industry.

On the other hand, as typically shown in the "Ruby City Matsue Project", there is another method to develop the ability of the provider side. In local cities in Japan, enterprises tend to depend on the orders from "Administration Sector" of their areas. It is obvious that these orders would not last for a long time in the future. Matsue City and Shimane Prefecture took the choice of this method, which not only uses open source technology but also adopts open source development style.

Thus, local governments in Japan can somehow lead regional-IT industry promotion policies, but the main constituents are local government administrations. Because that the local IT enterprises themselves have little motivation for improving their technological abilities which can expand their market. This is the one problematic point. And, the more acute point is that if they come up against the scramble for the market they wouldn't have the ability to overcome by themselves.

I think, while the competition between regions may accompany it, cooperation and collaboration of regions are also needed to expand their market, and to compete against the major IT vender enterprises. Though this may be a matter of the policy and the business, it is necessary to gaze at it as a research object.

# References

1. MIC Economic Research Institute: Current State and View of OSS Using IT Solution Market in Japan (2008)
2. The Ministry of Economy Trade and Industry of Japan: Specific Service Industry Investigation of Actual Conditions (2005-2008)
3. Nagasaki Prefecture: Promotion of Computerization from Purchaser Subject (2008)
4. Shimane Prefecture: Shimane Prefecture Commerce and Industry Labor Division (2005)
5. Shimane Prefecture Information Industry Association: Investigation Report Book of Software Industry in Shimane (2009)

# Comparing OpenBRR, QSOS, and OMM
# Assessment Models

Etiel Petrinja, Alberto Sillitti, and Giancarlo Succi

Free University of Bolzano, Italy

**Abstract.** The objective of this study was to investigate the quality and usability of three Free/Libre Open Source Software assessment models: the Open Business Readiness Rating (OpenBRR), the Qualification and Selection of Open Source software (QSOS), and the QualiPSo OpenSource Maturity Model (OMM). The study identified the positive and negative aspects of each of them. The models were used to assess two Free/Libre Open Source Software projects: Firefox and Chrome (Chromium). The study is based on a set of controlled experiments in which the participants performed the assessment using only one model each. The model used and the Free/Libre Open Source Software project assessed were randomly assigned to the participants. The experiment was conducted in a controlled environment with defined tasks to be performed in a given time interval. The results revealed that the three models provided comparable assessments for the two assessed projects. The main conclusion was that all the three models contain some questions and proposed answers that are not clear to the assessors, therefore should be rewritten or explained better. The critical aspects of each model were: Functionality and Quality for OpenBRR; Adoption, Administration/Monitoring, Copyright owners, and Browser for QSOS ; and Quality of the Test Plan, and the Technical Environment for OMM. Participants perceived the quality and usability of the three models of comparable level.

**Keywords:** FLOSS Assessment Model, Quality Criteria, Software Quality, FLOSS Development Process.

## 1 Introduction

The quality of Free/Libre Open Source Software (FLOSS) products is affected by many variables and it varies strongly in different products. Often, the adoption of a product is affected by the reputation of the producer rather than the real quality of the product itself. However, different indicators can provide hints on the quality of a FLOSS project, for example: the number of users, the longevity of the project, the documentation available on-line, etc. The list of possible indicators is limitless and besides the most well-known (number of product downloads, number of bugs reported, etc.) there are many others that can have different interpretations. Therefore, it is important to have a structured set of criteria to use to assess the quality of a FLOSS project. The most well-known set of criteria used to assess the quality of software development (usually Closed Source) is part of the CMMI model [1]. However, additional sets have been proposed in the last few years targeting FLOSS. Such models include:

P. Ågerfalk et al. (Eds.): OSS 2010, IFIP AICT 319, pp. 224–238, 2010.

- Open Source Maturity Model (OSMM) from Cap Gemini (2003) [3]
- Open Source Maturity Model (OSMM) from Navica (2004) [6]
- Methodology of Qualification and Selection of Open Source software (QSOS) (2004) [2]
- Open Business Readiness Rating (OpenBRR) (2005) [10]
- Open Business Quality Rating (Open BQR) (2007) [9]
- QualiPSo OpenSource Maturity Model (OMM) (2008) [7, 11]

The large plethora of available models witness the interest and the need of systematic approaches for the assessment of the quality of FLOSS projects.

The proposed assessment models provide a selected set of criteria with their interpretation and the description of how to use them. Besides the few mentioned criteria, there are several more indicators for the quality of the code, for the functionality, the usability, the testability, the documentation, the development process followed. Moreover, there are several ways to measure such characteristics. Therefore, it is essential to include in the assessment model a consistent subset of metrics that can be used for the assessment since not all of them can be used in all the cases. The proposers of a model have to take in consideration also different use cases for their model: a FLOSS developer, a FLOSS user, and/or a FLOSS integrator. All of them will probably have different expectations about the product and the development process. For these reasons, an assessment model must be flexible and be able to adapt to different use cases. An important aspect of the criteria included in the assessment model is the names of the criteria themselves and the wording of the related questions that are used to detail them.

Another important aspect in the evaluation of the quality of FLOSS is the development processes followed. Our opinion is that it is necessary to take in consideration both aspects of FLOSS: product and process. For example, the maintainability of the product is affected by both. FLOSS integrators may be interested in the documentation produced and if it is easy to use parts of it inside their other products. For such reasons, they will be interested in the process followed to develop the FLOSS product. In this case, the measurement of the FLOSS development process of the FLOSS project is important. Available FLOSS assessment models contain some aspects of the final product and some aspects of the development process. However, most of the models are focused on the assessment of the final product. Only the OMM model covers more in details the FLOSS development process, resembling to some extent the approach adopted in the CMMI. Nevertheless this difference of focus, we identified many commonalities between the analysed models, and we think that a comparison of three of them is reasonable.

We conducted this research comparing three similar models, partially to evaluate the OMM model that we developed but mainly to see how it is perceived by users in comparison with the other two models. Moreover, some of the results of this research related to OMM were useful for validation purposes of the model and its future improvement.

The research offers also a use case demonstration of the other FLOSS assessment models. We think that it is essential to verify how the proposed models can be used concretely and what are the perceptions of people and their confidence in the results obtained by using different models.

In the available literature, there are no comparisons of different use cases of available FLOSS assessment models. This research aims at (partially) filling this gap presenting some empirical data about the comparison of different models. We expect significant differences in the quality perception of models by users, however we must be aware that this difference depend also on the use cases adopted inside the experimentation. Some models are perceived better in the area of FLOSS communities, others are preferred by developers, and others by users.

This paper is structured as follows. After this first introductory section, we briefly present some related work. In the third section we describe our research design presenting in details the experimentation performed. The fourth section is the main part of the paper and presents research results. Afterwards, we present some threats to validity of the research conducted, and, finally, we present our conclusions.

## 2 Related Work

The OpenBRR and the QSOS models were partially validated by their developers and used in a small number of use cases with results available on their web portals [2, 10]. However, there are no empirical evaluations of the validity of the two models. Moreover, the number of use cases is limited and we noticed a quite steady number of reported FLOSS projects assessments on the web portals of the two models.

We conducted an initial validation of the OMM model as part of the QualiPSo project [8]. We involved all the partners of the project that are interested in the future use of OMM. The description of the initial validation process is available in reports of the project [8]. Inside those documents are available also results of the research presented in this paper with additional content that we were not able to present here due to space limitations. A key outcome of the research presented by authors lists the actions necessary to improve the OMM model. Such reports contains also an evaluation of critical elements identified mainly inside the OMM model but also inside the OpenBRR and the QSOS models.

In our knowledge, only a few researches have been published analysing and comparing available FLOSS assessment models. One of them was conducted by Deprez and Simons that compared the OpenBRR and the QSOS models [12]. They have done a rigorous comparison of both models and they identified advantages and disadvantages of both. The main difference between their approach and the one proposed in this paper is that we wanted to use the models with real FLOSS projects and try to find out what are the problems encountered by participants during the assessment process and collect their subjective perceptions related to the quality and usability of the models. Deprez and Simons proposed a detailed conceptual comparison of elements of the two models without conducting a real use case.

## 3 Research Design

Our plan was to conduct a controlled experiment [5, 4]. We managed to satisfy many requirements for a controlled experiment as: randomization of participants, and testing specimens, and the set up of a controlled environment, the detailed planning of the experimentation process, and others. However, we were not able to involve different types of participants, for example professional programmers. This can be a problem for the generalization of results. Anyway, we took different actions to mitigate these and few other threats to validity of our research and we present key one in section 5. A detailed description of the scope of the research and the methodology used are presented in the following subsections.

## 3.1  Scope

The research included three FLOSS assessment models: OpenBRR, QSOS, and OMM. For the research we used all questions of the three models unifying their presentation and structure. We did not use the on-line questionnaires provided by the two methodologies. We carefully, taking care not to loose any details of the two models, copied all questions in a uniform structure, with a comparable level of details. For the experimentation we choose two well known FLOSS projects: Firefox, and Chrome (Chromium). Google Chrome is not a FLOSS project but its development is tightly related to the Google Chromium FLOSS project. In this paper we will use generically the name Chrome. We assessed two FLOSS projects that are providing similar functionality, namely a web browser. Therefore, we were able to compare specific final FLOSS products and FLOSS development process characteristics.

## 3.2  Methodology

We planned to conduct a well structured and executed experimentation process according to the operational definition of the experimentation process stated in one of the most frequently cited papers about this subject [5]:

Controlled experiment in software engineering is: *"A randomized experiment or a quasi-experiment in which individuals or teams (the experimental units) conduct one or more software engineering tasks for the sake of comparing different populations, processes, methods, techniques, languages, or tools (the treatments)"*.

From the definition we see that an important aspect of a controlled experiment is randomization of:

- individuals participating to the experiment,
- the tasks that they will have to perform, and
- all the treatments that are included in the experiment.

We randomized most of the experimentation components. After we decided to use three assessment models (OpenBRR, QSOS, and OMM) we distributed them randomly between all participants. The only constrained we imposed was that the number of users using each model was the same. For the experimentation we decided to use two FLOSS projects: Firefox and Chrome. Partially we choose these two projects because during the previous two years most of our students were involved in university projects that used Firefox or Chrome as source of code, and other type of project data. We expected most of our participants will be therefore at least aware of the two projects. We distributed the two projects randomly between all participants, taking care to give the same number of each projects to participants using a specific model.

In this way we managed to have random participants using the same number of models on both FLOSS projects.

During the project planing process we addressed the following five aspects of the experimentation process:

1. **Object of the study – What is studied?**
   The object of the study was one of the three FLOSS assessment models. At the same time we were also interested in the components of the three models. Participants had to answer to all questions present in the model and they were also asked to express their opinion on each question.
2. **Purpose – What is the intention?**
   The purpose of the experiment was to predict the usability and precision of metrics inside the three models and the perception of the quality of the whole model. Based on the results, we wanted to know also which model better characterizes specific aspects of FLOSS.
3. **Quality focus – Which effect is studied?**
   Our quality focus was the completeness of specific parts of the model and the precision of results for specific parts of the models. How detailed the model is in specific areas of FLOSS and whether the answers from different participants were similar or they diverged.
4. **Perspective – Whose view?**
   Participants were students. We expected that they are mostly FLOSS users and a smaller percentage of FLOSS developers.
5. **Context – Where is the study conducted?**
   The experimentation environment was a university laboratory.

An important aspect of the experiment conducted was the environment where the experiment was conducted. We included in the experiment mostly free willing participants from our university. We managed to involve 26 participants coming from the last year of the software engineering Bachelor and Master programs at our university. The experiments were always conducted in the same laboratory room where participants were able to use a computer connected to the web. Participants were separated and each was using his own computer. They were not allowed to communicate during the experimentation process.

The experiment had three phases:

1. First participants received an initial questionnaire where they were asked to report contextual data (age, experience in programming, experience with the assessed project, experience with the assessment model, and others information). This phase lasted 20 minutes.
2. For the second phase we distributed the printed version of questions relative to the assessment model they had to use. We gave them two hours (120 minutes) to assess the FLOSS project that was assigned to them. They were able to browse web pages of the assessed project, search source code repositories, mailing lists, bug/issue management systems, and other web available sources to answer to questions that are part of the assessment model they used. We did not restricted their web access, they were allowed to search anywhere for information.

3. After the two hours, we asked them to finish the assessment process and we distributed a final questionnaire in which we asked them to describe their opinions of the quality of the model, the clarity of questions, the coverage of FLOSS aspects, and others. We gave them some possible answers to questions and allowed them to add also additional answers. The third phase lasted also 20 minutes.

Afterwords we collected results of all three phases and the experiment was concluded.

# 4  Results

We present first the contextual data about participants, then we present results of each assessment models used, their comparison, and at the end of this section we present opinions of the participants on the use of the three assessment models.

**Contextual Data**
We collected many contextual data during the first phase of the experiment. We present here only a few aspects we consider important for understanding the results of the experiment:

- the role that best describes the current position in the assessed FLOSS project,
- the number of FLOSS projects the participant is or were involved in, and
- their experience in the assesed project related tasks.

Some of the questions are personal, others are FLOSS specific, and some asks participants if they were already in contact with the assessed FLOSS project or the assessment model. If they were involved in the assessed project their assessment can influence the assessment process and can explain a better compilation of the assessment questionnaire.

Contextual question: **Role that best describes the position of the participant in the assessed FLOSS project**
From the three charts (Figure 1) we can see that almost all the participants have declared to have already used  Firefox or  Chrome browsers (nearly 100% of participants for both browsers; two participants did not choose any answer to this question). We do not know if they use them regularly or they have tried to use them just few times. Anyway, they are aware of the product and what it is used for. We can also see from the charts that few participants have contributed to the two FLOSS projects; they have declared to be testers, translators, or even active developers inside the Firefox or Chrome projects. We can see from the three charts that the number of developers is homogeneously  distributed in all three assessment models groups. From the third chart we can see that we have an equal number of FLOSS developers involved in both Firefox and Chrome projects using the OMM methodology. Another peculiarity of the group using OMM is that we have additionally also a small number of translators participating to the experiment.

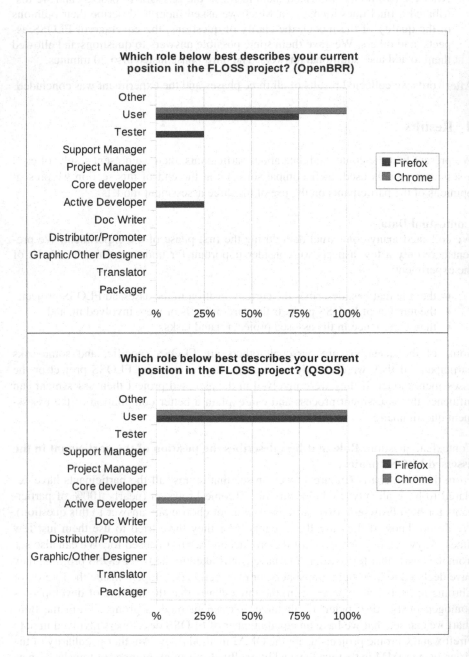

**Fig. 1.** Role of participants in the assessed FLOSS project

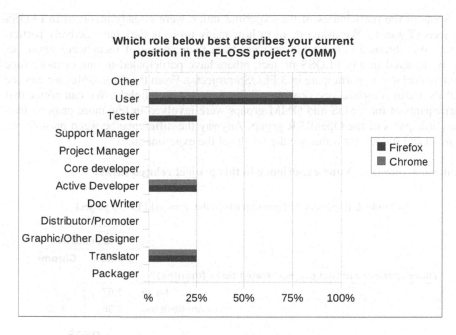

**Fig. 1.** (*Continued*)

Contextual question: **Number of FLOSS projects you are/were involved in?**

**Table 1.** Number of FLOSS projects participants were involved in

| | OpenBRR | |
|---|---|---|
| | **Firefox** | **Chrome** |
| **Number of FLOSS projects you are / were involved in?** | | |
| Mean | 0,67 | 0,75 |
| Standard deviation | 0,94 | 0,83 |
| | **QSOS** | |
| | **Firefox** | **Chrome** |
| **Number of FLOSS projects you are / were involved in?** | | |
| Mean | 1 | 0,5 |
| Standard deviation | 1,22 | 0,87 |
| | **OMM** | |
| | **Firefox** | **Chrome** |
| **Number of FLOSS projects you are / were involved in?** | | |
| Mean | 0,5 | 1,33 |
| Standard deviation | 0,5 | 1,25 |

Some of the participants of the experimentation were already involved in FLOSS projects (Table 1). We asked them in how many projects they have actively participated. We obtained comparable answers from them; some of them were never actively involved in any FLOSS project, others have participated in one or more (one student declared to participate in 3 FLOSS projects). From the three tables we can see that the mean is around one project and that it varies just slightly. We can notice that participants of the QSOS and OMM groups were involved in few more projects than the participants of the OpenBRR group. Anyway the difference between participants is small and it does not influence the results of the experiment.

Contextual question: **Your experience in this project related tasks?**

Table 2. Experience of participants in the assessed FLOSS project

| | OpenBRR | |
| --- | --- | --- |
| | Firefox | Chrome |
| **Your experience in this project related tasks (months)?** | | |
| Mean | 1,67 | 2 |
| Standard deviation | 2,36 | 1,63 |

| | QSOS | |
| --- | --- | --- |
| | Firefox | Chrome |
| **Your experience in this project related tasks (months)?** | | |
| Mean | 2,25 | 2 |
| Standard deviation | 2,49 | 2,12 |

| | OMM | |
| --- | --- | --- |
| | Firefox | Chrome |
| **Your experience in this project related tasks (months)?** | | |
| Mean | 2,25 | 4 |
| Standard deviation | 2,28 | 3,74 |

We wanted to know also if participants have actively contributed to the assessed FLOSS project. From the three tables (Table 2) we can see that participants were only marginally involved in the two assessed FLOSS projects. The variability between the three groups is small, therefore we can be confident that this aspect does not influence the results of the experimentation process.

**Results of assessing two FLOSS projects using three assessment models**
We provide here our interpretation for the results obtained by using each of the three methodologies; additionally, we present an overview of similar characteristics measured by the three methodologies. We calculated mean values and standard deviations for assessments done by participants for each quality characteristic. We expect that a low value of the standard deviation means a similar assessment result obtained by

**Table 3.** Use of OpenBRR

| | OpenBRR | | | |
| | Firefox | | Chrome | |
| | Mean | Standard deviation | Mean | Standard deviation |
|---|---|---|---|---|
| **Functionality** | 3,7 | 0,5 | 2,3 | 0,7 |
| **Usability** | 4,2 | 0,4 | 4,1 | 0,6 |
| **Quality** | 2,7 | 0,3 | 3,5 | 0,3 |
| **Security** | 2,7 | 0,7 | 2,3 | 0,7 |
| **Performance** | 3,0 | 0,0 | 3,1 | 0,5 |
| **Scalability** | 4,0 | 1,0 | 3,8 | 0,0 |
| **Architecture** | 4,3 | 0,5 | 2,3 | 0,2 |
| **Support** | 4,8 | 0,2 | 4,7 | 1,3 |
| **Documentation** | 4,2 | 1,2 | 3,1 | 0,5 |
| **Adoption** | 3,8 | 0,3 | 3,2 | 1,5 |
| **Community** | 3,8 | 0,8 | 4,3 | 0,8 |
| **Professionalism** | 2,7 | 0,2 | 2,7 | 0,6 |

different participants. This can confirm that the questions were clear, the people were able to find appropriate information on the web, and the threshold values were defined appropriately.

From the results presented in Table 3 we can notice that the Firefox project obtained better grades than the Chrome project; we can see this from most of the assessed criteria. The larger differences are on the Functionality and on the Architecture. Only two criteria obtained a higher grade for the Chrome project: Quality, and Community. The important information for us is the value of the standard deviation for different criteria. We can not identify a criteria that has a high standard deviation for both projects, therefore we can not be sure of the bad quality of a specific criteria. The criteria that were not assessed homogeneously for one or the other project (the standard deviation value is relatively large) were: Scalability, Documentation, Support, and Adoption.

The QSOS assessment methodology has a different number of thresholds for assessing specific criteria than the other two methodologies (Table 4). QSOS has just three different thresholds. This aspect changes the range of values of the standard deviation. Also smaller values of standard deviations compared with the other two methodologies represent considerable deviations of assessment values given by users. We can see from the table that the highest standard deviation values are in the following characteristics: Adoption, Administration/Monitoring, Copyright owners, and Browser features. Also by using the QSOS methodology we can see that the Firefox project graded slightly better than the Chrome Project; however, the differences between the two are smaller than in the case of the assessment using the OpenBRR model. From the table we see also that the QSOS methodology has four different granularity levels for the summary of results. This is different from the other two methodologies. We decided to present results for the third level of granularity that has a similar number and type of characteristics as the chosen level of granularity of the OpenBRR and OMM models.

**Table 4.** Use of QSOS

| | | | QSOS | | | |
|---|---|---|---|---|---|---|
| | | | Firefox | | Chrome | |
| | | | Mean | Standard deviation | Mean | Standard deviation |
| **Generic Section** | | | | | | |
| | **Intrinsic durability** | | | | | |
| | | Maturity | 2,5 | 0,5 | 2,4 | 0,1 |
| | | Adoption | 2,9 | 0,8 | 2,5 | 0,4 |
| | | Development leadership | 2,8 | 0,5 | 2,8 | 0,2 |
| | | Activity | 2,9 | 0,6 | 2,7 | 0,2 |
| | **Industrialized solution** | | | | | |
| | | Independence of developments | 2,8 | 0,1 | 2,5 | 0,5 |
| | | Services | 1,7 | 0,3 | 1,7 | 0,5 |
| | | Documentation | 3,0 | 0,1 | 2,3 | 0,5 |
| | | Quality Assurance | 2,9 | 0,2 | 2,5 | 0,4 |
| | | Packaging | 3,0 | 0,4 | 2,2 | 0,5 |
| | **Exploitability** | | | | | |
| | | Ease of use, ergonomics | 3,0 | 0,5 | 3,0 | 0,0 |
| | | Administration / Monitoring | 1,0 | 0,8 | 2,3 | 0,5 |
| | **Technical adaptability** | | | | | |
| | | Modularity | 2,9 | 0,1 | 2,0 | 0,0 |
| | | Code modification | 2,8 | 0,3 | 3,0 | 0,0 |
| | | Code extension | 2,4 | 0,6 | 2,3 | 0,5 |
| | **Strategy** | | | | | |
| | | License | 1,9 | 0,1 | 2,2 | 0,2 |
| | | Copyright owners | 1,4 | 1,0 | 1,7 | 0,5 |
| | | Modification of source code | 2,3 | 0,5 | 2,7 | 0,5 |
| | | Roadmap | 2,9 | 0,6 | 3,0 | 0,0 |
| | | Sponsor | 2,9 | 0,1 | 2,3 | 0,5 |
| | | Strategic independence | 2,8 | 0,5 | 2,0 | 0,0 |
| **Browser features** | | | 2,5 | 0,9 | 2,7 | 0,1 |
| **Accessibility features** | | | 2,4 | 0,6 | 2,4 | 0,3 |
| **Web technology support** | | | 2,9 | 0,4 | 2,2 | 0,1 |
| **JavaScript support** | | | 2,0 | 0,0 | 2,5 | 0,5 |
| **Protocol support** | | | 2,3 | 0,5 | 2,7 | 0,2 |
| **Image format support** | | | 2,5 | 0,4 | 2,7 | 0,4 |

From the mean values for different trustworthy elements (TWE is one of the characteristics measured inside the OMM model) composing the OMM model we can see in Table 5 that the Firefox project in comparison with the Chrome project obtained better grades. The difference of the quality of the two projects is even sharper than it appeared with the use of the other two methodologies. From our point of view, this difference between the two projects reasonable since the Firefox project exists a longer period and it has a larger community. There is only one TWE that is larger for the Chrome project than for the Firefox project and is: RASM. The level of RASM for Chrome is not high (2,1) but it is even lower for the Firefox project (1,8), showing that the Firefox product is not tested and/or presented appropriately on the project's website. The values of standard deviations calculated for OMM are lower than for the other two methodologies; keeping in consideration the 5 grades threshold levels. The higher standard deviations resulted for the Quality of the Test Plan (QTP) (Firefox project 0,6 and Chrome project 0,8) and the Technical Environment (ENV) (Firefox

**Table 5.** Use of OMM

| | OMM | | | |
| | Firefox | | Chrome | |
| | Mean | Standard deviation | Mean | Standard deviation |
|---|---|---|---|---|
| Product Documentation (PDOC) | 3,9 | 0,0 | 3,5 | 0,1 |
| Popularity of the SW Product (REP) | 2,7 | 0,1 | 2,4 | 0,4 |
| Use of established and Widespread Standards (STD) | 3,8 | 0,3 | 3,1 | 0,7 |
| Availability and Use of a Roadmap (RDMP) | 3,1 | 0,5 | 2,3 | 0,4 |
| Quality of the Test Plan (QTP) | 4,2 | 0,6 | 3,3 | 0,8 |
| Relationship between Stakeholders (STK) | 4,1 | 0,1 | 2,9 | 0,1 |
| Licenses (LCS) | 3,4 | 0,2 | 2,7 | 0,5 |
| Technical Environment (ENV) | 3,9 | 0,7 | 2,8 | 0,8 |
| Number of Commits and Bug Reports (DFCT) | 3,5 | 0,1 | 3,5 | 0,0 |
| Maintainability and Stability (MST) | 3,8 | 0,4 | 2,7 | 0,4 |
| Contributions to the FLOSS project from SW Companies (CONT) | 3,2 | 0,1 | 2,7 | 0,3 |
| Results of Assessment of the Product by 3rd Party Companies (RASM) | 1,8 | 0,3 | 2,1 | 0,1 |

project 0,7 and Chrome project 0,8). The use of the OMM methodology on both projects showed higher standard deviation values for QTP and ENV. We analysed more in details the two TWEs and identified the questions that obtained largely heterogeneous answers by different participants. We plan to propose some changes related to those TWEs in the newer version of OMM.

**Participants assessment of the three models**

In this section, we present answers to only two questions out of ten that we asked. We present only the answers that are important to better understand the use of the three methodologies.

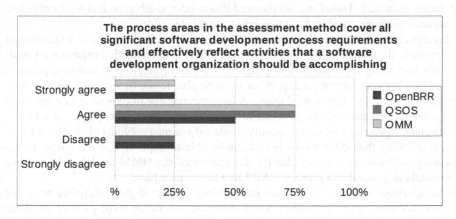

**Fig. 2.** Completeness of coverage of FLOSS process areas

As evident from Figure 2, participants perceived OMM as flexible to a large extent; 12% strongly agree that the model is flexible and a large 88% perceived it as flexible. We think that this result is a good indication of the modularity of the proposed model. Other two models obtained similar, just slightly lower values.

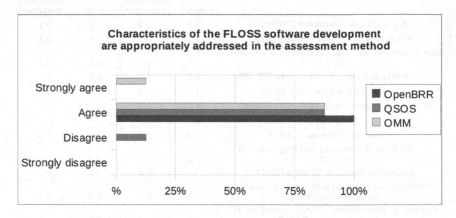

**Fig. 3.** The model addresses appropriately FLOSS characteristics

The OMM model scored best on the question about the quality of coverage of characteristics that were actually included in the model (Figure 3). A large majority of users of OMM agreed with the sentence and none disagreed with it. The other two methodologies scored also good as is evident from the chart, only the QSOS model obtained a 10% of negative answers.

## 5  Threats to Validity

The randomization of used assessment models, of assessed FLOSS projects, and participants was a key requirement for the experiment and a possible threat to validity of the results obtained. Therefore, we planned this aspect in advance and try to mitigate its negative elements as already presented in the third section.

An aspect that we want to improve is the number of participants of experiments coming from the industry. We plan to conduct additional controlled experiments with individuals coming from industry as programmers, and (integration) software projects managers. This is an important aspect for us to be able to see the usability of the three models in general for all potential users. A large percentage of experiments conducted with students can prevent the generalization of results. However, in the current experimentation process we wanted mainly to identify some problems of the three models and specially the OMM model in order to be able to improve it in the future. In the following iteration we want to identify the quality of the OMM model and the other two models in general and propose OMM to a larger user base.

Characteristics assessed by the three methodologies differ slightly, therefore not all of the characteristics can be compared. Small differences between some parts of the three models presented also a possible threat to validity of our conclusions. We managed to mitigate this threat by unifying the look and feel of the three models at the same time however not loosing any content part of the three assessment models. Since we were not able to remove all differences, we preferred to present in this document separately results obtained for all three models and provide an interpretation and comparison just for a limited subset of comparable characteristics.

## 6 Discussion and Conclusions

An important inconsistency in the used models was the naming of the characteristics assessed by each model. Although the three models contain characteristics that have sometimes the same name (for example license in OMM and QSOS) it is difficult or in some cases impossible to compare them because the internal questions are often (partially) different. For the conclusion we will present the comparison of one characteristic that is present in all three models (Documentation). The mean values obtained for the Firefox project using the three methodologies were: 4,2 (OpenBRR), 3,0 (QSOS), 3,9 (OMM). Taking in consideration that 3 is the highest value for QSOS we see that for this characteristic the three models obtained a similar value. The three values measure for the Chrome project were: 3,1 (OpenBRR), 2,3 (QSOS), and 3,5 (OMM). This three values are also quite similar (value 2 in QSOS is the intermediate value). Also the differences of the values for documentation of the Firefox and the Chrome projects are quite constant using all three models. For comparable characteristics we can see that the three models provide similar evaluation results for the two assessed projects.

Nevertheless we were mostly interested in the specific problems related to the OMM model, we identified also key problematic components of the OpenBRR and QSOS models. Based on the results of our experimentation process, we identify several problems related to each of the three assessment models. The Functionality and the Quality characteristics were assessed divergently by different participants using the OpenBRR model. The diverging values for the QSOS model were obtained for: Adoption, Administration/Monitoring, Copyright owners, and Browser characteristics. In the case of the OMM model the problematic characteristics were: Quality of the Test Plan (QTP), and the Technical Environment (ENV). Based on the final questionnaire filled during the experimentation process, we found out that most often the problems related to specific questions are caused by a not clear formulation of the question and in some cases the not clear understanding of the threshold value (available value for the answer) used by each model.

Based on the results of the experimentation process conducted we saw that OMM obtained at least as good results as OpenBRR and QSOS models. In few aspects (Figure 2 and Figure 3) it was perceived better then the other two models. With the experimentation we found out some elements of the model that have to be improved:

- the identification of misleading questions inside the QTP and ENV Trustworthy elements,
- simplification of the overall complexity of some questions,
- simplification of thresholds values (current answers are complex and to extensive),
- automation of the scoring mechanism for calculating Practices, Goals, and TWEs scores (in some case we identified errors in the calculated results in all three models); and
- creation of an easy to understand description of how to use the OMM model.

These action items will help us modify appropriately the OMM model in order to improve its quality and usability.

## Acknowledgement

The research was conducted in the scope of the QualiPSo project (FP-IST-034763). We are grateful to all QualiPSo partners and other participants to experimentation processes conducted.

## References

1. Amoroso, E., Watson, J., Marietta, M., Weiss, J.: A process-oriented methodology for assessing and improving software trustworthiness. In: Proceedings of the 2nd ACM Conference on Computer and communications security, pp. 39–50. ACM Press, New York (1994)
2. Origin, A.: Method for Qualification and Selection of Open Source Software (QSOS), http://www.qsos.org (Last visited: December 2009)
3. Duijnhouwer, F.-W., Widdows, C.: Capgemini Expert Letter Open Source Maturity Model, Capgemini (2003)
4. Hannay, J.E., Hansen, O., By Kampenes, V., Karahasanovic, A., Liborg, N., Rekdal, A.: A Survey of Controlled Experiments in Software Engineering. IEEE Trans. Softw. Eng. 31(9), 733–753 (2005), http://dx.doi.org/10.1109/TSE.2005.97
5. Kitchenham, B.A., Pfleeger, S.L., Pickard, L.M., Jones, P.W., Hoaglin, D.C., Emam, K.E., Rosenberg, J.: Preliminary guidelines for empirical research in software engineering. IEEE Trans. Softw. Eng. 28(8), 721–734 (2002), http://dx.doi.org/10.1109/TSE.2002.1027796
6. Navica Inc.: The Open Source Maturity Model is a vital tool for planning open source success, http://www.navicasoft.com/pages/osmm.htm (Last visit: December 2009)
7. Petrinja, E., Nambakam, R., Sillitti, A.: Introducing the OpenSource Maturity Model. In: Workshop on Emerging Trends in Free/Libre/Open Source Software Research and Development collocated with 31st International Conference on Software Engineering (ICSE 2009), Vancouver, Canada (2009)
8. Qualipso Consortium: QualiPSo - Quality Platform for Open Source Software, http://www.qualipso.org/index.php (Last visit December 2007)
9. Taibi, D., Lavazza, L., Morasca, S.: OpenBQR: A framework for the assessment of OSS. In: Open Source Software 2007, Limerick (June 2007)
10. Wasserman, M.P., Chan, C.: Business Readiness Rating Project, BRR Whitepaper 2005 RFC 1, http://www.openbrr.org/wiki/images/d/da/BRR_whitepaper_2005RFC1.pdf (Last visited: December 2009)
11. Wittmann, M., Nambakam, R., Ruffati, G., Oltolina, S., Petrinja, E., Ortega, F.: Deliverable A6.D1.6.3: CMM-like model for OSS, http://www.qualipso.org/sites/default/files/A6.D1.6.3CMM-LIKEMODELFOROSS.pdf (last visited December 2009)
12. Deprez, J.-C., Alexandre, S.: Comparing Assessment Methodologies for Free/Open Source Software: OpenBRR and QSOS, June 2008. LNCS, pp. 189–203. Springer, Berlin (2008)

# Joining and Socialization in Open Source Women's Groups: An Exploratory Study of *KDE-Women*

Yixin Qiu, Katherine J. Stewart, and Kathryn M. Bartol

University of Maryland, Van Munching Hall, College Park, MD 20742, USA
{yqiu,kstewart,kbartol}@rhsmith.umd.edu

**Abstract.** This research investigates how women's groups facilitate people's participation in the open source community by examining the joining activities in *KDE Women*. Leveraging literatures on group membership roles and socialization, and adopting a qualitative research method, a joining script of different kinds of participants was identified. It is found that members developed organizational and leadership skills and were engaged in defining group norms and values upon joining *KDE Women*. This study extends prior literature on socialization and provides better understanding of women's groups in open source.

**Keywords:** women's group, membership roles, joining, socialization, qualitative.

## 1 Introduction

Open source software (OSS) projects are increasingly recognized as an important domain of technological innovation. They constitute an important means of involvement in technical work and in informal networks that aid the acquisition and maintenance of technical expertise. Several studies have highlighted the importance of attracting and retaining participants needed for the success of OSS projects [10,16, 25,27]. Consistent with the general finding that women's rates of participation in IT related fields remain relatively low across both the United States and Europe [4,23,29], survey and anecdotal evidence have indicated that attracting and retaining women participants in OSS projects has been particularly challenging [13,28].

Recognizing the under-representation of women in OSS, several "grass roots" women's groups have formed to focus on facilitating the involvement of women in OSS. These are volunteer-led online groups that aim to provide a forum and resources for women participants in OSS projects, and ultimately seek to increase the number of female participants contributing to OSS. Examples of these groups include *LinuxChix, KDE Women, Ubuntu Women*, and *Debian Women*.

Prior research has focused on identifying barriers to women's participation in technical work (e.g., work-life conflict), and on understanding the impacts of diversity in work settings [1,4,5]. The FLOSSPOLS 2006 [24] presents some rather alarming findings about the gender issue in the open source community. They suggest that women are treated as either alien other or (in online contexts) are assumed to be male and thus made invisible. Women's contributions in areas other than coding receive less

P. Ågerfalk et al. (Eds.): OSS 2010, IFIP AICT 319, pp. 239–251, 2010.

valuation. Inflammatory talk and aggressive posturing exacerbates the confidence difficulties women tend to have [p.5].

Little research however, has explored the processes by which women actually do become involved in technical work, and, in particular, the role online grass roots women's groups may play in such involvement. An understanding of this is of critical importance for attracting women into the open source field because design guidelines could be drawn to make these online groups more effective in achieving their goals. Furthermore, people's interaction patterns found in these online women's open source groups will shed light on similar grass roots efforts in the offline settings and in other technology domains where lack of women's participation is an issue. This research intends to address the research gap by analyzing the actual group joining and socialization practices of one OSS women's group: *KDE Women*. KDE is a widely used desktop environment for UNIX workstations. *KDE Women* is a KDE sub-project, along with *KDE-devel*, *KDE-Accessibility*, *KDE-Artists*, *KDE-edu*, *KDE-Usability*, etc. The main goal of *KDE Women* is to "build an international KDE forum for women by providing a place where women can present what they already contribute to KDE and where women, who want to contribute, find a starting point." Through this project, they "actively want to contribute to the success of KDE" (http://women.kde.org/aboutus/announcement.php, accessed March, 2008).

According to Tilly [26], "joining" is a behavioral script that provides a structure for the activity of becoming a member of a collective action project. Prior research on joining patterns in OSS has identified patterns based on the amount and type of early activities of a member and demonstrated that certain kinds of joining scripts were associated with the later technical contribution of members [27]. Building on this definition, the study addresses two research questions. First, what joining and initial socialization activities are associated with participation in *KDE Women*? Second, do joining and socialization differ for people in different KDE participation phases and if so, how? Given the relatively limited prior works on female participation in open source and the exploratory nature of the study, we adopted a qualitative research method by combining the deductive and inductive coding method to answer these two research questions [11,17].

Von Krogh et al. [27] proposed that participants behaving according to a particular kind of joining script were more likely to be granted access to the developer community than participants who do not follow the joining script. Following on this work, we expect that identifying the joining scripts in OSS women's groups will be important to provide a window for further understanding the processes and practices through which a women's group facilitates women's participation in open source. Given the relatively broad mission of OSS women's groups, we take a broad view of participation, including both technical and non-technical activities (e.g., project documentation and general leadership activities).

We differentiate *KDE Women* participants into three categories based on their participation phases: those who have participated in other KDE groups before *KDE Women*, those who limit their participation to *KDE Women* only, and those who later engage in other KDE subgroups after *KDE Women*. Knowledge of people's joining patterns based on different participation phases will render a more nuanced understanding of the role of OSS women's groups. Potentially more targeted implications with respect to people's different motivations in joining OSS women's groups can be drawn from this investigation.

## 2 Theoretical Background

In order to understand the initial group joining behavior, we leverage literatures on group roles and group socialization of newcomers and combine this with an inductive analysis of the content unique to the group studied. In one of the seminal works on group membership roles, Benne and Sheats [6] identified three categories of functional member roles. 1) Group task roles. Members assuming these roles facilitate and coordinate group effort in the selection and definition of a common problem and in the solution of that problem [p.42]. Examples include initiator-contributor, information giver, information seeker, opinion giver, opinion seeker, coordinator, etc. 2) Group building and maintenance roles. Contributions of these roles are to alter or maintain the group way of working, to strengthen, regulate and perpetuate the group as a group [p.42] Examples include encourager, harmonizer, compromiser, etc. 3) Individual roles. The outcome of these roles is some individual goal which is not relevant either to the group task or to the functioning of the group as a group [p.43]. Examples in this category include aggressor, blocker, recognition-seeker, etc. While most research suggests that these group membership roles function in the offline face to face setting [15], Maloney-Krichmar & Preece [21] demonstrates that the group membership role schema by Benne and Sheats applies to members in the online setting as well. Through a study of an online health community, they found that such analysis can serve as a useful diagnostic tool in the online environment for determining if there are differences in the stated purpose of a group and the roles that members play [p.16].

Literatures on individuals' adaptation behavior in organization [2,12,14] identify different types of information people seek. Morrison [22] maintains that information seeking can facilitate newcomers' socialization process in terms of their task mastery, role clarification, acculturation, and social integration. The prior work shows that newcomers seek five types of information: (1) technical information, which is information about how to perform job tasks; (2) performance feedback, which is information about the appropriateness or correctness of fulfilling tasks; (3) referent information, which is information about job requirements and expected role behaviors; or what is expected of you in your job, (4) normative information, which is information about an organization's norms and values; or behaviors and attitudes that the firm values and expects, and (5) social feedback, which is information about how one's social behavior is being perceived and evaluated by others [p.174].

Based on these literatures we focus our qualitative investigation on understanding the initial roles that members play (task, group building or maintenance, or individual) and what information is sought or provided during members' initial interactions. Prior work on newcomer socialization has not examined the seeking and giving behaviors in a same context. Therefore, using the five types of seeking content and the membership role schema as a basis, we extended the coding to assess the "giving" and other actions associated with the different content of the socialization activities, in addition to the "seeking" action.

The next section discusses our methodological approach including the data collection approach and the analytical processes employed. This is followed by preliminary results from our analysis. We then discuss the implications from our findings.

## 3 Methodology

### 3.1 Data Collection

As mentioned earlier, multiple grass roots women's groups were identified. We selected *KDE Women* over others as a case to study women's joining and socialization in the open source world, because a relatively complete history of the group is available. As of May 2007, the group's activities tapered off: on average one message was added in each subsequent month. In spite of that, the group had been well established since its founding in 1998, and reached critical mass during the period from February 2001 to May 2005. A complete history of a well-developed group has advantages over other ongoing groups for capturing various joining and socialization patterns of people with different participation levels.

Because many of the activities and interactions within both *KDE Women* and the KDE OSS projects are conducted online through email lists, we looked into the archives of these lists as our main data source. We identified the email addresses of every person who has ever posted a message to the *KDE Women* mailing list and developed tools to download all messages in all threads in which these people have posted a message, both on the *KDE Women*'s list and other email lists associated with the KDE projects. There are 89 people who have posted on *KDE Women* at least once. Table 1 shows descriptive statistics about people's participation length in *KDE Women* ("Days

**Table 1.** Descriptive statistics of participation in *KDE Women* (KDEW) and KDE-affiliated projects (KDE)

|  | Days in KDEW | Days in KDE | Number of messages in KDEW | Number of messages in KDE |
|---|---|---|---|---|
| KDE then KDEW (N=28) | | | | |
| Average | 620.07 | 1868.11 | 8.96 | 760.96 |
| Max | 2125 | 3701 | 105 | 3912 |
| Min | 1 | 1 | 1 | 1 |
| KDEW only (N=51) | | | | |
| Average | 90.45 | n/a | 2.14 | n/a |
| Max | 2196 | n/a | 14 | n/a |
| Min | 1 | n/a | 1 | n/a |
| KDEW then KDE (N=10) | | | | |
| Average | 184.5 | 297.4 | 2.9 | 5.9 |
| Max | 1277 | 1381 | 11 | 18 |
| Min | 1 | 1 | 1 | 1 |

in KDEW") and other KDE-affiliated projects ("Days in KDE") as well as their posting activities, all of which were documented since people's first post in either *KDE Women* or other KDE projects to the date of data collection in July 2007. Based on the dates of these 89 people's first posts, we identified people at three participation phases.

### 3.2 Data Analysis

As mentioned earlier, a combination of deductive and inductive qualitative coding method was used to address the research questions. Every person's message(s) in their first thread posted on the mailing list of *KDE Women* were coded to capture the joining patterns. For the deductive coding of the joining activities, we utilized the group membership roles by Benne & Sheats [6] and the five types of information seeking behaviors for newcomers' socialization [22] as a coding scheme. In the meantime, we allow new meanings to emerge from the data, resulting in new categories.

Data were coded using NVivo 7.0. For the deductive coding, two researchers first independently coded the same subset of the data according to the coding scheme and then compared the results. After consensus was reached, each researcher coded half of the rest of the data. For the inductive coding, each researcher first developed new categories for a same cluster of data. These categories were compared and the interpretations of differences in the coding were reconciled. This process continued for another round and a condensed list of categories were finalized when a high level of agreement across researchers was reached and most messages were adequately coded using the existing set of categories.

## 4  Findings

### 4.1  Joining and Initial Socialization Activities on *KDE Women*

In this research, we are interested in exploring the joining behaviors of *KDE Women* participants. Adopting a qualitative approach, we developed a joining script of these people through combining the literatures on group membership roles, newcomer socialization, and inductive coding technique.

The first research question is on finding out the joining and initial socialization activities associated with participation in *KDE Women*. We found that the key group membership roles exhibited in the *KDE Women* are five types: information seeker, opinion seeker, information giver, opinion giver, and encourager. Except for "encourager", which is a group building and maintenance role, the others are task roles [6].

Socialization literatures indicate that newcomers usually seek five kinds of information: technical information, performance feedback, referent information, normative information, and social feedback. Building on this framework, our findings bring forth finer-grained dimensions and new meanings of these socialization activities in *KDE Women*. For *KDE Women* members, the technical component relates to using or developing specific software programs. The performance component includes group-related tasks, such as building, promoting or representing *KDE Women* in the KDE community in general. The referent dimension is concerned with members' seeking of job requirements and identity in *KDE Women*. The normative dimension elaborates on issues such as group goal setting, group legitimacy and composition, and technological ideology debate. Specifically, early *KDE Women* members collectively determined the group goal and mission statement (group goal setting). After the group was formed, later joined participants had heated debate on whether the *KDE Women* group was being sexist, and whether men should be allowed in the group (group legitimacy and composition). During the group discussions, the war between open source and windows was constantly brought up (technology ideology debate). And lastly, the social dimension was mainly concerned with people's communication behaviors where they expected others' response.

Linking these socialization activities with the membership roles creates a complete configuration of the joining script in *KDE Women*, as shown in Table 2. Cells with a number suggest the frequency of a particular type of joining script where members take on certain roles in specific socialization activities, therefore indicating the kind of behavioral structures people engage in when becoming a member of *KDE Women*. Grayed areas indicate no such behavioral structures were found. For example, the combination of "information seeker" and "technical" creates one kind of joining script, meaning that participants in *KDE Women* would seek technical related information. However, no joining script exists as a combination of "opinion seeker" and "technical", showing that no one seems to seek technical opinions while first joining the group. As shown in Table 2, there are several kinds of joining script. Firstly, people seek and give

**Table 2.** Joining script of *KDE Women* members

| Socialization Activities | Major group membership roles played | | | | |
| --- | --- | --- | --- | --- | --- |
| | Information Seeker | Opinion Seeker | Information Giver | Opinion Giver | Encourager |
| 1. Technical | 16 | | 5 | 1 | |
| 2. Performance 2.1 Group tasks | 7 | 6 | 1 | 3 | 2 |
| 3. Referent | 9 | | | | 1 |
| 4. Normative 4.1 Group goal | | | 2 | 3 | 8 |
| 4.2 Group legitimacy and composition | | | 2 | 10 | 2 |
| 4.3 Technological ideology | | | 3 | 3 | |
| 5. Social 5.1 Communication stimulation | 11 | | 2 | | 1 |

*Note.* Numbers indicate frequency of *KDE Women* members' joining script of that configuration. Grayed area: no joining script found.

information, as well as give opinions on technical issues. Secondly, they seek and give information, as well as seek and give opinions, and encourage each other on group tasks-related issues. Thirdly, for job requirement and membership identity issues, *KDE Women* members only seek for feedback and being encouraging about that, but don't engage in any giving activities. Fourthly, *KDE Women* members give information and opinions and provide encouragement on normative issues; and yet they don't seek anything on this dimension. Lastly, they seek social feedback and provide information or encouragement to be social, especially when they want to stimulate some conversations on the mailing list.

## 4.2 Joining Activities across People at Different Participation Phases

The second research question addresses whether joining activities differ for people in different KDE participation phases and if so, how? In Table 3, which is adapted from Table 2, "A" stands for people who participated in *KDE Women* after KDE (N=28), "B" stands for those who participated in *KDE Women* only (N=51), and "C" stands for those who participated in KDE after *KDE Women* (N=10). It shows where participants of *KDE Women* stand respectively in the joining script configuration. In other words, it demonstrates which of the three groups belong to which joining script configuration. For example, all three types of people seek technical information, type A and type B people seek opinions on group tasks, and only type B people give opinions on technical

**Table 3.** Joining script of *KDE Women* members at three different phases

| Socialization Activities | Major group membership roles played | | | | |
| --- | --- | --- | --- | --- | --- |
| | Information Seeker | Opinion Seeker | Information Giver | Opinion Giver | Encourager |
| 1. Technical | ABC | | AB | B | |
| 2. Performance  2.1 Group tasks | AB | AB | A | A | AB |
| 3. Referent | B | | | | B |
| 4. Normative  4.1 Group goal | | | A | AC | AB |
| 4.2 Group legitimacy and composition | | | AC | ABC | AB |
| 4.3 Technological ideology | | | B | AB | |
| 5. Social  5.1 Communication stimulation | ABC | | AB | | B |

*Note.* "A": people who participated in *KDE Women* after KDE (N=28); "B": people who participated in *KDE Women* only (N=51), "C": people who participated in KDE after *KDE Women* (N=10). Grayed area: no joining script found.

information. Below characteristics of these people's joining scripts on *KDE Women* will be presented. They are illustrated by sample quotes, and the corresponding codes representing joining scripts are included in the parentheses. These quotes are not meant to be exhaustive; rather, they are most representative of the joining activities of people at the three participation phases.

### 4.2.1  People Who Participated in *KDE Women* after KDE (Denoted as A in Table 3)

There are 28 people in this category. As demonstrated in Table 3, this group of people was very involved in group tasks issues and normative issues. They assumed all roles related to the former and all but information giver on the technological ideology issue to the latter. For example, this person was looking for volunteers to help with the *KDE Women* booth at the Linux Expo:

> *"Hello all, I am just in the process of organising the KDE booth at the Linux Expo's in London (3/4 July) and Birmingham (12/13 Sep)...I am looking for some kde-women representation at the booth. If you can help with the following, please get in touch..."* **(Performance / group task – information seeker)**

Another person gave suggestions as to what a PR article should cover about *KDE Women*:

> *"I think it's important to mention in the article that it's not only for women to participate. Rather it seems from the website that the focus is to get more women invovled with KDE and to address gender-specific issues in KDE."* **(Performance / group tasks – opinion giver)**

This person presented ideas on involving more women in KDE:

> *"What we want to do about this: getting women involved into KDE as an integrated part. Let them give the audience and the chance to change computing in a way they think it is right for them also....That also makes KDE the only project until now I know of that cares about women's needs generally in connection with the design and usage of graphical computer interfaces."* **(Normative / group goal – information giver)**

People in this category were also interested in technical issues: they were both information seeker and information giver on software usage or development. An example is as follows:

> *"I'm in desperate need of a program that:*
> *- has an interface to KOrganizer to keep track of dates*
> *- has an interface to MySQL to maintain a musicians and customers database*
> *- can do bookkeeping WITH the option of taking the VAT out by triple booking"*
> **(Technical – information seeker)**

In addition to the above joining activities, this group extended communication on the list in hope of receiving feedback from others by performing the information seeker and information giver roles. Interestingly, this group was not engaged in referent issues, probably because they were very clear about their membership identity, so it would be unnecessary for them to seek information on their responsibility for the group.

### 4.2.2  People Who Only Participated in *KDE Women* and Not Other KDE Projects (Denoted as B in Table 3)

In total 51 people belong to this category. Being the largest group on *KDE Women*, these people were very social, in the sense that they gave encouragement to other group members' ideas and opinions in several aspects, and wanted to receive feedback on their greetings to the list. For example, in terms of normative issues, while they didn't comment on group goals, they showed encouragement on that:

> *"Second, I great appreciation for this list and its members. It is so nice to know other women who are thrilled with similar interests. I've never been a room with another female programmer...Look forward to the list."* (**Normative / group goal -- encourager**)

This member ended the self-introduction with the hope to know more people on the group:

> *"I'm surprised at how much I have to say. I hope to get to know some of you-and learn from you."* (**Social / communication stimulation – information seeker**)

More interestingly, they were the only group that sought information regarding group membership identity. In other words, this group showed strong willingness to find a fit to contribute to the group:

> *"I'm not entirely sure how I can help, but I have a lot of skills and can probably help more than I know. I'm decently proficient in C... I mostly just like to help out."* (**Referent – information seeker**)

In addition to the above examples, this group of people showed interest in technical issues, and they were both seekers and givers in this regard. The following quote illustrates that a person encountering a technical difficulty was seeking help:

> *"I've seen articles saying that it can be done but I'm having trouble actually doing it. I need to compile a \*.so file for Solaris using a linux box...Has anybody done this?"* (**Technical – information seeker**)

This person expressed interest in working out some technical issues:

> *"I'm fairly familiar with php, and I wouldn't mind working with other kde-women to set something like this up"* (**Technical – information giver**)

### 4.2.3  People Who Participated in Other KDE Projects after *KDE Women* (Denoted as C in Table 3)

The smallest group in *KDE Women*, this category has 10 people. Their main joining scripts centered on normative issues. They were information giver and opinion giver on group legitimacy and composition, as well as opinion giver on group goals.

> *"I'd like to see KDE women as a forum to encourage to write (well documented) code... not just lending a helping hand with documentation and stuff. Don't misunderstand me: documentation is important. But in my sparetime I'd rather write my own stupid KDE program (I hope one day I will...) that possibly will never reach the quality of say Konqueror instead of writing documentation for a boy who liked coding but disliked documentation."* (**Normative / group goal – opinion giver**)

> *"As you can guess by the title I am against the group...Why bother, tell me something, are women \less capable then everybody else... Is there some kind of issue with women \ not being able to handle it on there own"* (**Normative / group legitimacy – opinion giver**)

In the meanwhile, they sought information on technical issues, as illustrated by the following example:

> *"I'm trying to install KDE3 next to KDE2 following the instructions on the page "How to get KDE2 and KDE3 (from cvs) working on the same machine". ... which 4 lines are meant here? All help welcome best regards"* (**Technical – information seeker**)

This group also sought social feedback from the list. While they were engaged in several seeking activities, they didn't seem to give back on those dimensions. Moreover, participants in this category did not show much concern about group related tasks, nor about their membership identity. They were also not engaged in much of the encourager role for any aspect of the joining activities.

## 5  Discussions

A comparison across these three categories of people indicates that there are two commonalities among them in terms of their joining and socialization activities. First, all three classes of members played the information seeker role on both technical and social dimensions. This suggests that people joined *KDE Women* in order to obtain answers to technical questions and receive feedback. A second similarity, which contrasts with prior literature, is that no one was engaged in the "seeking" activities regarding group normative information. Literatures on newcomers suggest the importance of adaptation to team expectations [7,8,9,20], and the impact of information seeking about organization's norms and values on individuals' social integration in the organization [22]. What we found instead, is that rather than seeking information on the group norms, members participated in shaping the group norms and values – they were information and opinion givers and encouragers in this respect. This finding well reflects the dynamics on the gender issues in the open source world. While the implicit consensus is that open source is a male-dominated environment, a group dedicated to involving more women still stirred a great deal of debate and encountered many opposing view points. While many participants embraced the idea of a women's group, others objected, arguing, for example, that it served to segregate women rather than engage them. The norms and values of *KDE Women* thus remained in flux throughout the life of the group, and served as a common point of interest to bring engage all different kinds of members in discussion.

People at the three participation phases also reveal apparent differences in their joining and socialization patterns. For the group performance or tasks related activities, only people who had experience with KDE before entering the *KDE Women*'s group were engaged in all aspects of the roles in this dimension. Through *KDE Women*, they were able to exercise organizational and leadership skills that, perhaps, they could not find the opportunity to develop in other KDE projects. People who never had any interactions with other KDE projects had diverse experience in *KDE Women*. Interestingly enough, only this group sought information regarding job requirements and membership identity, and showed encouragement to others on this dimension. Therefore, unlike those who were more experienced with KDE and took on group related tasks right away, these people, while very willing to contribute to *KDE Women*, were not quite sure how. Lastly, people who proceeded to other KDE projects from *KDE Women* started their participation on *KDE Women* through discussion on the normative topics of the group. Unlike the other two groups however, upon joining, they were not involved in any group task related activities. They were also less social: they did not give or share much information with others, nor did they show an encouraging attitude to other people. Overall, this group seems to have engaged in the one topic most unique to *KDE Women*: discussing the group itself, and targeted their task related participation to other KDE groups.

## 6  Conclusion and Future Research

In this research, we studied how women's groups facilitate people's participation in open source by investigating joining activities on *KDE Women*. Leveraging literatures on group roles and socialization, we developed an understanding of the joining scripts of different kinds of participants and thereby identified activities that people engage in to become a member of the group. In particular, we found two unique ways that *KDE Women* appeared to provide a place for participants to engage in ways that may not have been available to them in the broader KDE environment. These were (1) by providing an opportunity for members to develop their leadership skills and (2) by providing a place for members to participate in the development of an environment with norms and values consistent with their own vision. This study extends prior literature on socialization by including the giving side of members' activities on different dimensions. We also illustrated joining patterns of members at three participation phases, and thus highlighted the different ways in which people utilized women's group for their participation in open source.

The study remains a work in progress. As we continue with the research, we plan to expand our data coding and analysis in several ways. In particular, we plan to examine later threads of *KDE Women* members to understand their subsequent activities, and the outcomes of the various joining activities. We also plan to extend the study to other women's groups to gain a more comprehensive understanding of how these groups facilitate the participation of women and which of their practices are most successful and most likely to be adaptable to other contexts, such as the incorporation of women into professional technical roles.

**Acknowledgements.** We would like to thank Chang-han Jong and Lydia Chiu for excellent research assistance.

# References

1. Ahuja, M.K.: Women in the Information Technology Profession: A Literature Review, Synthesis and Research Agenda. European Journal of Information Systems 11(1), 20–34 (2002)
2. Ashford, S.J.: The Role of Feedback Seeking in Individual Adaptation: A Resource Perspective. Academy of Management Journal 29(3), 465–487 (1986)
3. Ashford, S.J., Taylor, M.S.: Adaptation to Work Transitions: An Integrative Approach. In: Ferris, G.R., Rowland, K.M. (eds.) Research in Personnel and Human Resource Management, pp. 1–39. JAI Press, Greenwich (1990)
4. Bartol, K.M., Aspray, W.: The Transition of Women from the Academic World to the IT Workplace: A Review of the Relevant Research. In: Cohoon, J.M., Aspray, W. (eds.) Women in Information Technology: Research on Underrepresentation, pp. 377–419. MIT Press, Cambridge (2006)
5. Bartol, K.M., Williamson, I.O., Langa, G.A.: Gender and Professional Commitment among IT Professionals: The Special Case of Female Newcomers to Organizations. In: Cohoon, J.M., Aspray, W. (eds.) Women in Information Technology: Research on Underrepresentation, pp. 421–438. MIT Press, Cambridge (2006)
6. Benne, K.D., Sheats, P.: Functional Roles of Group Members. Journal of Social Issues 4(2), 41–19 (1948)
7. Chatman, J.A.: Matching People and Organizations: Selection and Socialization in Public Accounting Firms. Administrative Science Quarterly 36(3), 459–484 (1991)
8. Chen, G., Klimoski, R.: The Impact of Expectations on Newcomers on Newcomer Performance in Teams as Mediated by Work Characteristics, Social exchanges, and Empowerment. Academy of Management Journal 46(5), 591–607 (2003)
9. Chen, G.: Newcomer Adaptation in Teams: Multilevel Antecedents and Outcomes. Academy of Management Journal 48(1), 101–116 (2005)
10. Crowston, K., Annabi, H., Howison, J.: Defining Open Source Software Project Success. Paper presented at the International Conference on Information Systems, Seattle, WA (2003)
11. Duriau, V.J., Reger, R.K., Pfarrer, M.D.: The Content Analysis of Content Analysis. Organizational Research Methods 10(1), 5–34 (2007)
12. Feldman, D.C.: A Contingency Theory of Socialization. Administrative Science Quarterly 21(3), 433–452 (1976)
13. Ghosh, R.A., Glott, R., Krieger, B., Robles, G.: Free/Libre Open Source Software: Survey and Study. In: Workshop on Advancing the Research Agenda on Free/Open Source Software, International Institute of Infonomics, University of Maastricht, Brussels, Netherlands (2002)

14. Graen, G.: Role Making Processes Within Complex Organizations. In: Dunnette, M.D. (ed.) Handbook of Industrial and Organizational Psychology, pp. 1201–1245. Rand McNally, Chicago (1976)
15. Hare, A.P.: Roles, Relationships, And Groups In Organizations: Some Conclusions And Recommendations. Small Group Research 34(2), 123–154 (2003)
16. Hars, A., Ou, S.: Working for Free? Motivations for Participating in Open Source Projects. International Journal of Electronic Commerce 6(3), 25–39 (2002)
17. Hennig-Thurau, T., Walsh, G.: Electronic Word-of-Mouth: Motives for and Consequences of Reading Customer Articulations on the Internet. International Journal of Electronic Commerce 8(2), 51–74 (2003)
18. Katz, R.: Time and Work: Toward an Integrative Perspective. In: Staw, B.M., Cummings, L.L. (eds.) Research in Organizational Behavior, pp. 81–127. JAI Press, Greenwich (1980)
19. Louis, M.R.: Surprise and Sense-making: What Newcomers Experience in Entering Unfamiliar Organizational Settings. Administrative Science Quarterly 25(2), 226–251 (1980)
20. Louis, M.R.: Newcomers as Lay Ethnographers: Acculturation During Socialization. In: Schneider, B. (ed.) Organizational Climates and Cultures, pp. 85–129. Jossey-Bass, San Francisco (1990)
21. Maloney-Krichmar, D., Preece., J.: A Multilevel Analysis of Sociability, Usability, and Community Dynamics in an Online Health Community. ACM Transactions on Computer-Human Interaction 12(2), 1–32 (2005)
22. Morrison, E.W.: Longitudinal Study of the Effects of Information Seeking on Newcomer Socialization. Journal of Applied Psychology 78(2), 173–183 (1993)
23. National Center for Educational Statistics, Table 46, http://www.nsf.gov/statistics/nsf04311/pdf/sectb.pdf (2004)
24. Nafus, D., Leach, J., Krieger, B.: Gender: Integrated Report of Findings. Free/Libre and Open Source Software: Policy Support FLOSSPOLS. Deliverable D 16, cover page (2006)
25. Roberts, J.A., Hann, I.-H., Slaughter, S.A.: Understanding the Motivations, Participation, and Performance of Open Source Software Developers: A Longitudinal Study of the Apache Projects. Management Science 52(7), 984–999 (2006)
26. Tilly, C.: Durable Inequality. University of California Press, Berkeley (1999)
27. Von Krogh, G., Spaeth, S., Lakhani, K.R.: Community, Joining, and Specialization in Open Source Software Innovation: a Case Study. Research Policy 32(7), 1217–1241 (2003)
28. Weiss, T.R.: Panel: Open-Source Needs More Women Developers. Computerworld (2005)
29. White House Council of Economic Advisors. Opportunities and Gender Pay Equity in New Economy Occupations. Available in the National Archives (2000)
30. Wu, C., Gerlach, J.H., Young, C.E.: An Empirical Analysis of Open Source Software Developers Motivations and Continuance Intentions. Information & Management 44(3), 253–262 (2007)

# Download Patterns and Releases in Open Source Software Projects: A Perfect Symbiosis?

Bruno Rossi, Barbara Russo, and Giancarlo Succi

CASE – Center for Applied Software Engineering
Free University of Bolzano-Bozen
Piazza Domenicani 3, 39100 Bolzano, Italy
{brrossi,brusso,gsucci}@unibz.it
http://www.case.unibz.it

**Abstract.** Software usage by end-users is one of the factors used to evaluate the success of software projects. In the context of open source software, there is no single and non-controversial measure of usage, though. Still, one of the most used and readily available measure is data about projects downloads. Nevertheless, download counts and averages do not convey as much information as the patterns in the original downloads time series. In this research, we propose a method to increase the expressiveness of mere download rates by considering download patterns against software releases. We apply experimentally our method to the most downloaded projects of SourceForge's history crawled through the FLOSSMole repository. Findings show that projects with similar usage can have indeed different levels of sensitivity to releases, revealing different behaviors of users. Future research will develop further the pattern recognition approach to automatically categorize open source projects according to their download patterns.

**Keywords:** Open source software projects, software releases, repository mining.

## 1 Introduction

Determining the success of software projects is very often non-trivial. There are many aspects to consider, and even the definition of success can depend from multiple point of views. Nevertheless, discerning the success of software projects is useful as researchers can evaluate approaches, methods, and processes that performed well - given a certain context. Furthermore, the availability of large data about open source software projects made such research more appealing. At the same time such task can be more difficult, as we miss some important in-context information that can be gathered only as insider of a development team. Reconstructing such information can be problematic when mining online repositories without strict contact with the original development team.

The definition of success is also not unique. One of the views of software projects' success is directly dependent on the users. Specifically, as reported in [2], in Information System (IS) research the success of a software system has been studied as directly dependent on system and information quality [4]. According to this view, system quality

P. Ågerfalk et al. (Eds.): OSS 2010, IFIP AICT 319, pp. 252–267, 2010.

impacts directly on software usage, and thus on users satisfaction. Deriving the success of a project is thus a question of considering a) the impact of the system quality on the users' usage level and b) the acceptance rate of the user. Once this has been determined, then it is relatively straightforward to associate successful projects to their development practices, development process characteristics, or even product features. Leaving aside software quality, the determination of criteria to measure the usage of software and relative users' satisfaction are relevant research problems.

If we focus on usage, and on commercial software, there is at least one single indicator that can be reliably used as an indication of usage: the number of copies sold on the market. It is difficult that applications bought do not translate in real usage of the application. For open source software, the situation is more fuzzy, as there are no unique indicators of usage. Different proposals have been made, like using the number of downloads [3], adopting software agents to monitor the software usage [2], tracking the inclusion in software distributions [2], using downloads time-series to detect the evolving users' community [10], or even use web search engines results to derive the popularity of projects [11].

Conversely, determining users' satisfaction is probably easier for open source software than for proprietary software. The large number of data available allows mining of repositories to audit mailing lists, forums, bug tracking reports, and so on. Also in this case, there is no single universal indicator of users' satisfaction and different proposals have been made: considering user ratings, opinions on mailing lists, or surveys [2] – among others.

In this work, we focus on software usage and specifically on download rates. Software usage needs to take into account different measures, but we believe that download rates have not been exploited to their full potential. Especially for open source software, downloads have been identified useful as a proxy of software usage. Recently, the interest is more on the patterns that have been detected rather than the mere download indexes (as for example [6] and [10] justify).

Our hypothesis is that it is not true that in all open source software projects download rates are in relation with software releases. More precisely, we believe that there are projects in which download rates are in relation with the application type (e.g. file sharing applications, where users want to have the latest application available, security fixes included) and others where users do not really care about the release date (e.g. applications installed and kept for longer time, like graphical utilities). From these hypotheses, it derives automatically that different patterns and - even more - download numbers can represent very different situations that are difficult to generalize. So, before reconstructing patterns to see whether projects were successful or not, we will need to know how much projects are sensitive to software releases. We have two research questions for this paper.

- RQ1. Are download patterns connected to releases in open source software projects?
- RQ2. If such relation exists, is the relation consistent in the same category of projects?

The paper is structured as follows. Section 2 presents background on deriving open source software usage and specifically on studies on the evaluation of usage by means of download rates. Section 3 discusses further the research questions by means of a problem statement. Section 4 presents a method for analyzing downloads time series. Section 5 proposes an experimental evaluation of the method by means of the highest ranked projects in *SourceForge's* history. The section includes experimental design, data collection, data filtering, evaluation of the method, findings and limitations. Section 6 is about conclusions and future works in prospective research.

## 2 Background

Different techniques have been proposed to derive software usage. If we specifically focus on download rates, this indicator has been found to have several advantages, like the fact that it is relatively easy to gather this kind of information from online repositories. Nevertheless, various disadvantages are also reported, like the issue that a single download may not really translate in software usage [2]. What is very often suggested is to use this measure with care [3].

There has been a move in recent years from considering mere download rates (we can report as an example [6]) towards analyzing time-series and emerging patterns. In this sense, there is a consensus now among all researchers that download averages, or totals, do not convey enough information to be used as independent or dependent variable in success prediction models. Time series of downloads convey a larger set of information. The problem is that in the open source scenario, with large datasets available, some kind of data compression or summarization is needed so that knowledge about single projects can be synthetically represented, summarized, and then visualized. We can report specific studies that are related to the current research (Table 1).

**Table 1.** Related Studies

| Paper | Study | Results |
|-------|-------|---------|
| [6] | Identification of classes of successful and unsuccessful projects according to download patterns | Six patterns of download rates identified. Justification of emergence of such patterns |
| [7] | Identification of successful open source projects by means of downloads numbers | Categorization of 122,065 projects in super, successful, and struggling. No evidence of Zipf's Law for the number of downloads for projects on *SourceForge* |
| [10] | Proposing a method to measure size of open source projects and use base based on downloads time series | Different types of users found according to adaptation of downloads to releases |

All these studies have in common the analysis of download rates. Research questions are different, as well as the experimental setting. In [6] and [7] the focus is on deriving successful projects, in [10] the focus is on deriving the use base from download patterns.

## 3 Problem Statement

To present the problem statement, we give a practical example by means of two well known software. Namely, we consider the most downloaded project of *SourceForge's* history - *eMule*[1] - and the *TCL*[2] application. This selection is not random, as those applications were selected in the categorization made in [6], so the interested reader can find the motivation more compelling. As can be seen, the two time series of downloads for the two projects are quite different (Fig. 1 and 2 ).

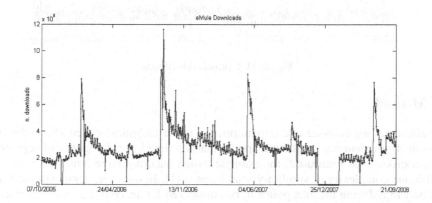

**Fig. 1.** eMule project downloads

What we see from the figures is that there are some short-time cyclic patterns in the download rates (maybe weekly) in both cases, but in the eMule case there are evident longer term cyclic patterns. What we ask ourselves is whether the latter type of patterns are related to software releases.

A simple approach would be to simply plot release dates on the same time series and evaluate manually the situation case by case. Instead, what we want to derive is an approach that allows to detect automatically - and without  visual inspection  - whether a time series is dependent on release dates. In the specific, if there are strong increases in download rates in coincidence with a software releases. Such method must also remove less important cyclic patterns in download rates. Considering the example in Fig. 2, we do not want to consider relevant low cyclic patterns that repeat weekly.

Thus, we present a proposed automated method that can be used applied to time series to evaluate the sensitivity to software releases. We apply it to downloads time series, but it can be potentially applied also to other types of time series (like the number of commits in time, for example). In the following, we use mostly the *TCL* project to explain the method.

---

[1] http://www.emule-project.net
[2] http://www.tcl.tk

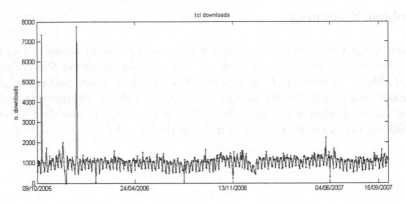

**Fig. 2.** TCL project downloads

# 4 Method

To eliminate non-relevant cyclic patterns, we use a technique called *Piecewise Aggregate Approximation (PAA)* that approximates a time series by means of segments. This approach has been used, for example, when handling large amounts of data to reduce the complexity of similarity search space [1]. In our case, segmentation helps not only in reducing the data points to be considered for analysis, but also in reducing short term cyclic patterns in the time series.

In the specific, in this research, we used a specialized form of *PAA* that uses a wavelet transform of the time series to decompose the segments of the time series [1, 8]. The wavelet used is the simplest form of wavelet: the *Haar* wavelet [8]. In Figure 3 we propose an example of *PAA* applied to a synthetic data set. The original time series is approximated by means of a wavelet that maintains similar patterns as the original. As can be seen directly from figures, the data loss is inversely proportional to the number of segments used for *PAA*. Conversely, more segments mean also keeping more short time periodic patterns.

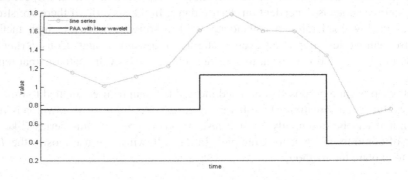

**Fig. 3.** Example of PAA by using Haar wavelets

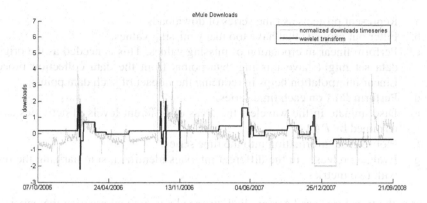

**Fig. 4.** eMule project wavelet transform

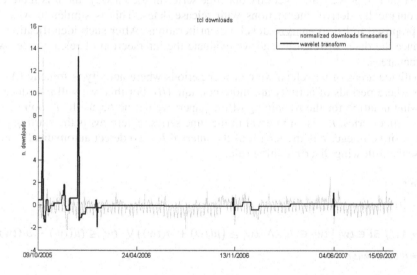

**Fig. 5.** TCL project wavelet transform

The whole theory of wavelets is far beyond scope for this paper, but an interesting overview can be found in [8]. Alternative and more sophisticate techniques for *PAA* can be found in [1]. To show the results of the technique, we applied it to the *eMule* and *TCL* projects (Fig. 4 and 5). After the application of the method, we have thus eliminated unwanted cyclic patterns from the time series, and got a simpler representation. We will then use this representation to detect the intersection of areas with releases automatically.

In detail, the whole approach is the following:

a.  Represent projects as time series of downloads;
b.  Filter out projects that have too many missing values;
c.  Perform linear interpolation of missing values. This is needed as the original data set might have missing data points from the data collection process. Linear interpolation helps in reducing the impact of such data points;
d.  Perform *PAA* on each time series;
e.  Discriminate in the wavelets the areas of different levels of activities as determined by *PAA*;
f.  Plot release information into the time series;
g.  Evaluate releases in the different intervals identified, summarizing the result with two metrics;

So, once the transform has been applied, we need a way to summarize the patterns of the original time series. For this, we divide the wavelet into different intervals according to the level of burstiness, identifying bursty intervals and more constant intervals. The reason is that we want to codify our time series in such a way that it is more easy to automatically derive intersections with release dates. This is similar to what has been proposed in [9] to analyze development iterations. After such identification, we introduce the dates of releases and we evaluate the intersection of release dates with different areas.

To divide areas of wavelets, we consider periods where activity is frenetic ($A$) and others where periods of activity are more constant ($B$). For this, we will introduce the following notation for the remaining of the paper: we define $tw_i$ as the $i^{th}$ point in the wavelet time series, $I_{k,\varepsilon}$ as an interval in the time series, where $tw_k$ is the starting point and a positive integer $\varepsilon$ is the length of the interval $I_{k,\varepsilon}$. To detect automatically areas, we use the following discriminating rule:

$$given\ \varepsilon > 0$$

$$I_{k,\varepsilon} = \left[ tw_{k\varepsilon+k}, tw_{(k+1)\varepsilon+k} \right]$$

$$I_{k,\varepsilon} \rightarrow A\ iff\ \exists i \in tw_i \mid tw_i \in I_{k,\varepsilon} \wedge\ tw_i \geq \left( \mu(tw) + \sigma(tw) \right) \vee\ tw_i \leq \left( \mu(tw) - \sigma(tw) \right)$$

$$I_{k,\varepsilon} \rightarrow B\ otherwise$$

We use this rule to classify between $A$, and $B$ periods. For convenience, we also define $J_A$ as each connected set of intervals $I_{k,\varepsilon}$ of type $A$. The same represents $J_B$ for $B$ periods. Fig. 6 shows the results of area mapping for the *TCL* project.

We next evaluate the relation between different area types and releases. We map thus how many releases happen for each project in specific areas. More releases in areas of type A mean that the project is more subject to a relation between releases and download rates. The simple approach that we used in current research is to use the intersection of releases and areas. If we apply this to *TCL* project, we can see that the project is scarcely sensitive to release dates (Fig. 7).

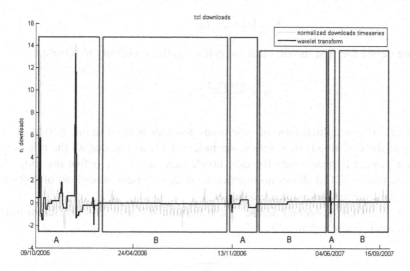

**Fig. 6.** TCL project with areas identified (spaces between areas are just to ease the interpretation of figure)

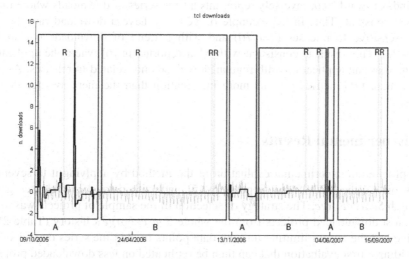

**Fig. 7.** TCL project with areas and release dates

To automate the process, and to allow for automatic use of the approach without examining the figures, we defined two metrics. One metric is about the sensitivity to releases, the other about the coverage of different areas. This information is needed as we can have the same level of sensitivity to releases but different level of burstiness of the time series. Without visual inspection, we will miss this relevant information. First we define the set of all releases of a project as:

$$R = \{r_1, r_2, \ldots, r_n\} \ .$$

Then we define the first metric - that we call $s$ - as the sensitivity to releases:

$$s = \frac{\sum [R \in J_A]}{\sum R} \ . \tag{1}$$

This metric is a weighted ratio of how many releases happen in one period of larger activity in the downloads time series. An index of 1.0 means that all the releases of a software project happen when the downloads time series are active the most. Conversely, a metric of 0.0 shows no reaction to releases: users' download of software is completely separated from release dates.

We then define the second metric as the amount of burstiness of different time series. We refer to this metric as $b$.

$$b = \frac{\sum |J_A|}{\sum (|J_A| \cup |J_B|)} \ . \tag{2}$$

An hypothetic index towards *1.0* means a time series of downloads completely bursty. An index towards *0.0*, inversely represents a time series of downloads where rates are almost constant. Thus in the example of *TCL*, we have a download rate that is not very sensitive to releases ($s=0.10$) and with a trend of downloads not so bursty ($b=0.31$). This result is consistent with what reported in [6] where the application is reported as an application with regular downloads not related to releases. As can be seen, these two metrics give us more information than the mere average download rates.

## 5  Experimental Results

We provide an experimental evaluation of the method by applying it to several projects out of the *FLOSSMole* repository [5] and with data about releases gathered through *SourceForge*. The strategy of selection of the sample of projects was to select the most downloaded projects from the whole *SourceForge's* history (Table 2). This choice had the aim of limiting missing data points in the time series on one side, and providing a first evaluation that can then be replicated on less downloaded projects. In this way, this first evaluation has the aim to gather findings from a well-known set of applications where downloads totals are quite consistent.

For each project, we gathered the time series of downloads, limiting the analysis to *1.000* data points. For the selection of the parameters, based on the sensitivity analysis reported in the next section, we selected the length of each interval $I_{k,\varepsilon}$ with $\varepsilon=30$. After collecting the data, we applied the whole approach to the dataset of each project.

**Table 2.** Most downloaded projects in Sourceforge's history

| Rank | Project Name | Type | Total Downloads | % Zeros TS 1000 |
|---|---|---|---|---|
| 1 | eMule | P2P Client | 510,493,881 | 0,00% |
| 2 | Azureus / Vuze | P2P Client | 455,284,828 | 0,30% |
| 3 | Ares Galaxy | P2P Client | 209,066,979 | 0,40% |
| 4 | 7-Zip | Compression Software | 76,806,020 | 12,30% |
| 5 | FileZilla | Client FTP | 71,295,059 | 34,40% |
| 6 | GTK+ and GIMP installers for Windows | Graphical tool | 64,212,148 | 24,30% |
| 7 | Audacity | Audio Editor | 64,051,083 | 4,40% |
| 8 | DC++ | P2P Client | 56,488,262 | 4,30% |
| 9 | PortableApps.com: Portable Software/USB | Utility | 55,713,765 | 18,90% |
| 10 | BitTorrent | File Sharing | 52,031,664 | 81,08% |
| 11 | Shareaza | P2P Client | 49,799,296 | 0,20% |
| 12 | VirtualDub | Video Editing | 47,835,670 | 10,20% |
| 13 | CDex | Digital Audio Ripper | 39,738,454 | 6,30% |
| 14 | Pidgin | Instant Messaging | 32,309,818 | 24,10% |
| 15 | aMSN | Instant Messaging | 31,175,716 | 6,30% |
| 16 | WinSCP | File Transfer Client | 29,681,313 | 6,10% |

## 5.1 Filtering

A first problem we had with the data set was how to handle zero counts in the time series. For this set of largely downloaded projects, our interpretation is that a zero in the download totals for one day means a missing point in the data collection process. The evolution of the download counts for such projects also justifies this view, with zeros interleaved to medium-high level download counts. Since our method supposes the application of interpolation, we wanted in any case to avoid its excessive application. For this reason, we filtered out from the sample the projects that had more than 10% zeros in the time series. Additionally, two projects were excluded from the sample. The *DC++* and the *CDex* projects were removed as we didn't have enough information about release dates.

## 5.2 Interpolation

For projects that were included in the sample, we interpolated linearly the missing points from previous and subsequent values in the time series. In this way, even with an approximation, we limited the impact of missing values from the data collection phase that could lead to an erroneous generation of different areas in the time series. In this experiment, we used a simple linear interpolation.

## 5.3 Sensitivity Analysis

There is one subjective choice when applying *PAA*: the selection of the number of segments to use. In our research, we considered monthly segments (segments of length *30*) and we believe this is as an appropriate number of segments according to our knowledge of the dataset, as we wanted to avoid weekly cyclic patterns. To support this decision, we performed a sensitivity analysis (Fig. 8), by calculating the Euclidean distance between the original time series and the wavelet. The analysis shows how *PAA* fits the original time-series according to different number of segments. With our selection of the parameters, we do not compress excessively the original representation of the dataset.

If we consider a higher number of segments, the fitting will improve going from monthly, to weekly segments, for example. By doing this, we will also introduce more cyclic patterns into the time series. So there is a trade-off in this sense. Our heuristic of selection for the estimation of the segments preferred to use monthly segments to reduce effects of weekly patterns in the time series.

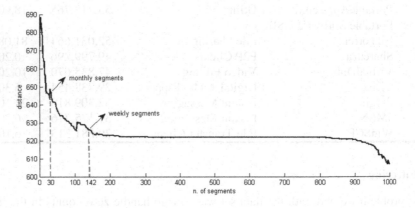

**Fig. 8.** TCL Project, distance between the wavelet and the time series according to number of segments (the lower value the better the fitting)

## 5.4 Results

After filtering and interpolation, we considered out of the initial 16 projects, just 7 projects (Table 3). The following categories were included: a) P2P Clients, b) Audio applications, c) Instant messaging, d) File Transfer Clients.

**Table 3.** Selected Projects

| Project Name | Type | Total Downloads |
|---|---|---|
| eMule | P2P Client | 510,493,881 |
| Azureus / Vuze | P2P Client | 455,284,828 |
| Ares Galaxy | P2P Client | 209,066,979 |
| Shareaza | P2P Client | 49,799,296 |
| Audacity | Audio(Audio Editor) | 64,051,083 |
| aMSN | Instant Messaging | 31,175,716 |
| WinSCP | File Transfer Client | 29,681,313 |

Then we applied to all projects the *PAA* technique, the derivation of areas in the time series, and the calculation of the metrics for sensitivity to releases and the level of burstiness. We report in the following the results.

For each project, we present the project name, the figure of the wavelet against releases, the parameters for sensitivity to releases, and burstiness of the wavelet as calculated by our approach (Table 4). The reader can see that in some cases, a high level of sensitivity to releases ($s$ parameter) can even be enforced by the fact that there are shorter areas of burstiness ($b$ parameter).

**Table 4.** Analysis of the Projects

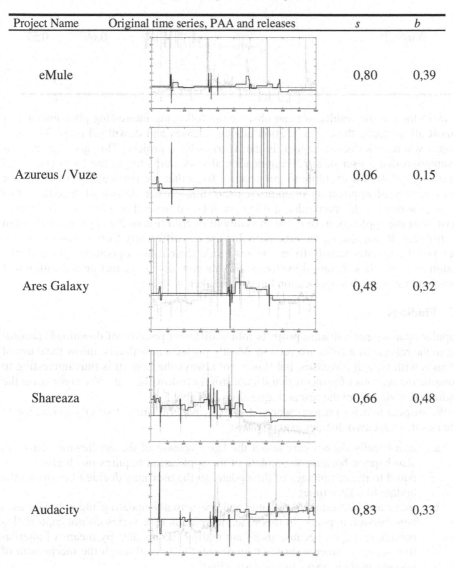

| Project Name | Original time series, PAA and releases | $s$ | $b$ |
| --- | --- | --- | --- |
| eMule | | 0,80 | 0,39 |
| Azureus / Vuze | | 0,06 | 0,15 |
| Ares Galaxy | | 0,48 | 0,32 |
| Shareaza | | 0,66 | 0,48 |
| Audacity | | 0,83 | 0,33 |

**Table 4.** (*Continued*)

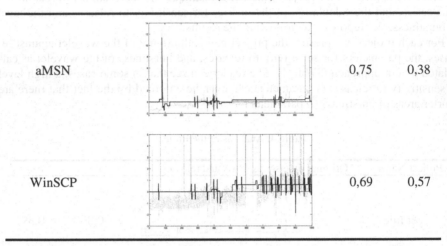

| | | |
|---|---|---|
| aMSN | 0,75 | 0,38 |
| WinSCP | 0,69 | 0,57 |

If we look at the results, we can observe the following interesting phenomena. For almost all projects, there is a relation among releases and download rates. The only project where this doesn't happen is the *Azureus/Vuze* project. This goes against our assumption that a user of a P2P application always wants to get the latest release as soon as possible, for example to get security fixes that are particularly important for this category of application or improvements like greater download speeds. If this doesn't happen for this particular application, it could mean that there specific characteristics of the application, or in the modality of distribution of the application that can be different. It can also be an indication that users – differently from the other cases – received the updates mostly from updates inside their Linux operating system distribution and not via software downloads. So this can also be in fact an indication that download rates for that application have to be taken with care.

## 5.5  Findings

Popular open source software projects follow different patterns of downloads according to the release of a software version. Mostly projects downloads follow the dates of releases with typical increases, but this is not always the case. It is thus interesting to examine the reasons of projects that do not strictly follow this rule. We summarize the findings deriving from the research questions in Table 5.

We suspect that for projects where download patterns are not strictly in relation to releases there are two distinct explanations:

a.  users really do not care about the latest release of the application. This can also happen because the update of the application requires much effort compared to the advantages of the update, so the user may decide to postpone the update to a later time;

b.  users are interested in updates and are actually updating the software as a new version appears. In this case, downloads time series do not capture this behavior, maybe because users are getting the updates by means of alternative sources (other websites than *SourceForge* or through the mechanism of updates in their own Linux distribution);

**Table 5.** Summary of the Findings

| Research Question | Finding(s) |
|---|---|
| RQ1. Are download patterns connected to releases in open source software projects? | We found that - in the majority of the projects analyzed - releases lead to an increase in download rates. In some cases, such behavior is less evident or even absent (e.g. Azureus). The explanation for this can be in the characteristics of users or the project features, but can also be an indication that download totals are not completely reliable for that specific application. |
| RQ2. If such relation exists, is the relation consistent in the same category of projects? | We found that the behavior is not consistent across all categories. Even in the limited set of categories we used, users respond in different ways to software releases even in the same category of applications. For example, in our sample, it is not true that users of P2P applications are more interested than other users in getting the latest release of the software. |

We argue thus that if we are in the a) case, downloads time-series can still be used as a somewhat reliable indicator of project's success in combination with other measures of usage and users' satisfaction. Conversely, if we are in the b) case, the evaluation of download rates must be complemented with additional information deriving – as an example – from projects' websites traffic, and/or search engines queries, like has been proposed in [11].

### 5.6 Limitations

The main limitation of the approach is about the definition of the parameters of *PAA* segmentation and areas definition. Although we provided the heuristic of selection and sensitivity of the model to the parameters when explaining the approach, it is clear that different parameters can lead to slightly different results. Specifically, the choice of the length of the interval $I_{k,\varepsilon}$ can give as result areas of different size to be used then in the metrics for calculation. Sensitivity analysis has been performed to reduce and limit this effect.

## 6   Conclusions and Future Works

We proposed a method to augment the expressiveness of downloads time series of open source software projects. We added information about the relation of projects' downloads to releases and defined two metrics. The metrics defined can give information about the responsiveness of the users to releases. This is a first step in research of automatic detection of patterns in downloads time series. Information from such patterns can then be used in models to detect projects' success.

We applied experimentally the method to a subset of projects in the *SourceForge* repository. We showed that codifying the downloads time series as two metrics conveys more information than using global metrics like average download rates or total download counts. As we have seen experimentally, even if projects have similar total download rates and counts, they can follow completely different download patterns. As such considering just those numbers can lead to wrong or biased conclusions. Furthermore, project downloads can be more or less related to software releases showing different behaviors from the point of view of users that can depend – and this will need to be validated in future research - on projects characteristics, application type or even modality of distribution.

Future research goes into two directions. One direction is to extend the approach to a larger data set, specifically focusing on projects' categories. The second direction is to investigate successful projects with an extension of the methodology developed in this paper.

**Acknowledgments.** We thank the creators and maintainers of the *FLOSSMole* repository for granting access and for their constant effort in providing a useful source of information about open source projects.

# References

1. Chakrabarti, K., Keogh, E., Mehrotra, S., Pazzani, M.: Locally adaptive dimensionality reduction for indexing large time series databases. ACM Trans. Database Syst. 27(2), 188–228 (2002)
2. Crowston, K., Annabi, H., Howison, J.: Defining Open Source Software Project Success. In: Crowston, K., Annabi, H., Howison, J. (eds.) Proceedings of the 24th International Conference on Information Systems (ICIS), pp. 327–340 (2003)
3. Crowston, K., Annabi, H., Howison, J., Masango, C.: Towards a portfolio of FLOSS project success measures. In: The 4th workshop on Open Source Software engineering, International Conference on Software Engineering (2004)
4. Delone, W.H., McLean, E.R.: The DeLone and McLean Model of Information Systems Success: A Ten-Year Update. J. Management of Information Systems 19, 9–30 (2003)
5. Howison, J., Conklin, M., Crowston, K.: FLOSSmole: A collaborative repository for FLOSS research data and analyses. International Journal of Information Technology and Web Engineering 1(3), 17–26 (2006)
6. Israeli, A., Feitelson, D.G.: Success of Open Source Projects: Patterns of Downloads and Releases with Time. In: IEEE International Conference Software Science, Technology, & Engineering, pp. 87–94 (2007)
7. Feitelson, D.G., Heller, G.Z., Schach, S.R.: An Empirically-Based Criterion for Determining the Success of an Open-Source Project. In: Proceedings of Australian Software Engineering Conference, pp. 363–368 (2006)
8. Li, T., Li, Q., Zhu, S., Ogihara, M.: A Survey on Wavelet Applications in Data Mining. SIGKDD Explor. Newsl. 4(2), 49–68 (2002)

9. Rossi, B., Russo, B., Succi, G.: Analysis of Open Source Software Development Iterations by means of Burst Detection Techniques. In: Proceedings of the 5th International Conference on Open Source Systems, pp. 83–93. Springer, Boston (2009)
10. Wiggins, A., Howison, J., Crowston, K.: Measuring Potential User Interest and Active User Base in FLOSS Projects. In: proceedings of the 5th International Conference on Open Source Systems, pp. 94–104 (2009)
11. Weiss, D.: Measuring Success of Open Source Projects using Web Search Engines. In: Scotto, M., Giancarlo, S. (eds.) Proceedings of the 1st International Conference on Open Source Systems, Genova, Italy, pp. 93–99 (2005)

# Modelling Failures Occurrences of Open Source Software with Reliability Growth

Bruno Rossi, Barbara Russo, and Giancarlo Succi

CASE – Center for Applied Software Engineering
Free University of Bolzano-Bozen
Piazza Domenicani 3, 39100 Bolzano, Italy
{brrossi,brusso,gsucci}@unibz.it
http://www.case.unibz.it

**Abstract.** Open Source Software (OSS) products are widely used although a general consensus on their quality is far to be reached. Providing results on OSS reliability - as quality indicator – contributes to shed some light on this issue and allows organizations to make informed decisions in adopting OSS products or in releasing their own OSS. In this paper, we use a classical technique of Software Reliability Growth to model failures occurrences across versions. We have collected data from the bug tracking systems of three OSS products, Mozilla Firefox, OpenSuse and OpenOffice.org. Our analysis aims at determining and discussing patterns of failure occurrences in the three OSS products to be used to predict reliability behaviour of future releases. Our findings indicate that in the three cases, failures occurrences follow a predetermined pattern, which shows: a) an initial stage in which the community learns the new version b) after this first period a rapid increase of the failure detection rate until c) very few failures are left and the discovery of a new failure discovery is rare. This is the stage in which the version can be considered reliable.

**Keywords:** Software failures, software reliability growth, open source software.

## 1 Introduction

Many of the open source projects do not have resources to dedicate to accurate testing or inspection so that the reliability of their products must rely on community's reports of failures. The reports are stored in the so-called bug tracking systems, are uploaded by the community, and moderated by internal members of the open source project. Reports are archived with various pieces of information including the date of upload and the description regarding the failure. What information can be collected from these repositories and how to mine them for reliability analysis is still an open issue ([6], [5]).

This paper proposes a method to mine bug repositories in order to determine patterns of failure occurrences that can be used to model reliability of past, current, and future versions of an open source product. In particular, this work discusses whether the traditional theory of software reliability growth can be readily applied to data coming from open source products. Our approach relies on an deep understanding of the bug tracking system used by each open source project and an accurate cleaning of

Ågerfalk et al. (Eds.): OSS 2010, IFIP AICT 319, pp. 268–280, 2010.

the data. Specifically, we have examined the user reports across versions of three open source projects, Mozilla Firefox[1], OpenOffice.org[2], and OpenSuse[3]. For each project and version, we have extracted information on the dates of issue opening and the number of open reports per day. We have fitted the traditional software reliability growth models with the time series of failures occurred per day for each version. We have repeated this procedure for major versions of the product and we have ranked the resulting best fit models for each version by a set of measures of model accuracy and reliability prediction. To understand whether there is a pattern in the failure occurrences of a product, we simply counted the number of times a model type is ranked at the top among all the versions and per measure of accuracy or prediction.

The paper is organized as follows. Section 2 consists of the background and some literature related to the work. This section briefly introduces to Software Reliability Growth Models and measures of accuracy and prediction. Section 3 illustrates the data sample and reviews the assumption we made in data collection and analysis. Section 4 introduces the method of regression across versions and pattern definition. Section 5, reports of the findings. Section 6 illustrates the limitation of the work and the future work. With section 7 we conclude.

## 2 Background

Modelling failure occurrences with Software Reliability Growth (SRG) is a classical approach in static analysis for software reliability ([2], [10], [11], [14]). In reliability growth, failure occurrences are assumed to grow with time such that the time to the next failure increases with time, too. This behavior is modeled with stochastic processes that describe the cumulative number of failures over time ([13]). The expected mean of the stochastic process defines a parametric function of time and represents the expected total number of failures at any instant of time t. Then the parameters of the expected mean are determined by non-linear regression on the actual dataset of cumulative number of failures over time.

The major challenges in SRG is to determine the mathematical expression of the expected mean either with a constructive procedure or with regression on predetermined families of time curves. While the former requires a deep understanding of the failures occurrence process in the operational environment of the software product ([10]), the latter focuses on an ex-post study shifting the complexity of the analysis to the data cleaning, model fitting, and interpretation, ([8], [9]). In this paper, we adopt the latter approach for the type of information we were able to gather with remote queries to data of the open source repositories.

Although this approach seems more feasible in our case, mining repositories for software reliability is not straightforward. To understand this we need to recall the meaning of software reliability. Software reliability refers to the capability of a software system to follow given specifications in a given interval of time and in given *operational* settings. A system experiences a failure when it deviates from this behaviour during its usage. As such, the best indicators for failures are users' reports on

---

[1] http://www.mozilla-europe.org/en/firefox
[2] http://www.openoffice.org
[3] http://www.opensuse.org/en

misbehaviour of the system during its usage. Gathering data on usage misbehaviour is hard as it depends on users' feedback, which is difficult to trace down. As a consequence, the majority of the classical works in SRG has used the same failures databases[4], not publicly accessible ones, or databases of defects discovered during internal testing ([10]) preventing proper conclusions on real system misbehaviour and replications in different applicative domains ([4]).

Nowadays, open source projects represents a new source of data for software reliability. They are as open to the everyone and provide enough information to perform analysis replications, but they may contain duplicate information, lack of complete information, and use localizations of standards and terminology. For example, in our case we were not able to find in the repositories information on deployment, customer usage profiles, and testing data that have been proved to be good predictor of failures occurrence rates and failures occurrence patterns for software systems prediction, ([2]). We also had to discard some open source projects that apparently had a large amount of users' report, but that at the end - after cleaning – data was too scarce to perform a rigorous analysis. Even with such drawbacks, we still believe that reliability analysis on open repositories is of great value as it is based on the real user's reports the effort is to understand whether the specific source provides suitable information.

Among several studies that analyzed open source software from the point of view of software reliability, we can cite ([12], [15], [18]). For example, Eclipse, Apache HTTP Server 2, Firefox, MPlayer OS X, and ClamWin Free Antivirus applications have been evaluated by means of several models, and the Weibull distribution has been found to adapt well in modelling simpler projects, although more complex models are claimed to be needed for Firefox and Eclipse [12]. In [15], authors are interested in evaluating the reliability of a complex system developed with a distributed approach, the Xfce desktop. Using decision support methods and non-homogeneous Poisson processes (NHPP), authors show how to model the complex interactions among   components for failure reliability prediction. In [18], several open source software projects have been analyzed and found to behave similarly to equivalent closed source applications. Also in this case, the Weibull distribution has been found to be a simple and effective way to represent software reliability growth.

In our research, after selecting the appropriate open source projects and cleaning the data our work provides an extension of the work of Li et al ([7]) and Succi et al. ([14]) combining the two approaches - software reliability growth across versions ([7]) and reliability growth and measures of accuracy and prediction ([14]) - and replicating the study on three different OSS. Mining repositories for software failures is a hot topic ([1], [2], [5], [6], [7], [8]), but at our knowledge it does not seem to be any research that has compared results across multiple releases and multiple OSS through multiple measures of accuracy and prediction.

## 2.1   Candidate Software Reliability Growth Models

We use the Software Reliability Growth Models (SRGMs) adopted in ([14]). SRGMs are stochastic processes described by a counting random variable determined by the

---

[4] Like the PROMISE Repository of Software Engineering Databases. School of Information Technology and Engineering, University of Ottawa, Canada http://promise.site.uottawa.ca/ ERepository/ or the new one http://promisedata.org/

cumulative number of failures over time ([13]). The expected mean at a given time t of these models is defined by a parametric curve in t, μ(t). The goal of software reliability growth is to determine the parameters of the expected mean. Many of the SRGMs we used are Poisson Processes whose mass probability is defined by a Poisson distribution. The basic model of them is the Musa exponential model ([11]) whose expected number of failures is defined by the parametric expression:

$$\mu(t)=a(1-e^{-bt}),\ a>0\ \text{and}\ b>0\ .\tag{1}$$

SRGMs can be S-shaped or concave depending whether they change concavity at least once. S-shaped models have an initial learning stage in which the failures detection rate, starting from very low values, shows a first increase and then a decrease that finally approaches to zero. Interpreting the model with time between failure occurrences, S-shaped models show an initial slow pace of failures occurrences, then a period in which failures are reported frequently until when relatively few failures remains in the code and failure discovery becomes hard. Example of S-shaped models are Weibull and Hossain Dahiya models. Concave models do not foresee such curve and reveal a good knowledge of the new released version. Failures are discovered soon after the release with a fast pace. Example of concave models are the Goel-Okumoto and Gompertz models.

In addition, models can be finite or infinite. Finite models have an horizontal asymptote and therefore a finite total expected number of failures are assumed in the product. The interested reader can find further details in the papers of Succi *et al.*, ([14]), and in the book of Lyu, ([10]).

## 2.2 The Measures of Accuracy and Prediction

The measures of accuracy and prediction capture the properties of a best-fit SRGM. We group the measures two by two by the attribute of the model we want to investigate. Table 1 introduces them indicating the references for further readings.

**Table 1.** Measures of accuracy and prediction used to rank the best fit models

| Model Attribute | Measure | Reference |
|---|---|---|
| Goodness of fit | Coefficient of Determination (R2) | [3] |
| | Akaike Information Criterion (AIC) | [7] |
| Precision of fit | Relative precision to fit (RPF) | [14] |
| | Coverage of fit (COF) | [16] |
| Forecasting ability | Predictive ability (PA) | [14] |
| | Accuracy of the final point (AFP) | [17] |

Goodness of fit and precision of fit refers to modelling the dataset, whereas forecasting ability defines the prediction nature of the model. Goodness of Fit expresses the ability of a mathematical model to fit a given set of data ($R^2$ and AIC). Precision of fit provides the extent (RPF) and the ability of the model to capture the data (COF) in the 95% confidence interval of the model. In general, these two measures are used in combination as they give complementary information on model precision. Measures of forecasting ability define the capability of a model to predict early in time (PA) or accurately (AFP) the final total number of failures.

# 3  The Dataset

We have selected three well-known OSS products: a web browser, Mozilla Firefox, an office suite, OpenOffice.org, and an operating system, OpenSuse. Each of these products maintains a bug tracking system open to the community based on Bugzilla[5]. We have chosen these three software projects also because their project's strategy for failures storage and the terminology used in Bugzilla is enough similar.

For each version found in the bug tracking system we have collected all the issues reported at our date of observation together with the date at which they were reported (date of opening). For each open source project, we have considered all the major versions until mid 2008 with more than 40 failures. For OpenOffice.org we were able to get 13 versions until 2006 (that we decided they were enough for the purpose of this analysis). Unfortunately, OpenSuse and Mozilla Firefox had not so many reports and versions at the time of our data collection and we had to limit the versions to five for OpenSuse and three for Mozilla Firefox.

Table 2 illustrates the dataset collected and the models used on the datasets.

**Table 2.** The Datasets

| Application | SRGMs type[6] | Versions | Time Window |
|---|---|---|---|
| Mozilla Firefox | Weibull, Goel Okumoto, Gompertz, Logistic, Goel Okumoto Sshaped, Hossain Dahiya | 1.5, 2.0.0.0 and 3.0.0.0 | 23.10.2006-8.6.2008 |
| OpenOffice.org | Goel-Okumoto, Goel-Okumoto S-shaped, Weibull, Hossain-Dahiya, Yamada | 1.0.0, 1.0.1, 1.0.2, 1.0.3, 1.1.0, 1.1.1, 1.1.2, 1.1.3, 1.1.4, 2.0.0, 2.0.1, 2.0.2, 2.0.3 | 16.9.2001-21.08.2006 |
| OpenSuse | Goel-Okumoto, Goel-Okumoto S-shaped, Weibull, Hossain-Dahiya, Gompertz, Logistic | 10.0,10.1,10.2,10. 3, 11.0 | 5.9.2005-1.6.2008 |

The Bugzilla repository allows to mine failures with sophisticated queries. A report in Bugzilla is called issue or bug and it may carry a large amount of information that needs to be pruned according to the research that is performed. To choose the view more appropriate for the research objective, one needs to understand in details the life cycle of an issue. In Fig. 1, we report the standard life cycle of a report submitted to a Bugzilla repository as described in the documentation[7].

Although the three projects we have chosen are all supported by a Bugzilla repository, we found out that their customization significantly varies. As such, starting from

---

[5] http://www.bugzilla.org
[6] The mathematical expression of the SRGMs model can be found in ([10]) or in ([13]).
[7] http://www.bugzilla.org/docs/tip/en/html/lifecycle.html

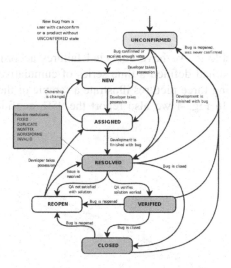

**Fig. 1.** Bugzilla life cycle of an Issue

the description in Fig. 1., we had to put some further effort in the homogenization of the terminology and the procedures used by the open source projects in moderating the issues.

### 3.1 Empirical Assumptions

In this section we discuss all the assumptions we have made on the data before starting our analysis.

Looking at the cumulative number of failures, we have noticed that in each version the rate of growth tends to zero for large values of time (Fig. 2, 3, and 4) indicating a bound for the total number of failures. For this reason, we have decided to include only finite models in our analysis, ([10]).

After a deep inspection of the repositories and of their documentation, we have decided to focus on those issues that were declared "bug" or "defect" excluding any issue that was called something like "enhancement," "feature-request," "task" or "patch". This would have guarantee that our analysis dealt with proper failures. For the same reason, we have considered only those issues that were reported as closed or fixed (according to the terminology of the single repository) after the release date of each version. Namely, reports before the release date were in general related testing release candidates and they were not expressing the reliability of the version.

For the same reason, we have cleaned the dataset from issues that were declared something like "duplicate," "won't fix," or "it works for me." Table 3 illustrates our choice.

**Table 3.** Chosen view of the Bugzilla repositories

| Issue Type | Status | Resolution | Platforms and operating systems |
|---|---|---|---|
| Defects/Bugs | Resolved, Verified, Closed, New, Started, Reopened | Fixed/Closed | All |

## 4  The Method

For each version of Table 2, we have collected failures according to their date of opening. In this way, we had defined a time series of cumulative number of failures per version. The following three pictures illustrate a sample of the time series and the plots of their SRGMs. In Fig. 4 we also report the 95% confidence interval of the Hossain Dahiya model.

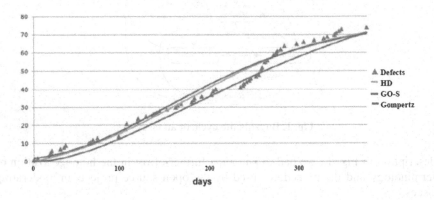

**Fig. 2.** Cumulative number of failures of Firefox version 1.0, the best-fit model (HD) and its 95% confidence interval. HD: Hossain Dahiya, GO-S: Goel Okumoto S-shaped, Gompertz.

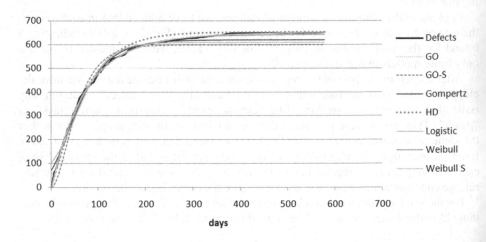

**Fig. 3.** Cumulative number of failures of OpenSuse version 10.0 and its best-fit models. GO: Goel Okumoto, GO-S: Goel Okumoto S-shaped, Gompertz, HD: Hossain Dahiya, Logistic, Weibull S: Weibull S-shaped, Weibull more S: Weibull S-shaped, Yamada.

For every OSS chosen, we have fitted the parametric expected mean of the SRGMs with the actual cumulative series of failures per each version of Table 2. As a result we have obtained a set of best fit models each corresponding to one type of SRGM in Table 2. At the end, we have obtained 18 SRGMs (6 each version) for Mozilla Firefox, 52 SRGMs (4 each version) for OpenOffice.org, and 25 SRGMs (5 each version) for OpenSuse. Fig. 2, 3, and 4 illustrate the time series and the curves corresponding to the best-fit models for one version of the three OSS. We have ranked the resulting best-fit models of every version by the measures of model accuracy and reliability prediction of Table 1.

A SRGM that outperforms for a given measure across versions of an OSS represents a pattern of reliability for that OSS. To determine the pattern we have simply counted for each measure of accuracy or prediction of Table 1, the number of time a best fit SRGM of a given type outperforms across the versions. In the following section, we discuss the existence and the type of pattern we have found.

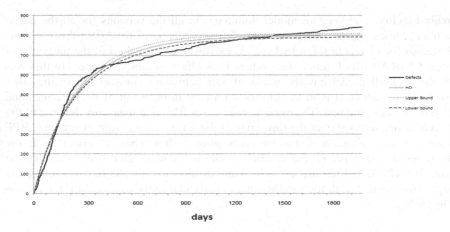

**Fig. 4.** Cumulative number of failures of OpenOffice.org version 1.0.1 and its best-fit models. HD: Hossain Dahiya.

## 5 Findings

In the following we report of the findings for the three OSS. The Yamada model is not included as its analysis does not report of significant results.

**OpenSuse.** For all the versions of OpenSuse, the Weibull model is the best one in all the measures but prediction ability. Thus, we can say that across versions it represents the data (Goodness of fit), it captures the majority of the data in small area of confidence (Precision of fit), and it is accurate in determine the final number of defects that partially determines forecasting ability. Namely it is not the best for prediction ability, thus we cannot use it to predict the total number of defects early in time. In fact, there are no SGRMs that can do it as no model outperforms across versions for predictive ability.

The predominance of the Weibull model confirms the findings in ([7]) extending the result to other measures of accuracy than the Akaike Information Criterion.

In Table 4 we report the rankings for version 11.0.

**Table 4.** Ranking of the best fit models for OpenSuse version 11.0

| Model Type | R Squared | AIC | CoF | RPF | AFP | PA |
|---|---|---|---|---|---|---|
| Weibull | **0.998** | **1.38** | **96%** | 3075.47 | **0,05** | 0.77 |
| Weibull S-shaped | 0.997 | 1.4 | **96%** | 4057.85 | 0.30 | 0.77 |
| Goel Okumoto | 0.995 | 1.49 | 82% | 792.31 | 1.91 | 0.77 |
| Gompertz | 0.993 | 1.6 | 18% | **144.02** | 3.15 | 0.75 |
| Logistic | 0.985 | 1.75 | 62% | 642.75 | 4.79 | 0.75 |
| Goel Okumoto S-shaped | 0.956 | 1.95 | 69% | 10838.81 | 7.51 | 0.79 |
| Hossain Dyohain | 0.935 | 2.05 | 2% | N/A | 9.64 | **0.71** |

**Mozilla Firefox.** The Weibull model dominates in all the versions for all the measures but accuracy of the final point and relative precision of fit. Again this confirms and extends the results in ([7]). In Table 5, we illustrate the values of the measures of version 1.5 of Mozilla Firefox. The values in boldface are the best ones. In this version, the Weibull model has the worst overall behaviour compared with the other versions. Nonetheless, it is in the top most ranking in three out of six measures and well performing in the rest. The exceptional case is RPF that measures the area of the 95% confidence interval over the time span of the series. As measure of accuracy RPF is subsidiary to CoF and it is used in combination with it. Thus, as models with low CoF are less relevant per se, the Weibull model is the most interesting, although the ratio CoF / RPF is not the highest. The Weibull model is not the best for prediction though. This is true for all the versions and in particular in the case of AFP for version 1.5 in Table 5.

**Table 5.** Ranking of the best fit models for Mozilla Firefox version 1.5

| Model Type | R Squared | AIC | CoF | RPF | AFP | PA |
|---|---|---|---|---|---|---|
| Weibull | **0.973** | 2.55 | **94%** | 35.52 | 0.04 | **0.62** |
| Goel Okumoto | **0.973** | **2.51** | 56% | 7 | **0.022** | **0.62** |
| Gompertz | 0.954 | 3.08 | 18% | **2.68** | 0.13 | N/A |
| Logistic | 0.934 | 3.45 | 25% | 6.43 | 0.15 | N/A |
| Goel Okumoto S-shaped | 0.956 | 3.02 | 37% | 4.52 | 0.17 | N/A |
| Hossain Dyohain | **0.973** | 2.54 | 80% | 13.02 | 0.024 | **0.62** |

**OpenOffice.org.** The Weibull model is again the best model across the measures of accuracy and prediction. In particular, for CoF the model outperforms 70% of the times across the versions.

The predominance of the Weibull model across so many versions as in OpenOffice.org is definitely significant.

**Table 6.** Ranking of the best fit models for OpenOffice.org version 2.0

| Model Type | R Squared | AIC | CoF | RPF | AFP | PA |
|---|---|---|---|---|---|---|
| Weibull | **0.998** | **3.08** | **0.95** | 27.09 | **0.001** | 47% |
| Goel Okumoto | 0.994 | 4.05 | 0.32 | 7.59 | 0.04 | 47% |
| Gompertz | 0.98 | 5.24 | 0.17 | **5.89** | 0.07 | 45% |
| Logistic | 0.965 | 8.85 | 0.24 | 23.86 | 0.09 | 45% |
| Goel Okumoto S-shaped | 0.946 | 6.21 | 0.23 | 14.35 | 0.12 | N/A |
| Hossain Dyohain | 0.991 | 4.42 | 0.34 | 11.44 | 0.054 | 46% |

Comparing the results for the three applications we can say that:

- For Goodness of fit (R square and AIC) the Weibull model confirms its superiority according to the work of Li et al. ([7]). The model well represents the data and the Weibull pattern can be used to represent future versions.
- For Precision of fit the 95% confidence interval of the Weibull model is the best in capturing data within its 95% confidence interval (CoF) although sometimes with a poor density (or equivalently in a large confidence interval, RPF). This is a measure of spread of data around the model also accounting for its variation in its confidence interval. Any significant variation of the model still represent the data with enough precision. This might turn to be useful when we discuss the time of occurrences in the open source repositories. The time reported in the repositories (calendar time) might not refer to the real time in which the user has been experiencing a failure. Some delay might have been occurred. As such, the result for Precision of fit even if it gives a positive answer for the Weibull model is not completely satisfactory and a Monte Carlo sensitivity analysis on the timing for each version will be matter of future work.
- For Predictive ability the Weibull model is definitely good to estimate the final total number of failures (AFP) but it cannot be used – as any other SRGM – for early prediction (PA). For this purpose, other approaches might be considered in combination ([1], [8], [9]). The lack of a pattern for PA further means that in the majority of the versions of the three OSS there is a low increase of the failure detection rate as in Fig. 2 (with or without a learning effect) so that it is impossible to predict the total number of failures of a version at the early stage of the failure reporting process. Thus, the case of Fig. 3 where a sudden and early increase of the detection rate appears, is less frequent across the versions of the three products Fig. 2 better represents the data as after 300 days the number of failures is still far to be near to the total final number of the version.

# 6  Limitations and Future Work

During our inspection, we have understood that the bug tracking system is regularly used by the internal team of the project. Internal team members know better the application so their reports might not represent typical end-user's reports of a failure. Although we have used some measure to limit this bias (section 4.1), as we could not differentiate between issues reported by the internal team and the rest of the world, we could not guarantee that the dataset is a dataset of failures reported only by end-users. In any case, as we have considered reports issued only after the release date, the reports of the internal team members refer to an operative period of the OSS and as such they contribute to some extant to the overall reliability of the OSS.

The date of report might be not exactly the date of failure discovery. There might have been some delay in reporting the issues and - depending on the repository – in the assignment of the date of opening during the moderation of the issue. This might create a noise in the timing and it will be matter of future research.

The number of versions, the number and types of SRGMs, and the time windows of the observations are different in the three OSS. This was due to some time constraints, the availability of the data in the repositories, and the missing values for the measures in Table 1. As we do not pretend to compare the three OSS, but we rather want to understand whether there is a pattern of reliability in each OSS, this difference is not crucial.

## 7 Conclusions

The goal of this paper was to present an approach to investigate reliability of OSS with software reliability growth. We used open on-line repositories to collect data of three different projects. We intensively cleaned the data we collected to limit the bias associated with the open nature of these repositories.

We found that the classical theory of software reliability growth is appropriate for such data and it is a good instrument to model failure occurrences across software versions.

We found that the Weibull model is the best model that fits the data across all the versions for each OSS (Goodness of Fit) with a low percentage of outliers (Precision of Fit). This confirms the results obtained in ([7]) for the Akaike Information Criterion and reveals a common pattern of software reliability for the three OSS. Namely, the Weibull model is the best SRGM that represents the failure occurrences in the three open source products. It is an example of S-shaped curve and as such it indicates an initial learning phase in which the community of end-users and reviewers of the open source project does not react promptly to new release. This slow pace at which failures are reported might originate from various causes like, for example, the unfamiliarity with the project or its complexity. Given the existence of candidate releases and intermediate versions one could expect that the community were ready to report of failures soon after the public release date. But the learning curve proves differently in three OSS.

Given the dominance of the Weibull model across the versions of an OSS we can assume that this type of model can be used for the future versions of the OSS. The open question is how to predict the parameters of this model without fitting the model on future data. In [8] the authors propose to use a combination of code and time measures. This approach will be matter of future research.

For OpenOffice.org and OpenSuse, the Weibull models can also be used for accuracy of the final point telling mangers when the version can be considered reliable as few failures remain to be discovered. As Fig. 2 shows, this does not hold for Mozilla

Firefox. This may suggest how the community reacts differently to new releases of Firefox showing a slow pace to report failures in the early days after the release.

Yet, SRGMs are not a good instrument to early prediction of the total number of failures within the operational life of an OSS. Other instruments like Bayesian models, Product/Process models or genetic algorithms might be explored in combination with predictive ability ([1], [8], [9]).

**Acknowledgments.** We thank Sufian Md. Abu, Stefan Mairhofer, Gvidas Dominiskaus for their help in data collection and cleaning. We also thank the projects Mozilla Firefox, OpenSuse, and OpenOffice.org for the data supplied for this study.

# References

1. Antoniol, G., Ayari, K., Di Penta, M., Khomh, F., Guéhéneuc, Y.: Is it a bug or an enhancement?: a text-based approach to classify change requests. In: CASCON 2008, vol. 23 (2008)
2. Bassin, K., Santhanam, P.: Use of software triggers to evaluate software process effectiveness and capture customer usage profiles. In: Eighth International Symposium on Software Reliability Engineering, Case Studies, pp. 103–114. IEEE Computer Society, Los Alamitos (1997)
3. Draper, N.R., Smith, H.: Applied Regression Analysis. Wiley- Interscience, Chichester (1998)
4. Fenton, N., Neil, M.: A Critique of Software Defect Prediction Models. IEEE Transactions on Software Engineering 25(5), 675–689 (1999)
5. Godfrey, M.W., Whitehead, J.: Proceedings of the 2009 6th IEEE International Working Conference on Mining Software Repositories, Vancouver Canada, May 16-17 (2009)
6. Li, P.L., Herbsleb, J., Shaw, M.: Forecasting field defect rates using a combined timebased and metrics-based approach: a case study of OpenBSD. In: 16th IEEE International Symposium on Software Reliability Engineering (ISSRE), pp. 10–19 (2005)
7. Li, P.L., Herbsleb, J., Shaw, M.: Finding Predictors of Field Defects for Open Source Software Systems in Commonly Available Data Sources: a Case Study of OpenBSD. In: 11th IEEE International Symposium on Software Metrics (2005)
8. Li, P.L., Shaw, M., Herbsleb, J., Ray, B., Santhanam, P.: Empirical evaluation of defect projection models for widely-deployed production software systems. In: Proceedings of the twelfth international symposium on Foundations of software engineering, pp. 263–272 (2004)
9. Li, P.L., Shaw, M., Herbsleb, J.: Selecting a defect prediction model for maintenance resource planning and software insurance. In: Proceedings of the Fifth International Workshop on Economics-driven Software Engineering Research, Oregon, USA (2003)
10. Lyu, M.R.: Handbook of Software Reliability Engineering. McGraw-Hill, New York (1996)
11. Musa, J.D., Iannino, A., Okumoto, K.: Software Reliability: Measurement, Prediction, Application. McGraw-Hill, New York (1989)
12. Rahmani, C., Siy, H., Azadmanesh, A.: An Experimental Analysis of Open Source Software Reliability. In: F2DA 2009 Workshop on 28th IEEE Symposium on Reliable Distributed Systems, Niagara Falls (2009)
13. Rigdon, S.E., Basu, A.P.: Statistical Methods for the Reliability of Repairable systems, p. 281. Wiley and Sons, Chichester (2000)

14. Succi, G., Pedrycz, W., Stefanovic, M., Russo, B.: An Investigation on the Occurrence of Service Requests in Commercial Software Applications. Empirical Software Engineering Journal 8(2), 197–215 (2003)
15. Tamura, Y., Yamada, S.: Comparison of Software Reliability Assessment Methods for Open Source Software. In: 11th International Conference on Parallel and Distributed Systems - Workshops (ICPADS 2005), vol. 2, pp. 488–492 (2005)
16. Wood, A.: Predicting software reliability. Computer 29(11), 69–77 (1996)
17. Yamada, S., Ohba, M., Osaki, S.: S-Shaped Reliability Growth Modeling for Software Error Detection. IEEE Transactions on Reliability, 475–484 (December 1983)
18. Zhou, Y., Davis, J.: Open Source Software Reliability Model: an empirical approach. In: International Conference on Software Engineering: Proceedings of the fifth workshop on Open Source Software Engineering, St. Louis, MO (2005)

# A Field Study on the Barriers in the Assimilation of Open Source Server Software

Kris Ven and Jan Verelst

Department of Management Information Systems,
University of Antwerp, Antwerp, Belgium
{kris.ven,jan.verelst}@ua.ac.be

**Abstract.** An increasing number of academic studies have been devoted to the organizational adoption of open source software (OSS). Most studies have either focused on determining which reasons influence the adoption of OSS, or which barriers prevent the adoption of OSS. To our knowledge, no prior study has been conducted to determine which barriers exist to the further adoption of OSS in organizations. Studies addressing this issue could provide more insight into whether organizations are effectively able to overcome the initial barriers to adoption, or whether new barriers arise when the organization expands its use of OSS. To this end, we conducted a qualitative field study involving 56 organizations that were asked to report on which barriers existed to their further adoption of OSS. The data was analyzed using a mixed methods approach by combining both qualitative and quantitative techniques. Results showed that the main barrier reported by organizations was a lack of internal and external knowledge. Furthermore, our results indicate that there was no relationship between the barriers reported by organizations and their extent of OSS usage. This indicates that these barriers remain important as organizations increase their assimilation of OSS.

**Keywords:** open source software, adoption, assimilation, knowledge, barriers.

## 1  Introduction

In the past few years, an increasing number of academic studies have been devoted to the organizational adoption of open source software (OSS). One line of research has investigated which reasons influence the adoption of OSS (see e.g., [32,8,14,18,16,15,31]). These studies have provided more insight into which factors have a positive impact on the adoption of OSS. Typically, such studies hypothesize that organizations that exhibit a number of favorable characteristics are more likely to exhibit a greater extent of adoption [6]. Hence, these studies consider both organizations with a low extent of adoption and organizations with a high extent of adoption. A second line of studies has investigated the non-adoption of OSS (see e.g., [20,12,5,13,19,10,11]). These studies are concerned with determining which barriers prevent organizations from adopting

P. Ågerfalk et al. (Eds.): OSS 2010, IFIP AICT 319, pp. 281–293, 2010.

OSS. These barriers can have an important negative influence on the organizational adoption of OSS. It is important to realize that adoption and non-adoption are indeed two distinct phenomena [21,9]. Gatignon and Robertson have noted that *"the variables accounting for rejection are somewhat different from those accounting for adoption; rejection is not the mirror image of adoption, but a different form of behavior."* [9, p. 42]. For example, it has been shown that even in the presence of several strong arguments in favor of the adoption of OSS, the existence of additional factors with a negative impact on adoption may result in non-adoption [13]. Previous studies that investigate the barriers to the adoption of OSS generally use a sample of organizations that have not adopted OSS (e.g., [10,11,13]).

To our knowledge, no prior study has been conducted to determine which barriers exist to the further adoption of OSS in organizations. However, we argue that it is interesting to consider which barriers exist for organizations that have already adopted OSS to some extent. These barriers may limit the ability of organizations to further increase their adoption of OSS. It seems unlikely that an organization that has adopted OSS to a limited extent will no longer experience any barriers when expanding its use of OSS. Instead, a more gradual process is likely to take place in which organizations slowly overcome the barriers to adoption over time. Hence, there is currently a lack of studies that consider the impact of barriers to the further adoption of OSS. Studies addressing this issue could provide more insight into whether organizations are effectively able to overcome the initial barriers to adoption, or whether new barriers arise when the organization expands its use of OSS. The results from such studies could allow to devise interventions that facilitate the further adoption of OSS in organizations.

In this study, we will address this gap in literature by investigating how the barriers reported by organizations evolve as organizations increase their use of OSS. To this end, a qualitative field study involving 56 organizations was conducted to determine which barriers existed to the further adoption of OSS. Our sample included organizations that did not adopt OSS, as well as organizations that have adopted OSS to at least some extent. Each respondent was asked to report which barriers they perceived to be present to the further adoption of OSS. We investigated whether the existence of these barriers was related to the extent to which the organization has adopted OSS. This will provide insight into whether these barriers are overcome by organizations as they increase their assimilation of OSS, and which additional barriers may arise later in the adoption process.

## 2    Methodology

Our study was conducted as a qualitative field study. A self-administered web survey was used to collect the data for our study. The scope of our survey was restricted to the use of *open source server software (OSSS)*. For the purpose of this survey, the term OSSS referred to a limited list of 7 OSS products consisting of Linux, BSD, Apache, Bind, Sendmail, Postfix and Samba. Respondents were instructed on each page of the survey that the term OSSS referred to this specific list of OSS products. Our sample consisted of Belgian organizations from different sectors and sizes. The target person in each organization was the IT decision maker, commonly the CIO or IT manager. We received a reply from 111 out of 153 organizations that were sent an invitation to participate, which corresponds to a response rate of 72.5%.

The main question in the survey that respondents needed to answer was the following: *"Which reasons prevent the (further) adoption of OSSS in your organization?"*. The question was open-ended, allowing organizations to provide a free format reply. This approach allowed us to obtain more in-depth information than by using a closed-ended question. In addition, organizations were also asked to report on their extent of Linux adoption and their degree of OSSS assimilation. The extent of Linux adoption was measured on a 7-point point Likert scale ranging from "no usage" to "to a very large extent". The assimilation of OSSS was measured by using the Guttman scale developed by Fichman and Kemerer [7] which was slightly reworded to fit the context of the adoption of OSSS (see also [32]).

A mixed methods approach was used to analyze our data, by using both qualitative and quantitative techniques [24,22]. In a first step, our data was analyzed by using qualitative techniques [17,2,27]. This analysis was performed by using NVivo 8. After importing the data in NVivo, all replies were coded to identify any barriers reported by respondents. In a first cycle, coding took an inductive approach by creating new codes as new reasons were identified in the responses [2,27]. After this first cycle, a list of 21 codes was obtained. In a second cycle, the codes were further aggregated into a hierarchy consisting of 8 main categories representing the barriers to the further adoption of OSSS [2,27].

In a second step, the qualitative data was converted into quantitative data. This process is commonly referred to in mixed methods research as *quantitizing* [24,23,3,28]. After quantitizing, statistical methods can be used to test hypotheses regarding the relationships between independent and dependent variables derived from both qualitative and quantitative data [28,3]. Combining qualitative and quantitative analysis techniques is an accepted approach in mixed methods research and can assist the researcher to more easily distinguish and communicate patterns within the qualitative data [24,23,3,28,17]. Each barrier that was coded in the qualitative analysis was converted into a dichotomous variable by using NVivo. Each case was assigned a value of one if the barrier was present, and zero otherwise. This resulted in a matrix in which the presence or absence of each code derived from the qualitative data was recorded for each case. This matrix is referred to as an *inter-respondent matrix* or a *case-by-variable matrix* [24,3]. This matrix was further extended with quantitative data concerning the degree of OSSS assimilation and the extent of Linux adoption. This matrix was then used to perform statistical analyses to further analyze the data [24,23,3,28]. These analyses were performed in SPSS 15.0.

# 3  Findings

A quick inspection of our data showed that 55 out of 111 respondents did not provide an answer to the question of which barriers (further) inhibit the use of OSSS. This was indicated by an empty string for the reply field. A missing reply could have two causes: (1) the respondent did not reply to the question, or (2) the respondent could not identify any barriers to the further adoption of OSSS. Unfortunately, we were not able to make a distinction between these two scenarios. All cases with empty responses were therefore coded as *unanswered*. These cases were discarded in the further analysis of our data. As a result, we obtained usable replies from 56 organizations in our sample. We will first present an overview of which barriers were reported by our respondents. Next, we will investigate whether these barriers are related to the assimilation of OSSS or the extent of Linux adoption.

## 3.1  Barriers to the Further Adoption of OSSS

A general overview of which barriers were mentioned by the organizations in our sample is presented in Table 1. This table contains the 8 barriers that were identified during the coding process. One barrier ("insufficient knowledge") is further divided into two subcategories ("internal" and "external" knowledge). The second column indicates how many organizations have reported each barrier.[1] The final column indicates the percentage of organizations that has reported each barrier. Since organizations could mention more than one barrier, the total of this column does not add up to 100%. This column represents the *frequency effect size* of each barrier [22]. We will now discuss each of these barriers in more detail.

**Insufficient Knowledge.** The most important barrier that was mentioned by almost a third of the organizations is a lack of knowledge about OSSS. We further distinguished between insufficient internal knowledge (i.e., knowledge that is held by employees within the organization) and external knowledge (i.e., knowledge that is offered by service providers). Our data shows that internal knowledge is the most important component. Many organizations reported that they lacked employees that are familiar with OSSS. One respondent mentioned that it was *"difficult to find in-house the right technical staff to maintain [OSSS]"*. In addition, the knowledge base within the organization may also restrict the opportunities to explore the use of OSSS. One respondent mentioned that the increased use of OSSS was difficult due to *"colleagues who are more proprietary-oriented"*. With respect to the availability of external knowledge, organizations were concerned that no commercial support was available for OSSS and that the internal IT staff would be responsible for supporting the software. One respondent noted

---

[1] Since one organization reported a lack of both internal and external knowledge, the number of organizations that reported insufficient knowledge is not the sum of the number of organizations that reported insufficient internal and external knowledge.

**Table 1.** Overview of Barriers Mentioned by Respondents

| Rank Barrier | Number of Organizations | Percentage of Organizations |
|---|---|---|
| 1. Insufficient Knowledge | 18 | 32.1% |
|     Insufficient internal knowledge | 13 | 23.2% |
|     Insufficient external knowledge | 6 | 10.7% |
| 2. No further barriers | 15 | 26.8% |
| 3. Limited functionality | 9 | 16.1% |
| 4. Management guidelines | 8 | 14.3% |
| 5. Insufficient resources | 7 | 12.5% |
| 6. Dependency on vendor | 3 | 5.4% |
| 7. Not a goal | 2 | 3.6% |
| 8. Satisfaction with proprietary software | 1 | 1.8% |
| Total organizations responding: | 56 | |

Note: percentages do not add up to 100% since organizations could mention more than one barrier.

that *"currently no suppliers offer/support OSSS"*. It was verified that this specific respondent was not interested in using OSSS in the organization, and can therefore be expected to have accumulated little knowledge about OSSS.

**No Further Barriers.** Some organizations explicitly mentioned that they could not identify any barriers to the further adoption of OSSS.[2] Although this is not a real barrier that inhibits the use of OSS, it is interesting to consider this factor as it identifies those organizations that have overcome all barriers to adoption. For one organization, the absence of any remaining barriers appeared to have paved the way to make OSSS the preferred technology for the organization, as the respondent wrote: *"We don't have any reason to prevent future use of OSSS! We have the intention to use OSSS as much as possible!"*.

**Limited Functionality.** Several organizations mentioned the limited functionality of OSSS as a major barrier to adoption. The most important remark in this regard was the poor interoperability with proprietary applications. For example, one respondent mentioned *"compatibility issues on file-level and application-level with Microsoft software"*. One organization from the financial sector also raised concerns with respect to the *"necessary security requirements in a financial environment"*.

**Management Guidelines.** The attitude of managers may also impede the use of OSSS. Several respondents expressed that managers in the organization were either *"unease about the use of OSS"* or simply *"not in favor of OSS"*. This

---

[2] Please note that we were not able to determine whether a blank reply meant that the respondent did not reply to the question, or that the respondent could not identify any barriers. It is therefore likely that our data slightly underestimates the number of respondents that could not identify any barriers.

appeared to be mainly due to a lack of trust in OSS. As one respondent mentioned: *"we work for a large organization who values the trusted vendors more than the cost savings of working with OSS"*. As a result, some organizations had a clear policy that did not allow the use of OSSS. The respondent in the financial organization mentioned above, for example, indicated that *"an in-house policy forbids the use of all OSSS"*. In organizations in which IT plays an important role, managers may also prefer to posit strict guidelines with respect to the use of IT. This was illustrated by one respondent as: *"the decision of the board to go with 'standardization', which seems to be Microsoft nowadays"*.

**Insufficient Resources.** Some organizations reported that a lack of resources (i.e., time, money, and human resources) inhibited the use of OSSS. For most organizations, the main issue in this respect seemed to be a lack of internal resources that are available to gain experience with emerging technologies. This was expressed by one respondent as: *"not enough manpower to try out new things"*. One respondent considered OSSS unsuitable for organizational use by saying that: *"we have no time for such hobbies"*. Another organization seemed to think that using OSSS still required much customization efforts by explaining that the use of OSSS would require *"too much investment to make tailor-made"*.

**Dependency on Vendor.** Some organizations mentioned that they could not adopt OSSS because they fully depended on a vendor for their software. Interestingly, this factor was mentioned by three public organizations, all so-called *Public Center for Social Welfare (PCSW)*. As expressed by one respondent, *"given the specific PCSW software, we are dependent on our software vendor"*. These organizations need to use specific software for their services, and only a limited number of vendors currently offer this software on the market.

**Not a Goal.** Two organizations mentioned that the further use of OSSS was not a goal in itself. One respondent indicated that the further use of OSSS was *"not a necessity"*. This seems to suggest that these organizations do not want to increase the use of OSSS per se, but rather decide on a project-per-project basis which solution (proprietary or OSS) is best suited.

**Satisfaction with Proprietary Software.** Finally, one organization mentioned that *"we are satisfied with our current proprietary software [operating system] and don't see the ROI or the major change in TCO to move to OSSS"*. If organizations are satifisfied with their existing systems, there may indeed be no compelling reason to consider the adoption of OSS.

### 3.2    Relationship with Assimilation of OSSS

Next, we determined whether the existence of certain barriers was related to the *assimilation stage* reached by the organization. The Guttman scale developed by Fichman and Kemerer [7] classifies organizations into 7 different assimilation stages. Organizations that are situated in stages 0 to 3 have not yet made a formal decision to adopt OSSS. They have progressed at most to the trial or

**Table 2.** Relationship between Assimilation and Barriers Reported

| Barrier | Number of organizations | | Percentage of organizations | | Fisher Exact Test |
|---|---|---|---|---|---|
| | NAD | AD | NAD | AD | |
| 1. Insufficient knowledge | 10 | 8 | 37.0% | 27.6% | .319 |
|    Internal knowledge | 7 | 6 | 25.9% | 20.7% | .441 |
|    External knowledge | 4 | 2 | 14.8% | 6.9% | .301 |
| 2. No further barriers | 5 | 10 | 18.5% | 34.5% | .148 |
| 3. Limited functionality | 4 | 5 | 14.8% | 17.2% | .547 |
| 4. Management guidelines | 3 | 5 | 11.1% | 17.2% | .395 |
| 5. Insufficient resources | 5 | 2 | 18.5% | 6.9% | .182 |
| 6. Dependency on vendor | 1 | 2 | 3.7% | 6.9% | .527 |
| 7. Not a goal | 2 | 0 | 7.4% | 0.0% | .228 |
| 8. Satisfaction with proprietary software | 1 | 0 | 3.7% | 0.0% | .482 |
|    Total number of organizations: | 27 | 29 | | | |

Note: percentages do not add up to 100% since organizations could mention more than one barrier.
Legend: NAD: non-adopter, AD: adopter.

evaluation stage, where the use of OSSS is still being considered. Organizations in the latter stages (4 to 6) have made a formal decision to adopt, and are using OSSS in a production environment. It therefore appears that stage 4 is a logical boundary between assimilation stages. For the purpose of this analysis, organizations that have not progressed beyond stage 3 (evaluation/trial) will be called *"non-adopters"*, while organizations that have reached at least stage 4 will be called *"adopters"*.

Table 2 shows how many adopters and non-adopters have mentioned each barrier. Similar to Table 1, we also calculated the *frequency effect size* for each barrier [22]. This effect size was calculated by dividing the number of non-adopters (or adopters) that mentioned a specific barrier by the total number of non-adopters (or adopters). These effect sizes are displayed in column 4 and 5 and indicate the relative importance of each barrier for each group.

In order to test whether there are statistically significant differences between the barriers reported by adopters and non-adopters, we constructed a 2 × 2 contingency table for each of the barriers by using the data from Table 2. The rows classified organizations according to whether they mentioned a specific barrier, while the columns distinguished between adopters and non-adopters. The *Fisher Exact Test* was used to test for significant relationships. The Fisher Exact Test can be used when the assumptions of the $\chi^2$-test are violated [26]. The use of this test was appropriate in this case given the use of a 2 × 2 table, the relatively small sample for each barrier, and an expected frequency below five for some cells [26]. The Fisher Exact Test returns the propability that exactly the same observed distribution or an even more disproportional distribution would be obtained by taking into account the row and column totals [26]. A one-tailed, directional test was performed since it is expected that adopters would report

**Table 3.** Relationship between Linux Adoption and Barriers Reported

| Barrier | Number of organizations NLU | LU | Percentage of organizations NLU | LU | Fisher Exact Test |
|---|---|---|---|---|---|
| 1. Insufficient knowledge | 13 | 5 | 39.4% | 21.7% | .499 |
| Internal knowledge | 9 | 4 | 27.3% | 17.4% | .220 |
| External knowledge | 5 | 1 | 15.2% | 4.3% | .665 |
| 2. No further barriers | 6 | 9 | 18.2% | 39.1% | .381 |
| 3. Limited functionality | 5 | 4 | 15.2% | 17.4% | .063 |
| 4. Management guidelines | 5 | 3 | 15.2% | 13.0% | .503 |
| 5. Insufficient resources | 6 | 1 | 18.2% | 4.3% | .418 |
| 6. Dependency on vendor | 0 | 3 | 0.0% | 13.0% | .512 |
| 7. Not a goal | 1 | 1 | 3.0% | 4.3% | .643 |
| 8. Satisfaction with proprietary software | 1 | 0 | 3.0% | 0.0% | .804 |
| Total number of organizations: | 33 | 23 | | | |

Note: percentages do not add up to 100% since organizations could mention more than one barrier.
Legend: NLU: non-Linux user, LU: Linux user.

fewer barriers. The output of the Fisher Exact Test for each barrier is shown in the final column in Table 2. As can be seen, none of these values are smaller than a critical level of $\alpha = .05$. Hence, we have to accept the null hypothesis that there is no significant difference between adopters and non-adopters with respect to any of the barriers.

We further investigated this relationship by considering whether organizations that mentioned a barrier exhibited a lower degree of OSSS assimilation than those organizations that did not report the barrier. To this end, a t-test for each barrier was conducted. None of these t-tests identified a significant difference at $\alpha = .05$.

### 3.3   Relationship with Extent of Linux Adoption

We also considered whether the extent to which the organization has adopted Linux had an influence on the barriers reported. The extent of Linux adoption was measured using a 7-point Likert scale, ranging from "*no usage*" to "*to a very large extent*". Based on this data, organizations were divided into two groups. The first group consisted of those organizations that did not make use of Linux, or only to a small extent (i.e., those who answered 1–3 on the 7-point Likert scale) and will be called "*non-Linux users*". The second group consisted of those organization who made use of Linux to at least a moderate extent (i.e., those who answered 4–7 on the 7-point Likert scale) and will be called "*Linux users*".

Table 3 shows a breakdown of the barriers reported by Linux and non-Linux users, as well as the frequency effect size for each barrier. Similar to before, the Fisher Exact Test was performed for each barrier to test whether the barriers reported differed based on the extent of Linux adoption. As can be seen in

Table 3, none of the Fisher statistics are smaller than the critical value of $\alpha = .05$. Hence, we have found no significant difference between Linux and non-Linux users with respect to the barriers reported.

We subsequently performed a series of t-tests to investigate whether organizations that reported a barrier exhibited a lower extent of Linux adoption. Results showed only one significant difference at $\alpha = .05$. It was shown that organizations that reported that they could not identify any further barriers to the adoption of OSSS exhibited a higher extent of Linux adoption ($t = -2.028$, $df = 54$, $p = .048$).

## 4    Discussion and Conclusion

Several of the barriers discussed above were also mentioned in previous research (see e.g., [10,20,11,12,25,5,19]). However, an important result of this study is that the most important barrier to the further adoption of OSSS was insufficient knowledge on OSSS. Our data suggests that the lack of internal knowledge is more important than a lack of external knowledge. This can be explained by the fact that it can be rather difficult to reorient the knowledge base of the organization. Previous research has shown that the skills of employees are often brand-specific [29], or that the IT staff may resist a change towards a new platform if it goes against their "vested interests" [30,33]. The use of service providers is more flexible since they can be hired on an ad-hoc basis to support the organization in the adoption of OSSS.

Overall, our results indicate that there are important knowledge barriers involved in the adoption of OSSS. Knowledge barriers may occur during the adoption of knowledge-intensive technologies [1]. To overcome these barriers, organizations must engage in a process of organizational learning [1,7]. With respect to the adoption of OSSS, organizations may have to invest considerable learning effort if they have primarily experience with proprietary software. The assimilation of OSSS then becomes a process of organizational learning.

These results are very consistent with our findings from previous quantitative research that investigated which reasons influence the assimilation of OSSS [32]. Results showed that the assimilation of OSSS was primarily influenced by the knowledge available to organizations, which suggested the presence of knowledge barriers [32]. In addition, our previous research also showed that the availability of internal knowledge had a more important impact on the assimilation of OSSS than the availability of external knowledge [32]. The most important factor influencing the assimilation of OSSS was the presence of boundary spanners in the organization [32]. Such boundary spanners are important in overcoming the knowledge barriers involved in the adoption of new technologies [4]. The findings of this study therefore provide further support for our previous conclusions. This is remarkable since the aim and research design of both studies was quite different.

We further used statistical techniques to analyze our qualitative data. Our results showed that organizations that have adopted Linux to a higher extent tend to report more often that they cannot identify any remaining barriers to the further adoption of OSSS. This suggests that barriers to adoption are only overcome in the final stages of the assimilation process. This illustrates the importance of also considering which barriers exist for organizations that have already adopted OSSS. In addition, we were not able to find any statistically significant indications that the barriers reported by organizations were related to their degree of OSSS assimilation or their extent of Linux adoption. This implies that these barriers remain an issue during the whole assimilation process. Hence, our data does not suggest that organizations are effectively able to overcome these barriers to adoption. This means, for example, that a lack of knowledge continues to be an important problem for organizations when increasing their assimilation of OSSS.

## 4.1  Contributions

The main contribution of this study is that it addresses a topic that has not been previously addressed in literature, namely the barries that exist to the further adoption of OSS. We provided an overview of which barriers exist to the adoption of OSSS by using a large-scale sample including both organizations that did not adopt OSSS, and organizations that have adopted OSSS to at least some extent. By investigating the relationship between the existence of these barriers and the degree to which the organization has adopted OSSS, it was shown that these barriers remain important in the whole assimilation process of OSSS. This highlights the importance of also considering the barriers that limit the further adoption of OSSS.

A second contribution is that we confirmed the results from our previous quantitative study. This is noteworthy since a very different approach was used in this paper. A first difference is that our previous study focused on determining the factors that influence the assimilation of OSSS. In our present study, we were concerned with which barriers inhibited the further use of OSSS. It has been noted in literature that adoption and non-adoption are two fundamentally different phenomena [21,9]. A second difference is that our previous study used quantitative techniques to analyze the data, while a combination of qualitative and quantitative techniques were used in this paper. The fact that the results from both studies using multiple methods are very consistent increases the (nomological) validity of our results.

An important practical implication of this study is that organizations should consider the adoption of OSSS to be a learning process. This learning process is required to overcome the knowledge barriers associated with the adoption of OSSS. This implies that organizations should invest sufficient internal resources to support this learning process, instead of relying exclusively on a service provider. By having sufficient internal knowledge, the assimilation of OSSS can be facilitated. Decision makers can also take initiatives to foster the acquisition and exploitation of internal knowledge. Since boundary spanners have found to be important in overcoming knowledge barriers, decision makers could try to stimulate the emergence of informal boundary spanners and seek their input during the adoption process. In addition, decision makers should be aware of the existence of these knowledge barriers when considering the adoption of OSSS. This means that they should not only consider the advantages that the adoption of OSSS could offer to the organization (e.g., lower cost, the availability of the source code, or the reduction of vendor lock-in), but should also take into account whether the organization has the ability to acquire the knowledge required to use OSSS.

## 4.2   Limitations and Future Research

This study has a few limitations that provide opportunities for future research. A first limitation is that we did not obtain the perception of each organization with respect to each of the barriers reported in Table 1. Instead, organizations were asked to report any perceived barriers in a free text field. This way, we only obtained those barriers that spontaneously came into the mind of the respondent, without further probing for their opinion on other barriers. This may have had an impact on our results. Future research may therefore take the list of barriers identified in this study as a starting point, and measure the perception of organizations towards each barrier. Finally, the external validity of our study is limited in the sense that the scope of our study was limited to Belgian organizations and OSSS. It would therefore be useful to replicate this study in other regions and by using a different set of OSS products (e.g., OSS desktop products such as OpenOffice.org) to see to which degree our results can be generalized.

## References

1. Attewell, P.: Technology diffusion and organizational learning: The case of business computing. Organization Science 3(1), 1–19 (1992)
2. Auerbach, C.F., Silverstein, L.B.: Qualitative Data: An Introduction to Coding and Analysis. Qualitative Studies in Psychology. New York University Press, New York (2003)
3. Bazeley, P.: The contribution of computer software to integrating qualitative and quantitative data analyses. Research in the schools 13(1), 64–74 (2006)
4. Cohen, W.M., Levinthal, D.A.: Absorptive capacity: A new perspective on learning and innovation. Administrative Science Quarterly 35(1), 128–152 (1990)
5. Dedrick, J., West, J.: Why firms adopt open source platforms: A grounded theory of innovation and standards adoption. In: King, J.L., Lyytinen, K. (eds.) Proceedings of the Workshop on Standard Making: A Critical Research Frontier for Information Systems, Seattle, WA, December 12–14, pp. 236–257 (2003)
6. Fichman, R.G.: The diffusion and assimilation of information technology innovations. In: Zmud, R. (ed.) Framing the Domains of IT Management: Projecting the Future Through the Past, pp. 105–128. Pinnaflex Educational Resources, Cincinnati (2000)
7. Fichman, R.G., Kemerer, C.F.: The assimilation of software process innovations: An organizational learning perspective. Management Science 43(10), 1345–1363 (1997)

8. Fitzgerald, B., Kenny, T.: Open source software in the trenches: Lessons from a large scale implementation. In: March, S.T., Massey, A., DeGross, J.I. (eds.) Proceedings of 24th International Conference on Information Systems (ICIS 2003), Seattle, WA, December 14–17, pp. 316–326. Association for Information Systems, Atlanta (2003)
9. Gatignon, H., Robertson, T.: Technology diffusion: An empirical test of competitive effects. Journal of Marketing 53(1), 35–49 (1989)
10. Goode, S.: Something for nothing: Management rejection of open source software in Australia's top firms. Information & Management 42(5), 669–681 (2005)
11. Holck, J., Larsen, M.H., Pedersen, M.K.: Identifying business barriers and enablers for the adoption of open source software. In: Proceedings of the 13th International Conference on Information Systems Development, Vilnius, Lithuania, September 9–11 (2004)
12. Holck, J., Larsen, M.H., Pedersen, M.K.: Managerial and technical barriers to the adoption of open source software. In: Franch, X., Port, D. (eds.) ICCBSS 2005. LNCS, vol. 3412, pp. 289–300. Springer, Heidelberg (2005)
13. Huysmans, P., Ven, K., Verelst, J.: Reasons for the non-adoption of openoffice. org in a data-intensive public administration. First Monday 13(10) (2008)
14. Larsen, M.H., Holck, J., Pedersen, M.K.: The challenges of open source software in IT adoption: Enterprise architecture versus total cost of ownership. In: Proceedings of the 27th Information Systems Research Seminar in Scandinavia (IRIS27), Falkenberg, Sweden, August 14–17 (2004)
15. Li, Y., Tan, C.H., Teo, H.H., Siow, A.: A human capital perspective of organizational intention to adopt open source software. In: Avison, D., Galletta, D., DeGross, J.I. (eds.) Proceeding of the 26th Annual International Conference on Information Systems (ICIS 2005), Las Vegas, NV, December 11–14, pp. 137–149. Association for Information Systems, Atlanta (2005)
16. Lundell, B., Lings, B., Lindqvist, E.: Perceptions and uptake of open source in Swedish organisations. In: Damiani, E., Fitzgerald, B., Scacchi, W., Scotto, M., Succi, G. (eds.) Open Source Systems, IFIP Working Group 2.13 Foundation on Open Source Software, Como, Italy, June 8–10, 2006. IFIP International Federation for Information Processing, vol. 203, pp. 155–163. Springer, Boston (2006)
17. Miles, M.B., Huberman, A.M.: Qualitative Data Analysis: An Expanded Sourcebook, 2nd edn. Sage Publications, Thousand Oaks (1994)
18. Miralles, F., Sieber, S., Valor, J.: An exploratory framework for assessing open source software adoption. Systèmes d'Information et Management 11(1), 85–111 (2006)
19. Morgado, G., van Leeuwen, M., Özel, B., Erkan, K.: Current status of F/OSS (2007),
http://www.tossad.org/content/download/1385/6894/file/tOSSad_D18_V2.3.pdf
20. Morgan, L., Finnegan, P.: How perceptions of open source software influence adoption: An exploratory study. In: Österle, H., Schelp, J., Winter, R. (eds.) Proceedings of the 15th European Conference on Information Systems (ECIS 2007), St. Gallen, Switzerland, June 7–9, pp. 973–984. University of St. Gallen, St. Gallen, Switzerland (2007)
21. Nabih, M.I., Bloem, S.G., Poiesz, T.B.: Conceptual issues in the study of innovation adoption behavior. Advances in Consumer Research 24(1), 190–196 (1997)
22. Onwuegbuzie, A.J.: Effect sizes in qualitative research: A prolegomenon. Quality and Quantity 37(4), 393–409 (2003)

23. Onwuegbuzie, A.J., Dickinson, W.B.: Mixed methods analysis and information visualization: Graphical display for effective communication of research results. The Qualitative Report 13(2), 204–225 (2008)
24. Onwuegbuzie, A.J., Teddlie, C.: A framework for analyzing data in mixed methods research. In: Tashakkori, A., Teddlie, C. (eds.) Handbook of Mixed Methods in Social and Behavioral Research, pp. 351–383. Sage Publications, Thousand Oaks (2003)
25. Paré, G., Wybo, M., Delannoy, C.: Barriers to open source software adoption in quebecs health care organizations. Journal of Medical Systems 33(1), 1–7 (2009)
26. Pett, M.A.: Nonparametric Statistics for Health Care Research: Statistics for Small Samples and Unusual Distributions. Sage Publications, Thousand Oaks (1997)
27. Saldaña, J.: The Coding Manual for Qualitative Researchers. Sage, Los Angeles (2009)
28. Sandelowski, M., Voils, C.I., Knafl, G.: On quantitizing. Journal of Mixed Methods Research 3(3), 208–222 (2009)
29. Shapiro, C., Varian, H.R.: Information Rules: A Strategic Guide to the Network Economy. Harvard Business School Press, Boston (1999)
30. Swanson, B.E.: Information systems innovation among organizations. Management Science 40(9), 1069–1092 (1994)
31. Ven, K., Verelst, J.: The organizational adoption of open source server software by Belgian organizations. In: Damiani, E., Fitzgerald, B., Scacchi, W., Scotto, M., Succi, G. (eds.) Open Source Systems, IFIP Working Group 2.13 Foundation on Open Source Software. IFIP International Federation for Information Processing, vol. 203, pp. 111–122. Springer, Boston (2006)
32. Ven, K., Verelst, J.: The organizational adoption of open source server software: A quantitative study. In: Golden, W., Acton, T., Conboy, K., van der Heijden, H., Tuunainen, V. (eds.) Proceedings of the 16th European Conference on Information Systems (ECIS 2008), Galway, Ireland, June 9–11, pp. 1430–1441 (2008)
33. Zmud, R.W.: Diffusion of modern software practices: Influence of centralization and formalization. Management Science 28(12), 1421–1431 (1982)

# Reclassifying Success and Tragedy in FLOSS Projects

Andrea Wiggins and Kevin Crowston

Syracuse University School of Information Studies
Hinds Hall, Syracuse NY 13244 USA
awiggins@syr.edu, crowston@syr.edu

**Abstract.** This paper presents the results of a replication of English & Schweik's 2007 paper classifying FLOSS projects according to their stage of growth and indicators of success. We recreated their analysis using a comparable data set from 2006. We also expanded upon the original results by analyzing data from an additional point in time and by applying different criteria for evaluating the rate of new software releases for sustainability of project activity. We discuss the points of convergence and divergence from the original work from these extensions of the classification and their implications for studying FLOSS development using archival data. The paper contributes new analysis of operationalizing success in FLOSS projects, with discussion of implications of the findings.

## 1 Introduction

Much of the empirical analysis of FLOSS has been undertaken using bespoke data sets laboriously created for a single analysis. However, over the last few years, research teams have developed several repositories of FLOSS data that provide reliable curated data about FLOSS projects (4; 7; 9). Use of data from these repositories relieves researchers of the need to spider and parse data from project repository sites, increasing productivity while also avoiding errors from problems in the data collection processes. These repositories of repositories (RoRs, (8)) are seeing increasing use by researchers. Much of the prior research that employed large-scale data sources should be possible to recreate and extend using the data from RoRs, allowing the research community to build more quickly on past work to refine theories and methods in FLOSS research.

In this paper, we adopt this approach in replicating English & Schweik's (3) classification of project success and failure in open source projects. In this paper, we do not engage in a detailed critique of the classification; rather our goal is methodological development in the area of large-scale analysis of archival data from FLOSS repositories. We note though that objective methods for identifying FLOSS project success and failure is a topic of interest for both researchers and practitioners. Researchers need to identify successful and failed projects to be able to investigate the potential causes of success or failure. Practitioners are interested in being able to evaluate success for several reasons: first, this gives an individual decision information with respect to whether to rely upon or become involved with a given software project; second, it gives software foundation decision-makers useful information for determining whether to admit projects or invest resources into developing them; and third, it provides an assessment of the health of projects in which individuals and larger organizations are currently engaged.

P. Ågerfalk et al. (Eds.): OSS 2010, IFIP AICT 319, pp. 294–307, 2010.

In the following section, we describe the classification developed by English and Schweik (3). We then outline the methodology we adopted to recreate their results using data from the Notre Dame SourceForge Research Data Archive (SRDA) (9). We discuss our results in relation to the original work, examine the outcomes of varying a single classification criterion and look at the changes to project classifications over time. We then reflect on the methodological challenges involved in replicating large-scale analysis of archival data on open source projects. Finally, we conclude with directions for future work.

## 2    Theory: Assessing Project Success

Crowston et al. (2) note that for FLOSS projects, success is a multi-dimensional construct that can be assessed from many perspectives. The original classification by English and Schweik presents a set of six classes of FLOSS projects, operationalizations for which are reproduced in Table 1 from Table 1 in (3). English and Schweik developed the original criteria for their classification based on interviews with FLOSS developers and the thresholds for the classification originated with their initial manual coding of a sample of projects. (The original paper provides the full rationale for their definitions.) The classification has two facets: the stage of the project, either initiation (I, first year of the project or up to three releases) or growth (G, subsequent to initiation thresholds); and the outcome, either success (S) or failure, which English and Schweik labeled as "tragedy" (T) in reference to the tragedy of the commons. In addition, projects might be labeled as in an indeterminate state (I) if success or failure cannot yet be determined. Projects were classified based on a number of factors, including age, releases and their timing, and downloads, which serves as a proxy measure for the creation of useful software. Finally, projects were labeled as unclassifiable if there is evidence that they may have distribution channels other than SourceForge, suggesting that the download or release count data are unreliable. The final column of Table 1 shows the operationalization we adopted in our reimplementation of the classification. In many cases, our criteria are identical, but we discovered a few necessary changes, as discussed below in the methods section.

### 2.1    Propositions

Turning to the substantive content of the paper, we present several propositions related to the three main areas of analysis, both methodological and theoretical.

1. First, we expect that our classification drawing on repository data will produce comparable results to those reported in the original work, possibly with some minor variations due to differences in sampling.
2. Second, we expect that the three variations in the classification criterion for the rate of releases, an indicator of project stability, will result in differences in classification, though these differences will be limited to specific classes, as not all classifications rely on this data.

**Table 1.** Six FLOSS success/tragedy classes and their methods of operationalization, from English & Schweik (2007)

| Class/ Abbreviation | Definition | Original Operationalization | Re-operationalization |
|---|---|---|---|
| Success, Initiation (SI) | Developers have produced a first release. | At least 1 release (Note: all projects in the growth stage are SI) | *Not explicitly classified: Sum of IG, SG and TG* |
| Tragedy, Initiation (TI) | Developers have not produced a first release and the project is abandoned. | 0 releases AND ≥ 1 year since SF project registration. | Same |
| Success, Growth (SG) | Project has achieved three meaningful releases of the software and the software is deemed useful for at least a few users. | 3 releases AND ≥ 6 months between [all] releases AND does not meet the download criteria for tragedy detailed in the TG description below. | ≥ 3 releases AND ≥ 6 months between most recent and third most recent release AND > 10 downloads |
| Tragedy, Growth (TG) | Project appears to be abandoned before producing 3 releases of a useful product OR has produced three or more releases in less than 6 months and is abandoned. | 1 or 2 releases AND ≥ 1 year since the last release at the time of data collection OR ≤ 10 downloads during a time period greater than 6 months starting from the date of the first release and ending at the data collection date OR 3 or more releases in less than 6 months AND ≥ 1 year since the last release | 1 or 2 releases AND ≥ 1 year since the most recent release OR 3 or more releases AND ≥ 1 year since most recent release OR ≤ 10 downloads[1] |
| Indeterminate, Initiation (II) | Project has yet to reveal a first public release but shows significant developer activity. | 0 releases AND < 1 year since SF registration | Same |
| Indeterminate, Growth (IG) | Project has not yet produced three releases but shows development activity OR has produced 3 releases or more in less than 6 months and it has been less than 1 year since the most recent release | 1 or 2 releases AND < 1 year since the most recent release OR 3 releases AND < 6 months between releases AND < 1 year since the most recent release | Same |

---

[1] Note: we used all-time downloads as this was operationally the same when combined with the release rate criterion.

3. Finally, we also expect to see change over time in the classification applied to individual projects, based on predictions about the potential next states of projects based on their current classification. For example, no project which has advanced to a growth stage (SG, TG, IG) can return to an initiation stage classification (TI, II). Projects that are II will not remain in that state for longer than a year, by definition, but may progress to any of the other classifications. In all cases where a project may become either a success or a tragedy in the next classification, we expect to see success less often than tragedy. As a result, as the number of projects grows, most classes should grow proportionally, but not all. We expect that over time, the number of tragedies will increase as a matter of accumulation of failures in the population as a whole, although we expect that the relative proportion of successes will remain stable. Further, we expect that the effects of time will lead to larger proportions of projects identified as being in the initiation stage, as the number of new projects grows, and this will in subsequent time periods lead to a slow increase in tragedies at the initiation stage.

# 3    Methods

The main features of the large-scale archival data analysis consisted of replication and extension of the original work. We applied an eScience strategy to conducting the analysis with respect to our choices of data sources and tools, and in the process created research artifacts in the form of processed data and analysis workflows that will support further extension of this work.

## 3.1    Replication

The replication of the original analysis required processing data from approximately the same time period to provide a suitable comparison. The extension of the work was intertwined with the replication, and involved preparing additional data for analysis of change over time along with the addition of new variations of the classification. As our goal with this research was methodological refinement, we selected and analyzed additional dates and variations to evaluate the performance of the classification, rather than to make a theoretically-informed evaluation of change to the community composition itself.

### 3.1.1    Data

In the original work by English and Schweik, the authors developed and tested their classification algorithm using FLOSSmole data from 2005, subsequently creating a sampling frame from the FLOSSmole project list of August 2006. They then spidered SourceForge around 16 October of 2006 for release information to augment the FLOSSmole data. However, the data spidered by English and Schweik do not include statistics for over 8,000 projects with incomplete data or that were deleted by SourceForge between August and September. The gap between collecting project data and release data might also invalidate a portion of the original analysis, as the period of up to two months between data collection times affects values for the thresholds for a project's active lifespan, achievement of maturity and release rate. For example, projects that made a first release in August or September could be misclassified as TG instead of IG.

The goal of our work was to demonstrate the use of a shared data repository for replication of the original research. Because of the limitations on the available data from FLOSSmole, as noted by English and Schweik, we selected SRDA as the data source, as it contained all of the necessary data for the classification at the time. In addition, we note that the SRDA data comes straight from SourceForge as a monthly database dump, which makes it an authoritative data source. FLOSSmole data are parsed from SourceForge HTML pages, and while the repository provides a reliable data source, its contents are one step removed from the original source. We analyzed data for October 2006 to match the data collection for releases in the original work; more specifically, we used the October 2006 release of SRDA data, which was captured on 23 October 2006.

### 3.1.2 Analysis

The analysis was replicated by careful examination of the original English and Schweik article, from which the requirements for data and processing were derived. A workflow for the data processing was developed in Taverna, a scientific workflow tool (1; 6). This approach made it easy to integrate data coming from several different sources, e.g., from the SRDA and a local cache of release data (previously retrieved from the SRDA), and to modularize the analysis steps. The analysis was implemented in several phases: a first phase to retrieve the basic data needed about project downloads and releases, followed by phase that ran simple tests on criteria to determine whether thresholds had been met, and finally a stage that classified the projects according to these tests.

For the replication of the classification, we found it challenging to determine the correct data selection or parameters for classification from the text of the English and Schweik article. Fortunately, the workflow implementation allowed us to remain flexible in our data selection and to parameterize the analysis. As we sought to reproduce the classification table in the published article in the format of a truth table to achieve completeness and exhaustiveness in the classification, we discovered that the classification as published was not complete. Some of the negative cases were not included in the published table, which is not to say that the authors did not consider these cases, perhaps because those combinations did not arise in the original data sample. For the most part, we were able to fill in the appropriate classifications for negative cases based on inheritance from other criteria for each class.

We also faced slight differences in the data available from the SRDA that required some changes to the operationalizations. For example, the SRDA did not have a convenient source for downloads within 6 months, though in the cases where that count was required, it was functionally equivalent to all-time downloads, which we used instead. The final column of Table 1 shows the operationalizations we developed and implemented in the workflows.

After debugging and verification of the workflow's performance with a subset of data, the final analysis was run with a full replicated data sample and compared to the original published research. In production, we eventually substituted SQL queries for the final two phases of the classification workflow due to significant economies in processing time. The analysis of the resulting data was conducted with R on the classified data stored in a SQL database. To encourage reuse of these data and analysis approaches, the workflows, SQL scripts and classification data will be made available to other researchers from our website, http://floss.syr.edu/.

As a result of using the SRDA data, the time required to retrieve and pre-process data were much lower than reported in original article. English and Schweik spent 22 days to spider the data for the analysis that were not already available in FLOSSmole. After optimizing our processes, we found that retrieving and preparing the data for classification requires less than 30 hours for data sets approaching 150,000 projects. In future efforts, this reduction in processing time will allow us to develop more granular analysis of changes to project classification over time by generating analysis-ready data for each monthly release of data from the SRDA.

## 3.2 Extending Analysis

In addition to replication, we wished to extend the prior work by examining stability of classification status over time, and evaluating the performance of two alternate means of operationalizing the rate of release.

### 3.2.1 Additional Dates

Our data processing times were significantly lower than reported for the original work, so we were able to analyze data for more than one date. Preparing data for analysis is the most time consuming part of the process, but by storing the prepared data, we are able to reuse it more readily with alternate classification schemes. In addition to the original date, we analyzed data from April 2006 to provide a short-term comparison that would help evaluate the stability of the classification over a relatively short period of time. We selected this time period because over a period of six months, there is opportunity for many projects to change status on several of the indicators, thereby affecting their classification status.

### 3.2.2 Implementing English and Schweik's Future Work

One of the more complex aspects of the original analysis is the qualification of release rate as an indicator of the sustainability of project activity. The assumption here is that projects which make releases too quickly cannot maintain the pace of activity; what this fails to take into account, however, is the wide diversity of release strategies employed by FLOSS projects. The original measure evaluates release rate by whether the project has had at least six months elapsed between first and last releases, which automatically privileges older projects rather than more stable projects. One solution, inspired by English and Schweik's discussion of these issues, is setting a threshold for the amount of the time between the most recent series of releases, rather than for all releases. A second option is to evaluate the average time between each release against a threshold, which applies the lifetime perspective of the original method, but seems likely to be more stable than the alternative that evaluates the time between only the most recent releases. We have implemented both of these variations, along with the original version, to evaluate the influence of this factor on project classification.

# 4  Results

We discuss the results of our analysis with three comparisons: comparison to the original published results, comparison of results from varying one classification criterion, and comparison of classification over time.

## 4.1  Comparison to Original Published Results

Our results for data from October 2006, using the same default values for the classification thresholds, are compared with the original results for the same time period in Table 2. We note that as percentages of the classified projects, our results are similar for the classes of II, IG, TG, and SG. They are also remarkably close in values for the number of "unclassifiable" projects.

Potential causes for variation in results could be discrepancies in the release and download data, which English and Schweik retrieved from different sources and at different times, as previously discussed. Variations in release data in particular would be problematic, as this could affect the determinations of whether or not the project is active, whether it has had enough releases, and whether or not the releases have occurred too quickly to be considered sustainable.

**Table 2.** Comparison of classification results to original results from English and Schweik

| Class | Original results | Replication Results | Difference |
|---|---|---|---|
| unclassifiable | 3,186 | 3,296 | +110 |
| II | 13,342 (12%) | 16,252 (14%) | +2,910 (+2%) |
| IG | 10,711 (10 %) | 12,991 (11%) | +2,280 (+1%) |
| TI | 37,320 (35%) | 36,507 (31%) | -813 (-4%) |
| TG | 30,592 (28 %) | 32,642 (28%) | +2,050 (+0%) |
| SG | 15,782 (15%) | 16,045 (14%) | +263 (-1%) |
| other | 8,422 | — | |
| *Total* | 119,355 | 117,733 (+ 9.6%) | |

Another variation between these results is that we produced no "other" classifications. We did not run into problems with differences between sampling frames and actual data that we were able to collect, as there was no delay between sampling and data collection. More specifically, we have not sampled so much as taken a census, as we have used all of the available data for each time period. The discrepancies in total numbers of projects, approximately 1,600 fewer in our sample, could also result from the deletions of inactive projects that the authors cited as a cause for the "other" projects; however, we were able to classify a larger number of projects overall. The date for the SourceForge dump upon which our analysis is based is slightly later in the month of October than the original analysis data collection time period (and two months later than the collection of project statistics), but in our case, we have no record for projects that were deleted from the SourceForge system. Overall, we consider the replication successful, as the greatest variation in classification by proportion is in the TI category, with a relative difference of just 4%.

## 4.2   Comparison of Release Rate Criteria

As discussed previously, we implemented three different variations for judging the sustainability of the rate at which a project is making releases. The original article called for a period of at least 6 months between the most recent release and the first in the window of three releases. English and Schweik mentioned examining the time between each release, and based on this idea, we implemented a density-based calculation of the time elapsed over the most recent three releases (Method Two). The final variation compares the average time between all releases in a project against a threshold (Method Three); notably, this is a more strict definition than the original and may merit a different threshold value. The original implementation allowed an average of three months between releases in the case of the minimum qualifying number of releases for evaluating success; the results are reported for a six-month threshold in Table 3.

**Table 3.** Classification outcomes from varying release lag measures for each time period, using a six month threshold

| 2006-10-23 | Method One | Method Two | Method Three |
|---|---|---|---|
| IG | 12,991 (11%) | 14,310 (12%) | 19,235 (16%) |
| II | 16,252 (14%) | 16,252 (14%) | 16,252 (14%) |
| SG | 16,045 (13%) | 15,426 (13%) | 3,143 (3%) |
| TG | 32,642 (28%) | 31,942 (27%) | 39,300 (33%) |
| TI | 36,507 (32%) | 36,507 (32%) | 36,507 (32%) |

It is clear that variations in this criterion result in a reclassification of SG projects as IG or TG projects. This occurs when the release lag evaluation comes out as "too fast" to be considered sustainable (IG), or the project has not made a release in the last year (TG). In all other ways, these projects are judged successful according to the other classification criteria—they have achieved "enough" releases and the software has been downloaded. This reclassification effect is exaggerated in the comparison between Method One and Three because the difference in calculation method suggests that a different threshold value should have been used.

While this change to the release rate criterion seems to have a small effect with the recent release density function (Method Two) and a larger effect with the averaging over all releases function (Method Three), comparison of the project-level classifications tells another story. We find that even at the six-month threshold, Methods One and Three are most consistent with respect to results; in every case, the changes to a project's classification is a transition from SG to IG or TG. However, applying Method Two yields changes from SG to IG as well as from IG to SG, and likewise with TG. In addition, more classifications are changed with Method Two than with Method Three, so the apparently smaller change in summary statistics masks a larger change in classification at the project level.

## 4.3   Comparison over Time

The comparison of classification results over time suggests interesting directions for future analysis at a more granular level. Table 4 compares the counts of projects at two points in time, April and October of 2006. The consistency in SG classifications over time demonstrates underlying regularity with respect to the proportion of projects which can meet the classification criteria for success. We also see relative stability in the IG and II classes. We also observe growth of the TG class over time as it gradually accumulates failures, as we would in fact expect.

Interestingly, the changes between these two periods show small increases in the proportions of most classifications, with a notable decrease in the frequency of the TI classification. This suggests that more projects are successful at making at least one release than odds would suggest. All other things equal, we might expect to small net increases across all categories, but this would not take into consideration the compound effects of change over time; the decrease in TI projects would suggest that there is a substantial number of projects which require more than one year to make their first release. Logically, these projects are most likely to become TG or IG projects.

**Table 4.** Classification outcomes from different time periods, using the original release rate criteria

| Class | 2006-04-21 | 2006-10-23 |
|---|---|---|
| IG | 12,166 (10.8%) | 12,991 (11.0%) |
| II | 13,592 (12.4%) | 16,252 (13.8%) |
| SG | 14,244 (12.7%) | 16,045 (13.6%) |
| TG | 28,777 (25.6%) | 32,642 (27.7%) |
| TI | 39,948 (35.5%) | 36,507 (31.0%) |
| unclassifiable | 3,343 (3.0%) | 3,296 (2.8%) |
| Total | 112,430 | 117,733 |

To describe the changes in project state between two points in time, we present the Markov model shown in Figure 1 that shows the percentage of projects that shift from one classification in April 2006 to a different one in October 2006. New projects are not included. Omitted from the diagram are the rates for projects remaining in the same classification: IG at 54%, II at 56%, SG at 100%, TG at 98%, and TI at 98%.

Initial observations from the model include the fact that once a project is classified a tragedy, it has a very low likelihood of escaping that classification. A TI project has a 1% likelihood of salvaging itself, while a TG project has a 2% chance–in both cases, these are not very good odds for survival. Likewise, once a project is labeled SG, there is no rescinding this title; this is inherent in the operationalization, because once the thresholds have been reached (adequate releases, downloads, an appropriate amount of time between releases), there is no criterion that will reverse them.

It comes as little surprise to see that the most common path from II, the default classification for a newly founded project, is to TI; for a new project, tragedy is four times more likely than moving on to growth in the IG class, with the potential for success. This confirms the common assertion that many projects are stillborn and never make a release, failing nearly immediately. IG projects are three times more likely to become tragedies than successes, but notably, this is the only route to success, and no project is an overnight success.

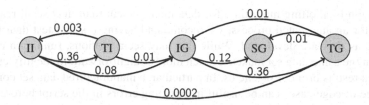

**Fig. 1.** Markov model showing changes in project classification over a six month period in 2006

## 5 Discussion

In this section, we discuss the implications of our findings and areas for future work. As the focus of this paper was primarily methodological, so are the implications. These are related to both the nature of the task for large-scale archival research, and to the substance of the task, classifying the successes and tragedies among FLOSS projects.

Large-scale analysis of FLOSS repositories sounds simple in conversation, but is never so straightforward in practice. This effort built on prior work that eased the development of data handling scripts and functionality, and yet demonstrated time and again that this style of analysis requires thorough knowledge of the data sources and analysis operations. Exceptions in the data source can wreak havoc with automated processing; while the data can simply be processed again, this can be very inefficient. Although it may only require 30 hours to analyze a snapshot of the entire SourceForge population, unexpected variations in source data realistically double or treble the time required to prepare data, once data verification procedures and troubleshooting are included, so permitting adequate time for development of the data handling is essential to this type of analysis. In the process of managing the flow of data between multiple workflows, or between workflows and other semi-automated processes, we also found that it is particularly helpful to use exactly the same names for variables in every location where they are used, from the database fields to the workflows and R scripts.

It is apparent that the challenges of working with the data sources have not changed significantly since the original classification by English and Schweik, although our tools and strategies are somewhat improved. We found that workflows were particularly good for dealing with retrieving and combining data from diverse sources when a single SQL query is not feasible. When preparing the data involve frequent and repeated format conversions (e.g., epoch times to SQL times and back again), sophisticated selection criteria (e.g., the most recent three releases), and careful handling of missing values, workflows are a better solution than most alternate procedures, particularly as they can be applied in precisely the same ways many times over, despite the complexity of the task.

One complication to analysis, as others have observed (5), is that the repository data structures represented in the SRDA can and do change over time. For example, all-time downloads data (used in the classification) are no longer available from the most recent monthly SRDA dumps, so a single analysis script will not work on data from different periods, complicating longitudinal analysis.

In addition to allotting more time for data management than may seem reasonable to expect for an automated process, we recommend having a set of test data for testing scripts as they are developed. While this may seem obvious, running a complex analysis on the same data set makes it much more straightforward to identify causes of unexpected results or inconsistencies. In particular, a manufactured data set containing a full range of edge cases can be useful in detecting errors in the script before scaling up analysis.

Although we have been tried to ensure that our algorithm matches the logic of the original paper as precisely as possible, we noted a number of cases that do not logically fit into the classifications that they are assigned. This kind of effect is difficult to trace back to its causes, which may include any of several issues such as bad data, incorrect implementation of the classification or a missing case in the classification. We suspect that some of the exceptions we have observed are cases that are illogical combinations of criteria that were not explicitly addressed in the original definition of the classifications. For example, the original classification does not cover projects that have downloads without releases (seemingly impossible), or how to classify a once-successful project that has long since fallen inactive. There would be no reason to expect that these configurations exist until they emerge as anomalies in the analysis, noticeable only through comparison to the expected results.

Finally, while our focus has been on methodological issues, our analysis does suggest some possible improvements to the English and Schweik classification. First, the influence of release rate and count thresholds is significant, and these are relatively dynamic measures compared to the other criteria, as our evaluation of variations in the release rate criterion demonstrated. Successes were reclassified as tragedies or as indeterminate; those which had not made a release in a year became tragedies, while those whose release rate was too swift became indeterminate. This suggests that additional testing and refinement of release-based criteria is primary task for improving the classification.

Second, as Figure 1 shows, a project classified as a success remains a success, even if it becomes inactive. This is a conservative classification choice, reflecting the reality that a successful project may not continue releasing new software indefinitely, but may enter a third stage of "retirement" in which it is still useful but is no longer under active development. This is a different state than one-time successes which are later abandoned and fall into disrepair; finding a way to distinguish between these cases would add significant nuance to our understanding of the lifecycle of open source project development.

## 5.1   Limitations

The results reported here are limited by the data sources and analysis methods, which are conversely also strengths in this analysis. This study is limited in generalizability in the same ways as the original work: neither result can be generalized to other repositories beyond SourceForge, and both are subject to flaws in the data sources. Both apply reductionist methods for operationalizing heuristics that are expected to indicate the development and success of FLOSS projects. While this mode of analysis has the advantage that it can be applied broadly, it also loses some face validity in broad application due to the existence of numerous individual examples (which may or may not be edge cases) that defy the assumptions embedded in the categorization of projects.

Limitations specific to this analysis include the change in data sources from the original, which introduces potential sources of error; however, we believe the SRDA data to be no less authoritative than combined FLOSSmole and bespoke data. Although we did not test additional variations on most of the parameters with the type of sensitivity analysis that large-scale analysis permits, doing so is simply a matter of choosing parameters and allotting processing time.

Finally, any classification of project success and failure will be challenging to validate empirically. English and Schweik developed their criteria for classification based on interviews with developers. However, once this classification has been applied to thousands of FLOSS projects, empirical validation becomes particularly challenging, as there is no established success metric against which to objectively evaluate the results. Feedback from the developer community on the results of the classification would provide a measure of validation; however, this method does not scale effectively. We note this limitation to both our work and the original classification as an opportunity for further development of the research on success in FLOSS projects.

## 5.2  Future Work

Replicating this classification using methods specifically oriented to further reuse of the data and analysis makes it feasible to consider a wide range of potential extensions to this work. More exhaustive testing of the threshold values is the most apparent direction for further refinement of the classification. In addition, taking advantage of the infrastructure to evaluate alternate measures of the classification criteria would permit the development of more sophisticated measurements. Just as we have tested variations on the release rate criterion, another possible variation could implement a function to adjust the threshold for downloads based on project lifespan or number of releases, which might better account for the current usefulness of the software.

While the operationalizations vary by the degree to which they capture the definitions, it is a nontrivial challenge to acquire and prepare data that would permit more accurate operationalization. For example, the definition of Indeterminate Initiation (II) is that the project has no public release but has significant developer activity; the operationalization is that the project has no releases and was founded less than a year before the data collection date. A more ideal operationalization would include explicit evidence of developer activity, such as communications or CVS activity. However, neither the original analysis nor our replication makes use of such data. The complexity of integrating additional data sources to produce such an analysis has been a barrier to developing more nuanced analysis, but our methodology and research artifacts can provide an extensible foundation for future work with more sophisticated measures.

As suggested by English and Schweik, incorporating data from CVS, email lists, and fora would create new potential for evaluating project activity. Rather than classifying projects as active or inactive based on having made a release within the last year, they could be classified based on the relative level of activity across a variety of channels for participation. While these data are available, incorporating them is tricky, and reshaping the classification criteria to make use of a greater variety of data would pose an interesting challenge. Refining a classification scheme such as this has the inherent problem that there is no objective way to determine what criteria are "best." We suggest that changes should be made based on improved congruence between definition and operationalization, and robustness of the measures to perturbation.

Finally, future work could more closely examine the shifts in classification over time. This effort would serve two goals: first, to further optimize the classification by identifying criteria that are more dynamic and potentially less representative, and second, to identify common developmental trajectories of FLOSS projects by charting their classifications across time. While the second goal is in many ways more attractive for researchers than the first, we note that refinement of the classification is key to generating valid results that can help us understand the implications of changes to project status over time.

## 6  Conclusions

In this paper, we replicated a classification of FLOSS project success and tragedy for all projects hosted on SourceForge at two points in time. The contributions of the work include the extension of the analysis, both in methods and in data analyzed. We extended the analysis to test three different methods of evaluating release rates; we tentatively suggest that the method that applies the average time between releases is the most stable and consistent with the intent of the analysis. We analyzed the data for two time periods, finding that the proportion of successful projects remained steady, while the number of tragedies appears to slowly increase over time in greater proportions than overall growth in sample size would predict. The implications of the work include recommendations for best practices for conducting large-scale analysis of archival FLOSS project data, and several suggestions for future work to develop the classification into a robust tool for research and practice.

**Acknowledgements.** The authors gratefully acknowledge James Howison's extensive contributions to the data and analysis infrastructure employed in this research. This research was partially supported by the United States National Science Foundation CRI Grant 07–08437.

## References

[1] Taverna project website, http://taverna.sourceforge.net/
[2] Crowston, K., Howison, J., Annabi, H.: Information systems success in free and open source software development: Theory and measures. Software Process Improvement and Practice 11(2), 123–148 (2006)
[3] English, R., Schweik, C.: Identifying success and tragedy of FLOSS commons: A preliminary classification of Sourceforge. net projects. In: Proceedings of the First International Workshop on Emerging Trends in FLOSS Research and Development, p. 11. IEEE Computer Society, Los Alamitos (2007)
[4] Howison, J., Conklin, M., Crowston, K.: FLOSSmole: A collaborative repository for FLOSS research data and analyses. International Journal of Information Technology and Web Engineering 1(3), 17–26 (2006)
[5] Howison, J., Crowston, K.: The perils and pitfalls of mining SourceForge. In: Proceedings of the International Workshop on Mining Software Repositories (MSR 2004), pp. 7–11 (2004)

[6] Hull, D., Wolstencroft, K., Stevens, R., Goble, C., Pocock, M.R., Li, P., Oinn, T.: Taverna: A tool for building and running workflows of services. Nucleic Acids Research 34(suppl. 2), W729–W732 (2006)

[7] Robles, G., Koch, S., Gonzalez-Barahona, J.: Remote analysis and measurement of libre software systems by means of the CVSAnalY tool. In: Proceedings of the 2nd ICSE Workshop on Remote Analysis and Measurement of Software Systems (RAMSS), Edinburg, Scotland, UK, pp. 51–55 (2004)

[8] Sowe, S., Angelis, L., Stamelos, I., Manolopoulos, Y.: Using Repository of Repositories (RoRs) to study the growth of F/OSS projects: A meta-analysis research approach. International Federation for Information Processing 234, 147 (2007)

[9] Van Antwerp, M., Madey, G.: Advances in the SourceForge Research Data Archive (SRDA). In: Fourth International Conference on Open Source Systems, IFIP 2.13 (WoPDaSD 2008), Milan, Italy (2008)

# Three Strategies for Open Source Deployment: Substitution, Innovation, and Knowledge Reuse

Jonathan P. Allen

University of San Francisco, 2130 Fulton St., CA 94117, USA
jpallen@usfca.edu

**Abstract.** As open source software adoption becomes mainstream, the question shifts from whether organizations should use open source, to how organizations can best deploy and use open source. Based on three distinct types of organizational outcomes for open source use, we propose three different strategies for deploying open source: a substitution strategy, an innovation strategy, and a knowledge reuse strategy. Limiting the deployment of open source to a substitution strategy can lead organizations to underestimate the strategic benefits of open source use.

## 1  Introduction:  Is Using Open Source Different?

As open source software adoption by organizations continues to grow, open source is increasingly perceived as a 'normal' option, rather than as a strange new technology requiring special justification and extraordinary precautions.  In a way, this is a victory for advocates who have fought to have open source software judged by the same criteria as 'normal' proprietary software (e.g.,[10]). The very term 'open source' itself was invented, in part, to downplay the differences between community-built and traditional software, making open source seem more familiar and acceptable for organizational use than 'free' (as in 'freedom') software [12].

However, in the rush to make open source software appear 'normal', there is a risk that the potentially unique benefits of open source might not be fully considered. Forcing a disruptive innovation to compete using existing performance criteria, rather than along new dimensions where it excels, often puts the disruptive innovation at a disadvantage [3]. There continue to be cases where using open source as a direct replacement for proprietary software is easily justified (e.g., [6]). However, it can be difficult to make the case for 'ripping out' established proprietary software that 'already works' and replacing it with an open source equivalent that the organization has no experience with.  In these situations, it would be helpful to have a clear argument for other performance dimensions along which open source use might be superior. It might also be helpful to think of these new performance dimensions not only in terms of justification, but also in terms of strategic use.  How would open source allow us to do things differently from proprietary software use?  What difference does using open source make?

The case for open source in software development has been made elsewhere (e.g., [7]), along with choices for open source development strategies.  Here, we focus on

Ågerfalk et al. (Eds.): OSS 2010, IFIP AICT 319, pp. 308–313, 2010.

the choices for user organizations that are not primarily in the business of software development.   Based on three different types of organizational outcomes, we offer three strategies for open source deployment in user organizations.

## 2   Open Source Deployment:  Three Types of Outcomes

Open source has been seen as a revolutionary, disruptive force for software development (e.g., [2]), but debate continues as to whether the organizational outcomes of open source use are similar to those of traditional proprietary software (e.g., [8]). We find it helpful to distinguish between the typical business benefits that come from using open source as a substitute for proprietary software, and the innovation and knowledge sharing benefits that are unique to open source software.

### 2.1   Substitution

Substitution takes place when open source software is used to replace the equivalent proprietary software.   Studies of open source adoption suggest that organizations are motivated by the desire to replace costly proprietary software with open source equivalents, providing similar functionality and performance (e.g., [4]).   Some go further and argue that organizations ignore the 'ideological' dimensions of open source–such as having the freedom access to source code–and only focus on practical benefits such as functionality and cost (e.g., [13]).

Table 1. Open source deployment: Three types of organizational outcomes

| Type | Activity | Example | Outcomes |
|------|----------|---------|----------|
| Substitution | Open source used to replace equivalent proprietary software. | Microsoft Office is replaced by OpenOffice. | Direct business benefits from software use. |
| Innovation | Open source used as a platform for creating new applications. | A new product promotion website uses WordPress. | Increased rate of innovation within organizations. |
| Knowledge reuse | Open source used as a platform for sharing new applications. | A new distribution of Drupal or Joomla is shared among not-for-profits. | Increased rate of innovation sharing between organizations. |

W expect that the main outcomes of open source use for substitution will be cost reduction and increased functionality.   A typical example of substitution can be seen in the case of an Irish hospital, searching for software that was "zero cost or as cheap as possible." ([6], p. 54)  The main outcomes reported in this case were a 6.5 million Euro initial purchase savings, and 12 million Euros in total savings over a 5 year period.   In other examples of substitution, pure cost savings are not as important as selecting the 'best technology' for the job, usually in terms of functionality, reliability, or security (e.g., [16]).

These evolutionary (rather than revolutionary) outcomes from open source software use would be similar to those expected from the use of proprietary software. An open source software package might provide the same or better organizational benefits—return on investment, functionality, security, or standards compliance—as a corresponding proprietary package. For example, a Windows server could be replaced with a Linux server because it provides better total cost of ownership, or Firefox could be chosen as a browser over Internet Explorer for improved security.

## 2.2 Innovation

The second outcome in our framework, innovation, takes place when open source software is used as a platform or foundation for creating new applications within an organization. Open source software is a 'generative system' [25] that allows organizations to create new applications by building on the freely available work of the community. The use of open source leads to increased innovation because of the leverage it provides, its accessibility for experimentation, and its adaptability due to source code access and modular design.

Open source can increase the rate of innovation by providing frameworks and libraries for programmers, such as when Django or Rails are used to develop new web applications quickly. Open source repositories can be used to share software across projects within an organization (e.g., [11]). Open source applications such as Drupal, WordPress, Joomla, or SugarCRM have modular architectures that facilitate innovative new applications with little or no custom programming. One example is a new website built by the City of San Francisco in a few weeks using the WordPress platform, instead of through the usual lengthy development process [1].

The unique aspects of open source licensing allow successful experiments to quickly spread throughout the organization, without having to be constrained by strict licensing terms and their associated costs. Organizations can commit serious resources only after an innovation has proven itself.

## 2.3 Knowledge Reuse

Knowledge reuse is the "sharing of best practices or helping others to solve common technical problems" ([9], p. 59). As software, open source facilitates knowledge reuse not only through shared repositories of knowledge about facts, but also by sharing procedural knowledge—code that runs business processes.

Knowledge reuse comes from the sharing of organizational expertise through open source software. One type of knowledge reuse comes from creating explicit partnerships or alliances to jointly develop open source business applications, such as the substance abuse treatment system developed in Maryland and Texas and now adopted by other states [15]. A different type of knowledge reuse comes from the creation of distributions, or versions of open source software that are pre-configured for specific business applications. For example, the CiviCRM project configures open source content management systems for the specific needs of not-for-profit organizations. Other open source projects allow users to easily create and share add-ons for specific business applications, such as plug-ins for WordPress sites.

Knowledge reuse can be seen as the most revolutionary, or disruptive, type of outcome from open source use in organizations. It might seem difficult to imagine that organizations would freely reveal their novel business applications to others. And yet, this is what the research on 'user-centric' or 'democratized' innovation implies will happen (e.g., [14]), if open source business software becomes widespread. The open innovation literature suggests that user organizations, not enterprise software vendors, might someday provide the majority of innovations, share them freely, and pool their work with other user organizations, as they do in surprisingly many other industries. This could lead to the free sharing of organizational innovations and best practices, through the use of open source software as platforms. Open source may be much more than low-cost software. It could be a mechanism for sharing and reusing organizational knowledge.

## 3 Three Strategies for Open Source Deployment

Because the types of outcomes for open source use are fundamentally different, we expect that achieving different outcomes will require different strategies.

The substitution strategy is probably the most commonly used today. The substitution strategy is to evaluate and adopt open source software in exactly the same way as proprietary software. The advantage of this approach is that it fits the way organizations already make decisions. The disadvantage is that evaluation and use might not take advantage of the unique strengths of open source software. The substitution strategy might force organizations to 'rip out' proprietary software that 'already works' for an unproven open source equivalent with roughly the same features. The open source package could have initial cost advantages, but the substitution strategy burdens the open source case with the switching costs.

**Table 2.** Strategies for open source deployment and use in organizations

| Strategy | Action Plan |
|---|---|
| Substitution strategy | Replace current software with cheaper and/or better open source equivalents. |
| Innovation strategy | Focus on new applications or needs that are not being addressed by proprietary software. |
| Knowledge reuse strategy | Collaborate with projects, or partners, who are already innovating with open source. |

A different strategy is to focus on business needs that are not currently being addressed by proprietary software. Open source software can be deployed and used without many of the usual cost and license considerations that limit proprietary software use. If there are many business processes that could be improved by using new applications, but are individually too small to justify a full-scale proprietary software acquisition project, then an innovation strategy might be effective. Once open source has been brought in for 'experimental' or 'prototype' projects, growing experience and expertise with open source could lead to wider deployment. In the banking industry, Linux servers at first were not sold as a direct substitute. But as the years went by, and Linux servers 'just worked', it was easier to make the case for using Linux more widely. Open source applications for business can follow a similar path of guerrilla first, mainstream afterwards.

The knowledge reuse strategy uses open source to find bodies of valuable knowledge (software, and people) that have already been created, and join that community in order to facilitate your organization's ability to reuse and refine that shared knowledge. In contrast with the innovation strategy, which involves deploying open source software that already exists, the knowledge reuse strategy is an attempt to improve a software project's usefulness to a user organization through contributions and community interactions. This strategy opens the possibility of obtaining the full benefits of open innovation. And it addresses the risk of not having enough influence on an essential software platform's future direction, if an organization does not contribute to its ongoing evolution (e.g., [5]).

## 4   Conclusion:   The Promise of Open Source

We expect that, like many new technologies, open source is mostly understood and used in the same ways as the technology that came before it. Open source use that substitutes for proprietary software can have a significant impact by changing cost structures, or by preventing any one competitor from controlling a technology standard. But open source use for innovation can make a dramatic difference as well. Within organizations, it gives IT departments the ability to create new business applications that would never be practical otherwise, possibly dramatically improving the performance of business tasks. Open source use for innovation also allows organizations to launch new products or services that would not have been possible with the license restrictions of proprietary software.

However, the most revolutionary potential for open source use is when organizations decide to jointly develop and deploy open platforms. The extension of democratized innovation [14], generative systems [17], and peer-production [2] to enterprise applications could result in an explosion of knowledge sharing and reuse around basic business processes. When sharing organizational knowledge through software becomes not just 'a nice thing to do', but actually the more efficient and effective way to operate, we will have reached an important cross-over point where freely-revealed software becomes the norm, rather than the exception; where the majority of business software innovations come from the business that use it, rather than from proprietary enterprise software vendors.

## References

[1] Allen, J.P.: Open source deployment at the city and county of San Francisco: From cost reduction to rapid innovation. In: Proceedings of the 43rd HICSS Conference, Kauai, USA (2010)
[2] Benkler, Y.: The Wealth of Networks: How Social Production Transforms Markets and Freedom. Yale University Press, New Haven (2006)
[3] Christensen, C.M., Overdorf, M.: Meeting the challenge of disruptive change. Harvard Business Review 78(2), 66–77 (2000)

[4] Dedrick, J., West, J.: An exploratory study into open source platform adoption. In: Proceedings of the 37th Hawaii International Conference on Systems Sciences, IEEE, Los Alamitos (2004)

[5] Enkel, E., Gassmann, O., Chesbrough, H.: Open R&D and open innovation: Exploring the phenomenon. R&D Management 39(4), 311–316 (2009)

[6] Fitzgerald, B., Kenny, T.: Developing an information systems infrastructure with open source software. IEEE Software 21(1), 50–55 (2004)

[7] Grand, S., von Krogh, G., Leonard, D., Swap., W.: Resource allocation beyond firm boundaries: A multi-level model for open source innovation. Long Range Planning 37, 591–610 (2004)

[8] Kessler, S., Alpar, P.: Customization of open source software in companies. In: Proceedings of the 5th IFIP WG 2.13 International Conference on Open Source Systems. Springer, Heidelberg (2009)

[9] Markus, M.L.: Toward a theory of knowledge reuse: Types of knowledge reuse situations and factors in reuse success. Journal of Management Information Systems 18(1), 57–93 (2001)

[10] Open Source for America (2009), Charter for Open Source for America, http://opensourceforamerica.org/charter (accessed August 23, 2009)

[11] Riehle, D., Ellenberger, J., Menahem, T., Mikhailovski, B., Natchetoi, Y., Naveh, B., Odenwald, T.: Open collaboration within corporations using software forges. IEEE Software 26(2), 52–58 (2009)

[12] Stallman, R.: Why "open source" misses the point of free software. Communications of the ACM 52(6), 31–33 (2009)

[13] Ven, J., Verelst, J.: The importance of external support in the adoption of open source server software. In: Proceedings of the 5th IFIP WG 2.13 International Conference on Open Source Systems. Springer, Heidelberg (2009)

[14] von Hippel, E.: Democratizing Innovation. MIT Press, Cambridge (2005)

[15] Wanser, D.: Crossing state lines to build better software. Behavioral Healthcare 28(7), 19–23 (2008)

[16] Wheatley, M.: The myths of open source. CIO Magazine (March 1, 2004)

[17] Zittrain, J.: The Future of the Internet–And How to Stop It. Yale University Press, New Haven (2008)

# Coordination Implications of Software Coupling in Open Source Projects

Chintan Amrit and Jos van Hillegersberg

IS&CM Department, University of Twente,
P.O. Box 217
7500 AE Enschede, The Netherlands
{c.amrit,j.vanhillegersberg}@utwente.nl

**Abstract.** The effect of software coupling on the quality of software has been studied quite widely since the seminal paper on software modularity by Parnas [1]. However, the effect of the increase in software coupling on the coordination of the developers has not been researched as much. In commercial software development environments there normally are coordination mechanisms in place to manage the coordination requirements due to software dependencies. But, in the case of Open Source software such coordination mechanisms are harder to implement, as the developers tend to rely solely on electronic means of communication. Hence, an understanding of the changing coordination requirements is essential to the management of an Open Source project. In this paper we study the effect of changes in software coupling on the coordination requirements in a case study of a popular Open Source project called JBoss.

**Keywords:** Software Coupling, Propagation Cost, Clustered Cost, Open Source, Coordination.

## 1 Introduction

Open Source developers generally rely on electronic means of communication, coordination in Open Source environments is difficult to achieve when compared to commercial software development. It is therefore essential for an Open Source project Manager to understand the changing coordination requirements in Open Source software in order to ensure successful coordination. While the coordination implication of software coupling has been suggested by various researchers [2-5], there has been little research done on the effect of the change in coupling on the coordination requirements of developers. Such research is especially important in the Open Source context, where the distributed and generally ad-hoc nature of development makes coordination of the development challenging.

MacCormack et al. [6] compare the architectures of Linux and Mozilla by comparing the pattern of distribution of their software coupling. They find that Linux had a more modular structure than the first version of Mozilla. While after a redesign the resulting architecture, Mozilla became more modular than the previous versions and even more modular than Linux. This result is in line with the view that in order to have a successfully coordinated Open Source project one needs to have a loosely

Ågerfalk et al. (Eds.): OSS 2010, IFIP AICT 319, pp. 314–321, 2010.

coupled and modular software [7]. Authors like O'Reilly [8] have claimed that Open Source software is inherently more modular than commercial software. Other authors have reasoned that Open Source software needs to be more modular so that the development process can be coordinated easily [7]. Paulson et al. [9], compare the coupling of Open Source projects (Apache, Linux and GCC) with three closed source projects. They do so, by comparing the growing versus the changing rate for software (as a tighter coupling will require more changes with each additional feature). Their results indicate that Open Source projects need more changes when new features are added. Hence, suggesting tighter coupling in Open Source projects than previously assumed. Parnas [1] described modularisation as a task assignment while Conway[2] analysed the relation between product architecture and the organizational structure. Since then, Conway's law [10] has come to denote the homomorphism between the product architecture (or software coupling [3]) and the organizational structure (or the communication between the software developers [3]). As the Open Source project gets developed, the software code evolves [11], and as a result the coordination requirements change [3]. As mentioned earlier, there has been little research done on the effect that the variation of software coupling has on the coordination requirements of the software developers. In this paper we try and fill this gap by analysing the effect of the changes in software coupling on the coordination requirements of the developers. We postulate that, if there is a sudden increase in the coupling of an Open Source system, then the coordination requirement among the developers' increases. Unless this coordination requirement is handled through communication, it could result in a coordination problem [12]. By conducting a case study of the of the JBoss application server, we observe the effect of the changes in coupling on the coordination of the project. The unique contribution of this paper lies in discussing the coordination implications of an increase in software coupling and then in demonstrating it through a case study that uses quantitative along with qualitative methods.

The rest of the paper is structured as follows; section 2 describes the Design Structure Matrices briefly along with the Clustered and Propagation Cost metrics used in this paper, section 3 describes the case study of JBoss, section 4 discusses the findings and finally section 5 concludes the paper.

## 2 Design Structure Matrix and Cost Metrics

In this section we describe the data structure and the metrics we use to study software coupling. Dependency Structure Matrices (DSM) have been used in engineering literature to represent the dependency between tasks, since the concept of the Design Structure Matrix was first proposed by Steward [13, 14]. A DSM highlights the inherent structure of a design by examining the dependencies between its component elements in a square matrix [13, 15]. Morelli et al. [16] describe a method to predict and measure coordination-type of communication within a product development organization. They compare predicted and actual communications in order to learn, to what extent an organizations communication patterns can be anticipated.

Sosa et al.[4] find a "strong tendency for design interactions and team interactions to be aligned," and show instances of misalignment are more likely to occur across organizational and system boundaries. Sullivan et al. [17] use DSMs to formally model (and value) the concept of information hiding, the principle proposed by Parnas to divide designs into modules [1]. Cataldo et al.[3] show how DSMs can be used to predict coordination in a software development organization and then they compare the predicted coordination DSM with the actual communication DSM. Sosa [5] builds on the DSM based method of Cataldo et al. [3] and provides a structured approach to identify the employees who need to interact and the software product interfaces they need to interact about. Amrit et al. [12, 18] take a similar approach and use DSMs to detect coordination problems in a software development environments.

We use the Software Dependency Matrix (the DSM of software dependencies) to calculate the Propagation Cost and Clustered Cost similar to what MacCormack et al. [6] do. Our unit of analysis is the source code file and we consider the function call dependencies among the files.

While the Propagation Cost assumes that the cost of dependencies between two elements are the same irrespective of where the elements lie (the path length between them), Clustered Cost assumes that the cost of dependency depends on whether the elements lie in the same cluster [6]. Together the Propagation and Clustered Cost measure both the number as well as the pattern of the software dependency [6]. In order to calculate the Propagation Cost, MacCormack et al. first raise their dependency matrix to successive powers of n and obtain the direct and indirect dependencies for successive path lengths [6]. They then obtain a Visibility Matrix by summing up all the successive powers of the dependency matrix. From the Visibility Matrix they calculate the "fan-in" and "fan-out" visibilities by summing along the columns or the rows and dividing the result with the total number of elements. As we consider undirected dependencies, we find the "fan-in" visibility to equal the "fan-out" visibility. The Propagation Cost measures the elements in the system that could be affected when a change is made to one element of the system (i.e. how the change *propagates*) [6].

Unlike the Propagation Cost, the Clustered Cost of an element depends on the location of the element (with respect to other elements). In order to measure the Clustered Cost, the DSM of the software call graph has to be first clustered. The clustering algorithm (described in [6]) tries to group all highly connected or dependent elements into one cluster. The clustering works by attaching a cost to each element, depending on where the element is located with respect to other elements (in the same vertical bus or in the same cluster)). The Clustered Cost of the software is then the summation of the individual Clustered Cost of the elements.

In the next section we describe the case study of the popular open source project JBoss. In the case study we describe how we apply the two metrics described in this section and the conclusions we draw from them.

## 3   Case Study of JBoss

JBoss project was started in 1999 by Marc Fleury who wanted to advance his research interests in middleware. JBoss Group LLC was incorporated in 2001 and JBoss became a corporation in 2004. After a few bids from big companies, JBoss was finally acquired by Red Hat in 2006. The JBoss Application Server is one of the main products of the JBoss project and is said to have pioneered the professional Open Source business model. JBoss has 79 listed developers and three project administrators of which one is the Chief Technical Officer (CTO) of JBoss.

The aim of the case study is to determine if there was a relation between the changes in the technical dependencies and the communication among the developers. For the technical dependencies, the JBoss Application Server (JBoss) source code was analysed over the period starting from May 2002 to December 2006 that covered the versions 3.0.0 to 4.0.3_sp1. We used a tool called TESNA [12] that uses DependencyFinder [19] to read the software code and create the DSMs. With the help of TESNA we could then calculate the Propagation and Clustered Cost based on the DSMs. The Lines of Code (KLOC) of the different versions of JBoss was also measured using the same tool.

To determine the communication patterns used by the developers, we analysed the Mailing List archive of JBoss. The JBoss Mailing List is used to discuss the development of the system, report bugs, coordinate the bug fixes, as well as discuss new features before and after the release of each version. An analysis of the different mediums of coordination in JBoss revealed that the Mailing List was the primary means of coordination. This is the case, as the usage of private means to communicate is considered unlikely, given the trend of openness in Open Source projects [20]. In order to find out the timeline around which developers discussed a particular release, we needed to first find out the coordination mechanisms used by the developers. We performed a qualitative analysis of the messages in the Mailing List archive where we read randomly selected mails (around each release) looking for coordination mechanisms as described in previous literature. The following post mailed on 28th of June shows how the management of each release was undertaken by one of the Project Leaders (Scott Stark in this case).

> Its about 36 hours until I'm planning on cutting the 3.0.1 release. Any changes you want in 3.0.1 should be in by Sat Jun 29 18:00:00 2002 GMT.
> xxxxxxxxxxxxxxxxxxxxxxxxx
> Scott Stark

This post also shows that the planning for a release was done around a month earlier to the release, as the release date for version 3.0.1 was on 6th August 2002. While the following post shows another instance of a post reporting a fix for a bug.

> Sender: d_jencks
> Logged In: YES
> user_id=60525
> I believe I have fixed this in HEAD. I'd appreciate verification before I backport it to 3.2, since it is a substantial refactoring of the ejb deployment/service lifecycle code. I'll close this after backporting to 3.2.

This post shows two important mechanisms; (i) the request for verification implying the coordination mechanism of code review as was described by Rigby et al. [21], (ii) the one which d_jenks refers to as "backport". By "backport" the author refers to making changes to the previous version well after the release (2002-08-27). This coordination mechanism coincides with what was observed by Yamauchi et al. [22], namely, a bias towards action first and coordination later. Given that the planning for the release and the coordination for the bugs in the release was conducted around a month before and a month after the release respectively, we decided to consider the messages related to a release over a three month window. Hence, the Mailing Lists were analysed from one month before each release to one month after each release, corresponding to the period of analysis of the JBoss code (i.e. from April 2002 to January 2007). We decided to consider all the messages in the three months window, as messages dealing with the coordination of the community for the following reasons:

1)    The threads containing more than one message is naturally a discussion thread implying coordination between messages

2)    Threads containing only one message were mostly announcements such as "Build Fixed" that warrants no further replies. However, such posts are also coordination alerts for the community to not worry about the compilation part of the particular version and to concentrate on other work.

Figure 1 describes the variation of the Propagation Cost of JBoss over the different versions, while Figure 2 denotes the variation of the Clustered Cost of JBoss over different versions. In both figures and particularly in Figure 1 we notice a sharp rise in the Clustered Cost for version 3.2.7. While the increase in the Propagation Cost is

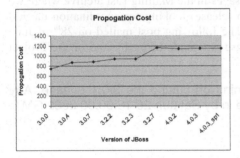
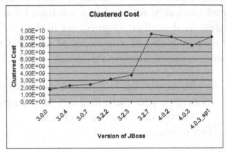

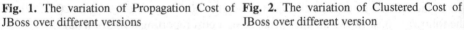

**Fig. 1.** The variation of Propagation Cost of JBoss over different versions

**Fig. 2.** The variation of Clustered Cost of JBoss over different version

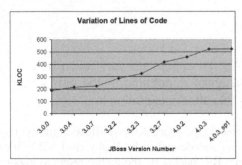
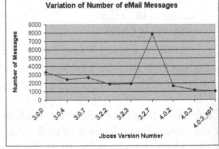

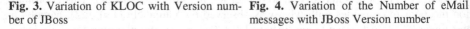

**Fig. 3.** Variation of KLOC with Version number of JBoss

**Fig. 4.** Variation of the Number of eMail messages with JBoss Version number

minor, the increase in the Clustered Cost for version 3.2.7 is quite marked. We calculated the KLOC (Lines of Code in thousands) of each of the versions to see how much code was actually added. Figure 3 shows the variation of KLOC over the different versions of JBoss. As can be seen from the figure, the trend is similar to the variation of coupling seen in Figures 1 and 2. The largest increase in KLOC, as evident from the slope of the graph in Figure 3, occurs for version 3.2.7. Clearly showing that for version 3.2.7 not only has the complexity of the code increased (with the increased coupling), but also the size.

This increase in modularity of the project causes an increase in the coordination requirement [3] and therefore require an increased amount of coordination to resolve the extra dependencies and features included for version 3.2.7.

Figure 4 describes the variation in the number of messages over the different versions of JBoss. We see a large increase in the number of messages for discussing the features and bugs for version 3.2.7. The increase in the number of messages is nearly 5000 and twice as much as the average number of messages (2650) discussing other versions.

## 4 Discussion and Conclusion

Though one needs to analyse the mails more closely to ascertain if they are indeed discussing the particular version, one can say with some confidence that this sharp increase in messages can be explained by the increased need for coordination. This increased need for coordination arises from the increased number of couplings and related features of JBoss in the release. Such an increase in the communication of the developers in the eMail List can indicate how the developers of JBoss satisfy the changing coordination needs for different versions and as a result remains a successful Open Source project. Had the coordination not increased to offset the increase in coupling and complexity of the software, we might have noticed a coordination problem as described by deSouza [23] and Amrit et. al [12].

In this paper we addressed the implications of coordination of an Open source project when the software coupling in the project changes. Clearly, the change in software coupling causes a change in the coordination requirements of the project as suggested by [2, 3, 6]. Unless this increase in the coordination requirement is compensated by an increase in communication related to the coordination, (as in the JBoss case study) one can expect consequences to the software quality of the project. Hence, this research has implications for the Open Source project manager. As such a manager has to be aware of the increased coordination requirement arising from changes in the project's software coupling.

The contribution of the research in this paper is twofold; (i) a discussion on the coordination implications of an increase in software coupling and (ii) the case study demonstrating the coordination implication using appropriate metrics like Propagation, Clustered Cost, KLOC and number of Mailing List messages. The email archive of JBoss also reveals two particular coordination mechanisms used to coordinate the development of JBoss, namely code reviews [21] and post-release coordination [22]. Future work can look at why the clustered and propagation cost differed in describing the coordination requirements in this case. Also, future work could look into different perspectives of comparing the effect of other technical dependencies on social coordination in Open Source projects. We are also studying the effect the change of coupling has on the health of the Open Source project.

# References

1. Parnas, D.L.: On the criteria to be used in decomposing systems into modules. Commun. ACM 15, 1053–1058 (1972)
2. Conway, M.: How do Committees Invent. Datamation 14, 28–31 (1968)
3. Cataldo, M., Wagstrom, P., Herbsleb, J.D., Carley, K.M.: Identification of coordination requirements: implications for the Design of collaboration and awareness tools. In: Proceedings of the 2006 20th anniversary conference on Computer supported cooperative work. ACM Press, Banff (2006)
4. Sosa, M.E., Eppinger, S.D., Rowles, C.M.: The Misalignment of Product Architecture and Organizational Structure in Complex Product Development. J Manage. Sci. 50, 1674–1689 (2004)
5. Sosa, M.E.: A structured approach to predicting and managing technical interactions in software development. Research in Engineering Design 19, 47–70 (2008)
6. MacCormack, A., Rusnak, J., Baldwin, C.Y.: Exploring the structure of complex software designs: An empirical study of open source and proprietary code. Management Science 52, 1015–1030 (2006)
7. Mockus, A., Fielding, R.O.Y.T., Herbsleb, J.D.: Two Case Studies of Open Source Software Development: Apache and Mozilla. ACM Transactions on Software Engineering and Methodology 11, 309–346 (2002)
8. O'Reilly, T.: Lessons from open-source software development. Commun. ACM 42, 32–37 (1999)
9. Paulson, J.W., Succi, G., Eberlein, A.: An Empirical Study of Open-Source and Closed-Source Software Products. IEEE Transactions On Software Engineering, 246–256 (2004)
10. Herbsleb, J.D., Grinter, R.E.: Architectures, Coordination, and Distance: Conway's Law and Beyond. IEEE Software 16, 63–70 (1999a)
11. Koch, S.: Software evolution in open source projects—a large-scale investigation. Journal of Software Maintenance and Evolution: Research and Practice 19, 361–382 (2007)
12. Amrit, C., van Hillegersberg, J.: Detecting Coordination Problems in Collaborative Software Development Environments. Information Systems Management 25, 57–70 (2008)
13. Steward, D.: The design structure system: a method for managing the design of complex systems. IEEE Transactions on Engineering Management 28, 71–74 (1981)
14. Steward, D.V.: Partitioning and tearing systems of equations. SIAM J. Numer. Anal. 2, 345–365 (1965)
15. Eppinger, S.D., Whitney, D.E., Smith, R.P., Gebala, D.A.: A model-based method for organizing tasks in product development. Research in Engineering Design 6, 1–13 (1994)
16. Morelli, M.D., Eppinger, S.D., Gulati, R.K.: Predicting technical communication in product development organizations. IEEE Transactions on Engineering Management 42, 215–222 (1995)
17. Sullivan, K.J., Griswold, W.G., Cai, Y., Hallen, B.: The structure and value of modularity in software design. In: Proceedings of the 8th European software engineering conference held jointly with 9th ACM SIGSOFT international symposium on Foundations of software engineering, pp. 99–108. ACM Press, Vienna (2001)

18. Amrit, C., Van Hillegersberg, J.: Exploring the Impact of Socio-Technical Core-Periphery Structures in Open Source Software Development. Journal of Information Technology (forthcoming, 2010)
19. Tessier, J.: Dependency Finder (Retrieved on March 1st 2009)
20. Raymond, E.: The Cathedral and the Bazaar. Knowledge, Technology, and Policy 12, 23–49 (1999)
21. Rigby, P.C., German, D.M., Storey, M.A.: Open source software peer review practices: a case study of the apache server. In: Proceedings of the 13th international conference on Software engineering, pp. 541–550 (2008)
22. Yamauchi, Y., Yokozawa, M., Shinohara, T., Ishida, T.: Collaboration with Lean Media: how open-source software succeeds. In: Proceedings of the 2000 ACM conference on Computer supported cooperative work, pp. 329–338 (2000)
23. de Souza, C.R.B., Redmiles, D., Cheng, L.-T., Millen, D., Patterson, J.: Sometimes you need to see through walls: a field study of application programming interfaces. In: CSCW '04: Proceedings of the 2004 ACM conference on Computer supported cooperative work, New York, NY, USA, pp. 63–71 (2004)

# Industry Regulation through Open Source Software: A Strategic Ownership Proposal

Jean-Lucien Hardy

Eurocontrol
jl.hardy@eurocontrol.int

**Abstract.** This paper is about a twofold proposal submitted to the scrutiny of the OSS scientific community. It is first argued that OSS should be considered a means to establish an industry regulation. The motivation of this first proposal is the need for harmonization of the supply chain in certain industrial sectors. The Air Traffic Management industry (ATM) is the only case considered in this paper. However, it is assumed that the regulatory advantage of OSS is not specific to that industry. The second proposal is about how to establish such a regulation through OSS. It is argued that the legal ownership of the OSS product should be assigned to a public organization, preferably to an organization that would be dedicated to monitor and promote the evolution of that product. The motivation for these proposals is based on the analysis of possible scenarios of OSS ownership in the case of ATM. Perspectives concerning the preliminary implementation of the proposals are introduced.

**Keywords:** Industry Regulation, Open Source Software, OSS, Intellectual Property Rights, IPR, Ownership, Secondary Software Sector, Air Traffic, ATM, ATC.

## 1 Introduction

In this introduction, the role of the scientific community in OSS adoption will first be discussed. Second, the strategic importance of legal ownership of OSS products is emphasized. Third, the absence of a public regulatory role concerning OSS is pointed out. Fourth, the context of ATM is introduced as a target case.

### 1.1 The Role of the OSS Scientific Community

Science is built on research results, but can only progress with development and foresight. The role of the scientific community is not only to do research on existing phenomena, but also to advise decision makers about appropriate deployment choices, based on clear concepts and hypotheses. OSS is not a fundamental science. Like the global warming issue, OSS has a strong social component and depends heavily on managerial or political decisions. The role of the scientific community is critical in terms of advices to decision makers involved in long term planning. A scientific approach must guide the decisions concerning the strategic use of OSS.

Ågerfalk et al. (Eds.): OSS 2010, IFIP AICT 319, pp. 322–329, 2010.

Scientific advisers should discuss their advices first within the scientific community, before presenting them to decision makers. Such an open work-flow is particularly important when the domain is related to public issues and public governance. OSS is such a domain intrinsically connected to public issues.

A scientific conference about OSS is an opportunity to emphasize, discuss, refine and publish hypothetical scenarios for decision makers, in order to focus further research on concepts and hypotheses useful beyond the academic world.

## 1.2  The Ownership of OSS Products

In legal discussions concerning the adoption of OSS, there is generally more emphasis on the choice of an appropriate license than on the choice of appropriate ownership of IPR (Intellectual Property Rights). However, the choice of ownership is a prerequisite to the choice of an appropriate license.

Informal discussions with people unfamiliar with the OSS domain tend to show that many of them are aware of a special kind of software license for OSS without understanding that there is a legal owner behind this license. The word "public" in the name of the most popular OSS license (GPL stands for "General Public License") contributes to such a misunderstanding of which the consequences are numerous. First, it must be counter-explained to people who discover OSS that a product distributed under GPL terms does not belong to the public. Second, the unnecessary multiplication of OSS licenses may well be motivated by the hidden pride to emphasize the existence of a product owner erroneously considered "public". Third, the perceived lack of an OSS owner who could be held responsible or accountable for problems is used as an excuse for not adopting OSS products [1].

The power of the IPR owner is critical in terms of business, as shown by the case of MySQL and OpenOffice.org, two popular OSS products whose ownership was recently transferred as an asset through the acquisition of Sun by Oracle. Oracle Corporation is now in a strong position to promote or jeopardize the OSS spirit concerning these products. There is actually no legal obligation for Oracle to consider the public interest in its strategy and support to OpenOffice.org and MySQL. Hopefully, it will be to its corporate and commercial advantage to promote the public interest.

In this paper, it is argued that the ownership of OSS products by public organizations could be the leverage of a new regulatory role.

## 1.3  The Absence of a Public Regulatory Role Concerning OSS

In terms of IPR, there are tremendous regulatory efforts built around legal patents. This regulation is transcribed into legal procedures administered by strong public organizations (patent offices). However, there is nothing similar for OSS. A quick look at the literature confirms the absence of a governmental regulatory role concerning OSS. The word "regulation" is not mentioned in any title of the hundreds of papers presented to the previous IFIP conferences on OSS. The interaction between industry regulation and OSS does not seem to have yet been investigated.

In this paper, it is argued that regulation of an industry through OSS could be effective in the interest of all players in that industry, and first of all the customers.

## 1.4 The ATM Supply Chain

Air Traffic Management covers a spectrum of activities including real-time Air Traffic Control (ATC), planning air traffic flow, design of 3D routes for air traffic, environmental concerns in air traffic, taxation of air traffic, etc.

From a conceptual point of view, it has been recently pointed out in a research panel [2], that an ATM supply chain exists, organized on a continental basis. In Europe, the ATM supply chain includes:

- the so-called "ATM industry" selling the ATM systems,
- the so-called ANSPs (Air Navigation Service Providers) mainly providing ATC services,
- many ATM sub-contracting firms supplying particular hardware and software equipment, and
- the public administrations (civil and military) contributing to operations, research, co-ordination, regulation and taxation. EUROCONTROL is such a public organization at the European level.

Most ATM activities are software intensive. Software tools are used for operations, as well as for research and innovation. In terms of software categorization, ATM is part of the secondary software sector [1,3]. Most ATM software is currently proprietary.

In this paper, the ATM supply chain is considered a study case for a new kind of regulation based on OSS.

## 2  Failure of OSS Adoption in ATM

There are no formal research results about the adoption of OSS in ATM. However, from the beginning of this century there were some serious initiatives by EURO-CONTROL to increase awareness and to study the potential of OSS in ATM [4,5,6]. ATM was also considered a case study in the European CALIBRE project. Subsequently, the author was interviewed in the context of a formal academic research concerning the adoption of OSS in the secondary software sector [1].

A general consensus concerning the potential of OSS in ATM emerged from various events and conferences. One of the main arguments is that OSS could help harmonize ATM systems in Europe. For the last two decades, the need for this harmonization has been frequently expressed. Recently, some thorough performance studies have highlighted the important costs of fragmentation of the ATM solutions in Europe [7].

Considering the concept of an ATM supply chain [2,8], OSS appears to be the right harmonizing technology, as it improves software co-operation and interoperability.

When he visited EUROCONTROL for a first seminar in October 2009, Rishab Ghosh also emphasized benefits in terms of sustainability [9], since OSS offers better guarantees than proprietary software, especially in a niche market like ATM. Sustainability seems to be necessary to guarantee harmonization over time.

Other technical and business benefits, as well as drawbacks, have been reported for the European secondary software sector [1,3]. Ideally, there should be a systematic research in terms of Pareto analysis to determine what the most critical drawback is. However, this research could hardly be based on strong facts, but rather on statistics based on experts' opinions that would be artificially consolidated for the sake of a paper. In the present paper, a Pareto analysis is based on rational reasoning built on top of the study by Lorraine Morgan and Patrick Finnegan [1], and applied specifically to the ATM case.

This study reports that the business drawbacks appeared to pose a bigger challenge for OSS than their technical counterparts. The lack of backing support from the company and the lack of ownership are mentioned first. The other business drawbacks found in the same study seem to be derived from the lack of backing: insufficient marketing, training, competencies and access to code.

In terms of top decision making for the ATM domain, it seems that the harmonization and sustainability benefits expected from OSS in ATM cannot explain the lack of managerial backing.

Therefore, the hypothesis considered in the present paper is that the lack of relevant ownership is the most critical drawback from which the others are derived, including the lack of backing support. In terms of a Pareto analysis, it is useful to guide the decision making with a proposal to overcome this major drawback.

# 3 Seven Counter-Productive Scenarios for OSS Ownership

Various scenarios of OSS adoption can be considered depending on who is the IPR owner of the OSS product(s). Considering the classical limiting factor analyzed by G.A. Miller [10], seven scenarios are discussed here concerning ATM.

## 3.1 OSS Owned by a Small Company

It is possible for a small company that owns a software product to publish it in OSS mode, inviting anyone to join a community to develop and improve that product. There was a recent example of such a scenario in ATM (www.albatross.aero). For a profit oriented company, the ownership can be motivated by financial speculation on the value of the IPR enhanced by the value of the OSS community that develops the OSS product. However, such an underlying speculative objective could hardly serve the efficiency of the ATM supply chain, because it cannot be shared by other partners. Therefore, an OSS tool driven by financial speculation would probably increase fragmentation in ATM.

## 3.2 OSS Owned by a Major Player in the Industry

In May 2006, there was a CALIBRE meeting in Spain where two major telecommunication players, Vodafone and Telefonica, explained their OSS initiatives: since the differentiating power of the software was rather low – Telefonica said 5% – they decided to go open source in order to enlarge their supplier basis. It appears that the two companies were creating their own OSS projects, as neither of them expressed a willingness to join the OSS initiative of the other. Telefonica and Vodafone gave the impression of using OSS to compete in attracting more suppliers.

If the same scenario happens between major ATM companies, it would not be a solution for the whole ATM supply chain, because the duplication of the OSS community would not contribute to harmonization.

### 3.3 OSS Owned by an OSS Foundation

The copyright of the ADA development environment GNAT was created by a grant from the US Air Force to the New York University. The copyright of GNAT was assigned from the New York University to the Free Software Foundation (FSF). Such a scenario prevents speculation and favors fair competition [11]. Now, what if the IPR of ATM tools were owned by an OSS foundation, like the FSF? Such a scenario would probably not work well, because the ATM domain is a specific niche, and because the FSF is not part of that niche. Therefore, such a foundation could not manage the use of the OSS ownership in the interest of the ATM world. For example, if there were a demand for a commercial license, a disconnected OSS foundation would have neither the means nor the competence to negotiate this demand for the benefit of the ATM niche and its specialized players.

### 3.4 OSS Owned by Cooperatives of Users

Some ATM software tools have their own informal community of users. It could be easy to create *ad hoc* legal bodies on top of these communities and to assign them the OSS copyright. The drawback of such a scenario would be the addition of an administrative layer with a need for co-ordination, and therefore yet another source of fragmentation in the ATM sector. Given the volatility of such small legal bodies, the sustainability criteria would not be fulfilled either.

### 3.5 OSS Owned by a National Public Administration

A country in EUROPE could decide to use OSS for its ATM sector. It would make sense, especially for countries that have a long tradition in public administration of ATM. This would facilitate fair procurement for the development and maintenance of OSS tools while avoiding the risk of speculation. However, such a national scenario would reinforce the existing geographical fragmentation of the European ATM.

### 3.6 OSS Owned by a Continental Public Organization

EUROCONTROL, a leading public organization in European ATM, is the owner of a portfolio of software tools used for either operations or R&D (Research and Development). Recently, there were two attempts to publish R&D tools in OSS terms (nogozone.sourceforge .net and atv3d.sourcegorge.net). However, such a publication cannot be extrapolated for commercial software tools, because EUROCONTROL is a public body that is not supposed to act in the role of a Pan European software house disturbing competition in the ATM market.

### 3.7 OSS Owned by a Global Public Organization

ATM is also in the scope of worldwide organizations. ICAO is a specialized organization of the UN dealing with civil aviation. IATA and CANSO are trade organizations for airlines and ANSPs. Assigning the OSS ownership to such bodies would have the advantage of providing global harmonization. However, performance studies indicate that the improvement needs are not the same in Europe and the US [7]. Therefore an attempt to reach a global harmonization through OSS tools is not a relevant scenario. It seems useful to keep different ATM solutions based on continental specificities.

# 4 Attaching the OSS Ownership to an Industry Regulatory Role

The seven scenarios of OSS ownership are not productive because the owner does not have a role that justifies its ownership on behalf of the OSS community.

Therefore, the question is whether it is possible to find such an appropriate role. From the seven scenarios, it appears that such a role must include protection against speculative or competitive use of OSS ownership, and governance in the niche domain of the OSS tools for the best interest of the public, e.g. aircraft passengers.

Considering that European ATM needs harmonization (and not fragmentation) of its supply chain, a regulatory role seems to be needed [8]. Since OSS technology has intrinsic advantages in establishing harmonization, the OSS ownership appears to be a prerequisite or an advantage for the creation of a regulatory role based on OSS.

Such a role must be assigned to a public organization at the level of the situation to be improved and harmonized. To tackle the fragmentation of the European ATM, the European Commission and EUROCONTROL are two candidate institutions, but EUROCONTROL is probably better suited for a specific OSS role in ATM given its mission that covers technical aspects of ATM.

To promote its regulatory role, the organization should not be involved in competitive activities such as software development or software services. These activities belong to the private companies involved in the regulated supply chain.

By holding the IPR ownership, a public regulatory body avoids speculative or competitive counter-productive effects. It could also tune the licensing scheme towards harmonization. For example, a dual licensing scheme could be used on demand. The reply would be a regulatory process which allows integration of OSS products with proprietary products. In such a process, the productivity of the supply chain would be optimized in terms of quality improvements and/or in terms of royalties for the OSS community. A preliminary step towards such a dual licensing scheme is implemented by the two ATM tools launched in OSS. For these R&D tools, the door is left open for a commercial license with a counterpart for the OSS community (nogozone.sourceforge .net and atv3d .sourceforge.net).

Figure 1 summarizes the hypothetical proposal. The harmonization and sustainability need of the whole industry determines the choice of technology (OSS) and the main attribute of the regulatory role (IPR). A key issue is that the regulatory role is also impacted by the choice of technology. Therefore, technology and regulation have a positive and correlated impact on the efficiency of supply chain.

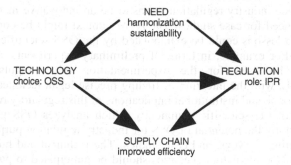

**Fig. 1.** Technology and regulation contributing to the efficiency of the supply chain

# 5  Conclusion and Future Perspectives

This paper describes a twofold strategic proposal. First, it is argued that OSS should be used as a tool for establishing industry regulation when needed. In particular, such a need exists in the case of a complex industrial supply chain using software products: the regulated approach through OSS is a way to achieve sustainable harmonization. The second argument is that an industry regulation through OSS is possible when the ownership of the OSS products is assigned to the regulatory body, preferably an organization with the necessary domain specific background to monitor and guide the evolution of the OSS products in the interest of the public.

In the particular case of the ATM sector in Europe, EUROCONTROL seems to be in the right position to play such a regulatory role. In a paper concerning the regulation of European ATM [8], Hervé Dumez and Alain Jeunemaître point out "the need to create new tools for thinking European regional regulation and European infrastructure management." The present proposal of an ATM regulatory role through OSS could be an attempt to fulfill this need.

By extrapolation from the ATM case, the proposal could go a step further towards public OSS. The ownership of OSS should be placed in the hands of public regulatory bodies, especially when the need for harmonization becomes greater than the need for innovation. Private companies that own a software product in the phase of commoditization and no longer commercially differentiating [12] should not just offer an OSS license for their product. They could consider donating or selling the ownership of that software product to a public entity in order to facilitate *de facto* standardization around that product. On the contrary, the ownership of a software tool by a private company, even in the case of an OSS product, is the evidence of a competitive advantage that inevitably hampers contributions, co-operation and harmonization around that tool.

In the end, the validity of the hypothetical strategic proposal will only be established if it is adopted by top decision makers who manage private or public organizations. Based on the present proposal, they could immediately apply the principle of Alan Kay: "The best way to predict the future is to invent it." However, the best approach for the interest of a large public might be in a continuous back-and-forth between innovation and research.

Therefore, some preliminary steps would be useful before a decision can be made concerning the implementation of industry regulation through OSS. First, it would be useful to survey the OSS ownership and to hear the opinions of OSS specialists, in order to predict if the regulation hypothesis is valid *a priori*, how it should be implemented, and how its value (efficiency and effectiveness) should be evaluated *a posteriori*. Second, since industry regulation seems to be an innovative matter in the OSS arena, there is a need for case studies. The ATM context might be considered a case study from which lessons could be extrapolated by the OSS scientific community to other industries. For example, in terms of preliminary experiments, a few software products could be used for the experimentation of a regulatory role by EUROCONTROL, with the intention of finding precise legal, technical, economical, social, organizational, and institutional implications of this regulatory role.

It is the role of the scientific community which analyzes OSS phenomena and communities to study the potential of OSS for industry regulation purposes, based on conceptual modeling, surveys, and case studies. The technical and business benefits and drawbacks of the regulatory scenarios should be anticipated to pave the way for future OSS policies and decisions.

**Disclaimer and Acknowledgment.** The author is a civil servant of EUROCONTROL, but the present proposal is a personal viewpoint that does not reflect the official views and policies of EUROCONTROL on the matter. The author is also contributing to the OSS2010 conference as a founding member of the IFIP Working Group 2.13 on Open Source Software. He is grateful to the two anonymous OSS2010 reviewers for their challenging comments on the first version of this paper, to his friend John Seifarth of Words-and-Wires-sprl (www.waw.be) for keen comments on the final version, and to his wife Jung Yeon Kim for correcting English mistakes despite an imminent extension of the family. A special dedication goes to Professor Dieudonné Leclercq of the University of Liège who introduced the passion for software into my life in 1975.

# References

1. Morgan, L., Finnegan, P.: Benefits and Drawbacks of Open Source Software: an Exploratory Study of Secondary Software Firms. In: Feller, J., Fitzgerald, B., Scacchi, W., Sillitti, A. (eds.) Open Source Development, Adoption and Innovation. IFIP, vol. 234, pp. 307–312. Springer, Boston (2007)
2. Jeunemaître, A.: Panel "Innovation in ATM". In: EUROCONTROL 8th Innovative Workshop and Exhibition (2009)
3. Ågerfalk, P.J., Deverell, A., Fitzgerald, A., Morgan, L.: Assessing the Role of Open Source Software in the European Secondary Software Sector: A Voice from Industry. In: Scotto, M., Succi, G. (eds.) First International Conference on Open Source Systems (OSS 2005), pp. 82–87 (2005)
4. Hardy, J.-L., Bourgois, M.: Open Source Implications for EUROCONTROL (OSIFE). In: 3rd EUROCONTROL Innovative Research Workshop (2004)
5. Bourgois, M., Hardy, J.-L., O'Flaherty, J., Seifarth, J. (eds.): Potential of OSS in ATM (2005), http://www.oss-in-atm.info
6. Hardy, J.-L., Bourgois, M.: Exploring the potential of OSS in Air Traffic Management. In: Darniani, E., Fitzgerald, B., Scacchi, W., Scotto, M., Succi, G. (eds.) Open Source Systems. IFIP, vol. 203, pp. 173–179. Springer, Boston (2006)
7. EUROCONTROL and FAA: U.S./Europe Comparison of ATM-related Operational Performance (2009),
   http://www.eurocontrol.int/prc/gallery/content/public/Docs/
   US_Europe_comparison_of_ATM_related_operational_
   performance.pdf
8. Dumez, H., Jeunemaître, A.: Restructuring Regulation in Europe. The Case of Air Traffic Services. Concurrences 1, 1–3 (2008)
9. Ghosh, R.A.: Free Software: How Does it Work. In: European Broadcasting Union Seminar (2007),
   http://www.slideshare.net/kamaelian/free-software-how-does-
   it-work
10. Miller, G.A.: The magical number seven, plus or minus two: Some limits on our capacity for processing information. Psychological Review 63(2), 81–97 (1956)
11. Gasperoni, F.: COTS, FLOSS, and Market Freedom in Safety-Centric Industries. In: [5] (2005)
12. Van der Linden, F., Lundell, B., Marttiin, P.: Commodification of Industrial Software: A Case for Open Source. IEEE Software 26(4), 77–83 (2009)

# Proposal for Solving Incompatibility Problems between Open-Source and Proprietary Web Browsers

Jun Iio[1], Hiroyuki Shimizu[1], Hisayoshi Sasaki[2], and Akihiro Matsumoto[2]

[1] Research Center for Information Technology, Mitsubishi Research Institute, Inc.
{iiojun,hshimizu}@mri.co.jp
[2] Gluegent, Inc.
{sasaki,matsumoto}@gluegent.com

**Abstract.** We conducted demonstration experiments to promote the use of open-source software on desktops as a part of the Open-Source Software-Utilization Development Program for the education sector. In these experiments, we found some barriers to popularizing the use of open-source software in end-user desktop applications. In this paper, we report some typical problems of Web-browser compatibility, which are considered to be obstacles for promoting open-source software on desktops. We also introduce a tool that we developed to help developers avoid such pitfalls while designing Web applications.

## 1 Introduction

In the late 1990s when open-source software was not considered as a commodity, open-source server software was used much more widely than open-source desktop applications. Recently, the open-source desktop applications have also become popular, and some well-known applications such as OpenOffice.org, Firefox, and Thunderbird are gaining market share and awareness on the Internet. According to the report from the open source application foundation (OSAF)[1], the use of Linux on desktops will first become popular among high-end users and IT engineers, then its use by routine workers (such as call-center operators) will follow. Ultimately, even office workers will use open-source software on their desktops.

## 2 Background

In order to accelerate the spread of open-source applications on desktops, we conducted a series of demonstration experiments from 2004 to 2008 as part of the Open-Source Software-Utilization Development Program for the education sector, which was sponsored by the Information-Technology Promotion Agency Japan (IPA) and the Center for Educational Computing (CEC). In these experiments, Linux desktop computers were deployed at more than 50 schools, and

P. Ågerfalk et al. (Eds.): OSS 2010, IFIP AICT 319, pp. 330–335, 2010.

students used Linux every day in the classroom. As a results of these experiments, we demonstrated that there were almost no problems using open-source desktop computers instead of the existing proprietary computers for IT-driven classes. In addition, we showed the possibility of reducing management cost of IT systems by adopting open-source software.

On the other hand, our study based on those experiments revealed many problems that must be solved in order to replace proprietary software with open-source software on desktop computers. One of the key issues in the use of open-source applications in the field of education is the ability to handle established course material for e-learning systems.

Many e-learning systems are constructed as Web applications. In general, Web browsers are used as client software, and viewing course material via the Internet requires a Web browser. However, much educational content is designed to be viewed by a particular Web browser (many e-learning systems are built to run on Microsoft Internet Explorer (MSIE) and they are tested only on that browser), In addition, many Web pages are consistently produced with defective content that cannot be properly viewed by open-source browsers.

Obviously, the trend toward such inappropriate Web content exists not just in the e-learning field, but also in general Web-based applications in other fields. Therefore, a study to identify browser-incompatibility trends was required; it was essential to clarify the practical problems and solutions in order to popularize the use of open-source browsers.

## 3   Browser Dependency Problems

Following are the typical types of browser-incompatibility problems.

**Embedding of Video and Other Objects.** This problem is caused by a difference in the handling of two tags: `<embed>` and `<object>`. To create a document that contains such an object and that is rendered correctly in every browser, the object must be embedded in the document using both tags.

**Requirements of Particular Plug-ins.** Shockwave, XVL Player / Viewer, and certain other plug-ins using Active-X technology cannot be handled in the Linux.

**Unsupported Functions in Japanese documents.** Text is written in two directions. MSIE implements the product-specific extension "writing-mode: tb-rl" in its style-sheet definition to enable vertical writing mode. However, Mozilla Firefox does not support this function.

**Layout Fragmentation by Line-Spacing Misalignment.** Some documents specify line spacing using a property of the tag `<p>`. This is valid only in MSIE, and not valid in Firefox. According to HTML specification, using this tag property is inappropriate, and the `<div>` tag should be used to adjust line spacing instead.

**Garbled Characters.** In the representation of Japanese text, there are four sets of character codes. Especially in Java applications, not assigning explicit character codes for the text can cause garbled characters to occur.

**Difference in the Versions of Java Runtime Environment (JRE).**
Some systems cannot run as expected because of issues with Java applet initialization.

**Defects in Different JavaScript (ECMAscript) Implementations.**
Because of differences in the implementation of JavaScript in MSIE and Firefox, many malfunctioning scripts, which have been tested only on MSIE, have been encountered.

### 3.1 Solutions for This Problem

These problems are serious, because Internet has achieved an important position in the infrastructure of the information society.

We investigated the problems that exist in the Internet space and their effect on Internet users. The result of our investigations [2] revealed that approximately 170 incompatibility problems potentially exist in the Internet. Some of other problems were categorized as fatal errors. For example, some data would never get displayed in a particular browser. Based on our findings, a recommendation document [3] was published.

As a solution to this difficult situation, we designed a tool for Web designers, Web developers, and Web application programmers to make the development of multi-browser Web application quick and easy. The key advantage of this tool is the information it offers to Web developers to avoid issues affecting interoperability. The "Pirka'r" system, which is introduced in this paper, was developed and released to the Internet as an open-source software. Pirka'r version 1.0 was released at the end of September 2009. It is now available under the Apache License, Version 2.0, from http://pirkar.ashikunep.org/.

## 4 System Overview

In this section, we provide an overview of Pirka'r and its ecosystem.

### 4.1 Overview of Pirka'r

Pirka'r is an integrated development environment for Web designers and Web application developers consists of two parts: a client running on the user's workstation, and a verification server. The verification server can be installed either on the workstation or on a separate PC.

The Pirka'r front-end was developed as an application in Eclipse Rich Client Platform (RCP). Following are the main functions available in Pirka'r:

- Verification against standards.
- Assessment of multi-browser interoperability.
- Multi-browser previewing.
- Automatic rendering.
- Multi-functional editing.
- Auto-downloading of a set of Web pages.

The Pirka'r user can verify whether the cascading stylesheet (CSS) definition in his Web content complies with standards. In addition, HTML/CSS/JavaScript can be examined in regard to browser-dependent descriptions.

The target Web page can be displayed in multiple browser panes. This function relieves the designers of the annoyance of handling multiple browsers and checking rendering results in each browser.

Web content written by the multi-functional editor is rendered in the multi-browser preview panes in real time. Designers are freed from the need to reload manually. The multi-functional editor is a HTML/CSS/JavaScript editor, which has convenient functions such as code completion, syntax highlighting, and grammatical checking.

## 4.2 The Pirka'r Ecosystem

Figure 1 shows an overview of the Pirka'r ecosystem.

**Fig. 1.** Overview of the Pirka'r ecosystem

The function that verifies interoperability problems on target Web content uses a verification engine running separately with the Pirka'r client. The verification engine provides the results of verification processes, which are based on verification data provided from the verification data-management server.

A single management server that manages verification data can deliver the set of verification data to the verification engines that are installed locally on any network.

If a designer detects a new interoperability-discrepancy problem, he can report the problem via the reporting form provided. When we receive the report, committers in our community attempt to create new verification data that contains a check script, as well as a series of documents that show how to fix the problem.

### 4.3  Verification Script

The verification script is written in JavaScript. It searches for defects in the target Web content, traversing over its Document Object Model (DOM).

If the problem is not too complex, the verification script is quite simple. We have already prepared more than 100 scripts to check existing problems, and stored them in the master database. We are now asking committers who belong to the Seaser Foundation to update verification scripts. Seaser Foundation is one of the most famous organizations in Japan associated with Web application development.

Committers can register new verification data using the submission form of the verification data-management server, which is also managed by the Seaser Foundation. A set of verification data consists of a name, category (HTML, CSS, or JavaScript), language, severity level, reason for the problem, proposals to solve this problem, verification script, examples of descriptions, and attached files (if any).

## 5  Related Works

A study on interoperability-discrepancy problems associated with Web standards was also carried out by the standard and certification working group of the North-East Asia Open-Source Software promotion forum. In their group activities, information on problems was shared by delegates from China, Japan, and Korea [4]. Furthermore, solutions to some problems were discussed in the collaborative work of the members [5].

A number of Web services can provide screenshots of many types of Web browsers. Moreover, a variety of studies, such as analyses of Web service interoperability [6] and browser-compatibility testing methods [7], have been conducted from the standpoint of software engineering.

Several studies on the standardization of Web content have been conducted over the past few years. Result of these studies too are useful for our work. Peter K. conducted massive studies of U.S. government Web sites [8] and the People's Republic of China government Web sites [9], using the W3C Validator. However, we consider his approach to be practically insufficient, because problems such as those discussed in this paper may not be detected by simply checking syntax errors.

## 6  Conclusions

We studied the interoperability-discrepancy problems of Web content for more than three years. Our activities resulted in the development of Pirkar, which helps Web designers easily create multi-browser Web applications.

Recently, browser vendors have become increasingly aware of standards. If all Internet users were able to access Web sites using the latest browsers, there would be no problem. However, it cannot be ignored that there are still many legacy users, who continue to use old-fashioned browsers that have interoperability problems, as mentioned in this paper. This is a strong motivation for developing the Pirka'r tool, which supports Web designers. All of the functions provided by Pirka'r are language-independent, so this tool can be used by Web developers worldwide. We will work in the future to promote this activity.

## Acknowledgements

This study and Pirka'r development were conducted as part of the Open-Source Software-Utilization Development Program, supported by the Information Technology Promotion Agency, Japan.

## References

1. Decrem, B.: Desktop Linux Technology & Market Overview. Open Source Application Foundation (July 2003)
2. IPA, Research on the Improvement of Web Contents Compatibility Conducive to the Widespread Use of OSS Desktops, Resarch Report (2007),
   http://www.ipa.go.jp/software/open/ossc/download/Web_Research_En.pdf
3. IPA, Research on the Improvement of Web Contents Compatibility Conducive to the Widespread Use of OSS Desktops, Written recommendations (2007),
   http://www.ipa.go.jp/software/open/ossc/download/
   Web_Recommendations_En.pdf
4. NEA-OSS Promotion Forum WG3, Information Technology – Report of Web Interoperability, WG3 SWG2 N054, TR00003 (2007)
5. NEA-OSS Promotion Forum WG3, Information Technology – Solution of Web Interoperability Discrepancy, WG3 SWG2 N063, TR00004 (2008)
6. De Antonellis, V., Melciori, M., Plebani, P.: An Approach to Web Service Compatibility in Cooperative Processes. In: Proceedings of SAINT 2003 Workshops, pp. 95–100 (2003)
7. Xu, L., Xu, B., Nie, C., Chen, H., Yang, H.: A Browser Compatibility Testing Method Based on Combinatorial Testing. In: Cueva Lovelle, J.M., Rodríguez, B.M.G., Gayo, J.E.L., Ruiz, M.d.P.P., Aguilar, L.J. (eds.) ICWE 2003. LNCS, vol. 2722, pp. 310–313. Springer, Heidelberg (2004)
8. Peter, K.: Government Web standards usage: USA – standards-schmandards (2005),
   http://www.standards-schmandards.com/2005/government-web-
   standards-usage-usa
9. Peter, K.: Government Web standards usage: People's Republic of China – standards-schmandards (2006),
   http://www.standards-schmandards.com/2006/gvmt-standards-prc

# FLOSS Communities: Analyzing Evolvability and Robustness from an Industrial Perspective*

Daniel Izquierdo-Cortazar[1], Jesús M. González-Barahona[1], Gregorio Robles[1], Jean-Christophe Deprez[2], and Vincent Auvray[3]

[1] GSyC/LibreSoft, Universidad Rey Juan Carlos, Mostoles, Madrid
{dizquierdo,jgb,grex}@libresoft.es
[2] Centre of Excellence in Information and Communication Technologies, Charleroi, Belgium
jean-christophe.deprez@cetic.be
[3] PEPITe, Liège, Belgium
v.auvray@pepite.be

**Abstract.** Plenty of companies try to access Free/Libre/Open Source Software (FLOSS) products, but they find a lack of documentation and responsiveness from the libre software community. But not all of the communities have the same capacity to answer questions. Even more, most of these communities are driven by volunteers which in most of the cases work on their spare time. Thus, how active and reliable is a community and how can we measure their risks in terms of quality of the community is a main issue to be resolved. Trying to determine how a community runs and look for their weaknesses is a way to improve themselves and, also, a way to obtain trustworthiness from an enterprise point of view. In order to have a statistical basement, around 1400 FLOSS projects have been studied to create thresholds which will help to determine a project's current status compared with this initial set of FLOSS communities.

**Keywords:** Libre software communities, quality models, data mining.

## 1 Introduction

QualOSS[1] (Quality of Open Source Software) is a research project focused on the assessment of the quality of FLOSS (free, libre, open source software) endeavor. A FLOSS endeavor is composed of a set of community members (or contributors), a set of work products including code, a set of development processes

* This work has been funded in part by the European Commission, under the FLOSS-METRICS (FP6-IST-5-033547), QUALOSS (FP6-IST-5-033547) and QUALIPSO (FP6-IST-034763) projects, and by the Spanish CICyT, project SobreSalto (TIN2007-66172).

[1] The QualOSS project is coordinated by CETIC, and includes also University of Namur, Universidad Rey Juan Carlos, Fraunhofer IESE, Zea Partners, UNU-MERIT, AdaCore and PEPITe. The work described in this paper has been performed, or coordinated, mainly by the GSyC/LibreSoft group at Universidad Rey Juan Carlos. More info about the project: http://qualoss.org/

P. Ågerfalk et al. (Eds.): OSS 2010, IFIP AICT 319, pp. 336–341, 2010.
© IFIP International Federation for Information Processing 2010

followed by the community to produce work products, and a set of tools used to support the endeavor, to produce work products and to run the FLOSS software component [2]. In other words, a FLOSS endeavor is really like an enterprise working on FLOSS development projects. In turn, the exact goal of QualOSS aims at assessing the robustness and evolvability of FLOSS endeavors.

When acquiring software, enterprises are not only interested to know about the product and its quality but also interested in who produced that product and its reputability. For traditional enterprises, reputability can be check based on financial strength of the software provider. However, for the FLOSS world, we must find other ways to determine if a FLOSS endeavor (or a FLOSS project) is serious. This can be done by studying the behavior of a FLOSS community. In particular, a FLOSS community should behave in a manner to convince potential FLOSS integrators from industry that it is dependable.

## 2    Related Research

In the area of FLOSS, several models have been proposed as well, such as Open-BRR[2] or QSoS[3]. They consider metrics in several realms relevant to FLOSS development and maintenance, ranging from product to process or community metrics. In addition, some of them lack the needed benchmarking to fine-tune the methodologies proposed, and in some cases do not consider some important aspects of FLOSS development or maintenance [1]. Finally it is necessary to show that in the case of OpenBRR the number of metrics associated to the community side are just two metrics. On the other hand, QSoS provides four metrics related to activity over the source code.

QualOSS is aimed to fill this gap [6], and specifically in terms of community assess to provide a methodology whose metrics and indicators are semi-automatically retrieved and all based on a theoretical framework. Thus, results are influenced by the existing methodologies and their lack of information regarding communities, what we think that it is a key factor to take into account in software maintenance process, as well as interviews with FLOSS integrators.

## 3    Methodology

For achieving its aims, QualOSS started by applying a Goal-Question-Metric methodology, from which relevant goals, questions, and finally metrics and indicators were derived. The computation of metrics measurements and aggregation for answering questions of the QualOSS quality model was partially automated with several tools.

**Interviews and GQM**
The first step of the QualOSS methodology consisted of gathering the current state of the art on the topic, from a theoretical and empirical point of view [3].

---

[2] http://www.openbrr.org
[3] http://www.qsos.org

In addition, some companies were interviewed to know about their needs, direct or indirect, regarding the quality of software. Those interviews were held with the goal of identifying the needs from an industrial point of view. Focusing on the community side, companies with a business model specificallybased on FLOSS are usually not directly worried to use products still in their preproduction state and with no stable releases. They highlighted the importance of the surrounding community and the support it may provide. Therefore these companies do not hesitate to interact with the community, sharing technical and non-technical goals.

Using the Goal-Question-metric approach, a goal is defined by an issue, a context, a point of view and the object to analyze. In QualOSS, the issues consist of identifying the risk to collaborate with a community. The context assumes that an enterprise considers integrating a FLOSS component and collaborates fully with the existing FLOSS community. The point of view represents the role of people in the enterprise who are concerned about the community issue. And finally, the object to analyze is the FLOSS endeavor itself.

Then, to polish the quality focus it was necessary to create a quality model based on the ISO 9126 and merging it with the criteria proposed by the companies. From the information above, it is then possible to refine a set of goals related to community. For each goal, a set of questions determine how to verify if the goal is fulfill. Those questions are: First: "how can we measure community robustness?". Second: "how can we measure community evolvability?".

Both questions were further elaborated into several sub-questions, with the aim of characterizing the object of measurement.

### Size and Regeneration Adequacy:

- **Definition**: The degree to which the size evolution and regeneration of a FLOSS community happens at an adequate rate to maintain a sustainable community size.
- **Questions**:
  - **sra2**- New code contributors evolution.
  - **sra3**- New non-code contributors evolution.
  - **sra4**- New core contributors evolution.
  - **sra5**- Evolution of core members who stopped contributing.
  - **sra6**- Balance between new core contributors and those who left the project.
  - **sra7**- Average longevity of committers to the FLOSS endeavor
  - **sra9**- Number of code contributors submitting changes in major releases.

### Interactivity and Workload Adequacy:

- **Definition**: The degree to which the community interacts adequately and partition the workload among FLOSS community members adequately to maintain a community cohesion and motivation.

- **Questions:**

  - **iwa1-** Is the number of events adequate (to show a lively community)?
  - **iwa2-** Is the number of code commits adequate?
  - **iwa4-** Are there sub-groups disconnected or are active community members serving as bridges between these sub groups?
  - **iwa5-** Is there enough community supporting a FLOSS desired version?
  - **iwa7-** Is the current team concerned about the entire source code?

It should be noticed that there are missing questions. Since this analysis is focused on the source code management system (the most usual data source found in FLOSSMetrics databases), those metrics related to other data sources (and so, their questions) have been removed.

**Indicators**

For each metric defined above (each metric answers one question), there is a high-level risk indicator. This indicator measures one aspect of the risk taken by a company engaging in a full FLOSS collaboration. The QualOSS project has defined the following four color-coded risk levels, in order of decreasing risk: black, red, yellow and green. However, it should be noticed that indicators should not be trusted blindly. In particular, they should not be considered separately, but rather jointly to form an overall risk picture associated to a project.

In this paper, we adopt a data-driven approach to define indicators. Using the notion of quantile, we search for a partition of a metric's values into a given number of intervals with equal probability.

To compute our metrics and estimate our indicators, we use data collected by the FLOSSMetrics [5][4] project. Many of the open-source projects considered by FLOSSMetrics appear to be very small and are thus not representative of our population of interest. Indeed, we are assessing the risks associated to a full FLOSS collaboration with projects. This precludes very small projects for which, from a business perspective, a fork should be more appropriate.

# 4   Threats to Validity

As it was said, during the extraction data from the FLOSSMetrics databases, it did not provide full of data regarding other data sources except for the source code management system. Thus, what it is presented in this paper is just based on those metrics which are retrieved specifically from that data source.

It is also necessary to deal with the fact that projects stored in FLOSSMetrics database do not represent the whole population of FLOSS projects.

Finally, depending on the policy, projects may have a small set of developers who are in charge of committing all the changes. This will skew the results produced by our methodology.

---

[4] http://flossmetrics.org

# 5   Results for Illustration

In first place, an approach using the slope as our metric and then defining the indicators showed a lack of information since most of the slopes were pretty similar. We used another approach based on the slope, but modifying the way the indicator is calculated. In fact, those indicators are not dependable of a given statistical approach and some other may be used to improve the indicator accuracy. Projects presented for illustration are well known projects and even when the Evince community is more active than the Nautilus's or HTTPD1.3's, all of them show a similar activity in terms of commits per committer, handled files and other metrics.

**Table 1.** Illustration in some projects using a linear model

| Pred. diff. | evolution | evince | nautilus | httpd1.3 |
|---|---|---|---|---|
| sra7 | 8.6044 Y | 5.0272 B | 7.4484 R | 15.7206 G |
| iwa4 | 0.3666 G | 0.4216 G | 0.3168 G | 0.0923 G |
| iwa5 | 374772.4 B | 48282.8 R | 246991.9 B | - |
| iwa7 | 0.145 G | 0.1904 Y | 0.0491 Y | 6.0e-4 B |
| $\Delta_{90}$ sra2$_{90}$ | 0 Y | 0 Y | 0 Y | 0 Y |
| $\Delta_{90}$ sra3$_{90}$ | 0 Y | 0 Y | 0 Y | 0 Y |
| $\Delta_{90}$ sra9$_{90}$ | 0 Y | 0 Y | 0 Y | 0 Y |
| $\Delta_{90}$ iwa1$_{90}$ | -0.23327969 R | -0.0312513 R | -0.20157016 R | -0.62113349 R |
| $\Delta_{90}$ iwa2$_{90}$ | -1.03448107 R | 0.31998312 Y | -1.60891953 R | -0.76343814 R |
| $\Delta$ sra4 | 1 G | 2 G | 0 Y | 0 Y |
| $\Delta$ sra5 | 1 B | 1 B | 2 B | 0 R |
| $\Delta$ sra6 | -1 R | 0 Y | -1 R | 0 Y |

# 6   Conclusions and Further Work

We have presented a way to estimate "quality" from an industrial perspective based on statistical analysis of hundreds of FLOSS projects. FLOSS communities are key actors in the software evolution and maintenance process and better understanding their behavior through their life will improve the make decision process.

For instance, checking the tendency by means of the methodology explained in this paper, we are able to know if a community is growing, or if the number of core committers is decreasing over and over. Perhaps we are interested in a specific product and we know that some of its weaknesses are motivated because of a really high turnover of developers.

Thus, we can check how reliable are the FLOSS communities looking at the values for the given set of metrics. As it was mentioned, indicators define relative, and not absolute, risks. A low risk does not mean that the metric value is good, this is just good compared to other projects. In this case, this is useful if we are interested in guessing the activity of the community, the general tendency and how it behaves compared to some other set of projects.

As further work, we should say that indicators must be polished by using new data from FLOSSMetrics (current status of the Melquiades database[5] shows an increase of 600 projects since results were retrieved for this paper). There are other publicly available data sources which may be checked in order to add more accuracy to this analysis. Specifically OSSMole[6] or Ohloh[7] are some examples which have been used for academic purposes. The FLOSSMetrics project is currently adding data from bug tracking systems what means that some other indicators will be added using the statistical approach defined here.

We also need to include advanced aspects of community behaviors and how they can be integrated in the QualOSS methodology. For instance, community-driven projects show interesting interactions among participants [8,4,7].

# References

1. Deprez, J.-C., Alexandre, S.: Comparing assessment methodologies for free/open source software: OpenBRR & QSOS. In: Jedlitschka, A., Salo, O. (eds.) PROFES 2008. LNCS, vol. 5089, pp. 189–203. Springer, Heidelberg (2008)
2. Deprez, J.-C., Fleurial-Monfils, F., Ciolkowski, M., Soto, M.: Defining software evolvability from a free/open-source software. In: Proceedings of the Third International IEEE Workshop on Software Evolvability, October 2007, pp. 29–35. IEEE Press, Los Alamitos (2007)
3. Deprez, J.-C., Ruiz, J., Herraiz, I.: Evaluation report on existing tools and existing f/oss repositories. Technical report, QualOSS Consortium (2007)
4. González-Barahona, J.M., López-Fernández, L., Robles, G.: Community structure of modules in the apache project. In: Proceedings of the 4th Workshop on Open Source Software Engineering, Edinburg, Scotland, UK (2004)
5. Herraiz, I., Izquierdo-Cortazar, D., Rivas-Hernández, F.: Flossmetrics: Free/libre/open source software metrics. In: CSMR, pp. 281–284 (2009)
6. Izquierdo-Cortazar, D., Robles, G., González-Barahona, J.M., Deprez, J.-C.: Assessing floss communities: An experience report from the qualoss project. In: OSS, p. 364 (2009)
7. Madey, G., Freeh, V., Tynan, R.: Modeling the Free/Open Source software community: A quantitative investigation. In: Koch, S. (ed.) Free/Open Source Software Development, pp. 203–221. Idea Group Publishing, Hershey (2004)
8. Mockus, A., Fielding, R.T., Herbsleb, J.D.: Two case studies of Open Source software development: Apache and Mozilla. ACM Transactions on Software Engineering and Methodology 11(3), 309–346 (2002)

---

[5] http://melquiades.flossmetrics.org
[6] http://ossmole.sourceforge.net/
[7] http://www.ohloh.net/

# BULB: Onion-Based Measuring of OSS Communities

Terhi Kilamo, Timo Aaltonen, and Teemu J. Heinimäki

Tampere University of Technology
firstname.lastname@tut.fi

**Abstract.** Up to date information on the associated developer community plays a key role when a company working with open source software makes business decisions. Although methods for getting such information have been developed, decisions are often based on scarce information. In this paper a measuring model for open source communities, BULB, is introduced. BULB provides a way of collecting relevant information and relates it to the well-known onion model of open source communities.

## 1   Introduction

A company working with open source software (OSS) is often dependent on the developing community. Especially, when a product or service is sold the connection is obvious. In order to make business decisions, the need for up to date information about the community is clear. For example, the size of the community should be known, the activity of developers is interesting, and how easy penetrating into the community is should be found out.

Currently getting this kind of information is hard. Precise models for such have been developed. For example, *social network analysis (SNA)* [3] has been suggested to get a strict view to the community. SNA analyses a mathematical graph, where nodes are the members of the community and arcs model relationships between them. Different kinds of surveys can be given as another example of community information digging.

The industry does not seem to use such advanced methods today. On the contrary, to our knowlegde the business decisions are often based on simple models, and, it is not unusual that the only two sources of information are the number of messages in discussion forums and the number of downloads.

In order to be adopted in the industry, metrics need to be instantly meaningful. We propose is a measuring model *BULB*, which relates the measurements to the well-known onion model [7] of OSS communities. The onion model is commonly accepted and it is easy to grasp, so BULB conforms to the prerequisites. The measurements are based on robots digging continuously information from various sources, like the discussion forum, the version control and the bug repository, i.e. the framework conforms to the rest of the conditions.

The rest of the paper is structured as follows. In Section 2 ways to measure open source communities are discussed. The BULB model for measuring open

P. Ågerfalk et al. (Eds.): OSS 2010, IFIP AICT 319, pp. 342–347, 2010.

source communities is introduced Section 3 and applying it to an industrial community is given Section 4. The paper in concluded in section 5

## 2    Measuring Open Source Communities

The community behind an open source project is a key component that effects the success of the project. Information on the nature of the community is needed in order to make informed decisions on adopting open source software and to aid running a successful business based on open source components. Several different approaches have been suggested to provide support in the decision making ranging from easy to get to more extensive analysis.

**Social Network Analysis.** Any open source project can be seen as a social network of developers. The developers are linked to each other through different kinds of relationships that are created and maintained in OSS projects mainly by computer-enabled channels. The OSS community is thus seen as a graph with the developers as the nodes and the social relationships between developers as the edges. Social network analysis (SNA) [3,10] can be used to study the community and its structure.

**Business Readiness Rating.** Business Readiness Rating (BRR) [2] proposes a method for assessing open source software. The goal is to get a rating on the open source software through four steps (1. quick assessment, 2. target usage assessment, 3. data collection and processing and 4. data translation) As BRR itself admits that phase three is the most time consuming and yields best results for mature projects, its value is somewhat limited to eliminating bad candidates. It also seems apparent that BRR is no longer being developed further at the moment.

**Simple metrics.** One way to evaluate the open source project is to measure some publicly available data that are easy to access and measure. Very simple metrics, such as the amount of downloads or the daily amount of discussion on email lists or the project discussion forum are used. Naturally the level of activity in the community shows if the community is still alive, but says very little on the product or the sustainability of the community on the whole.

**Software use.** Software use is naturally a popular metric albeit one that is difficult to measure reliably in the case of open source software. Numbers of downloads alone do not reliably tell about adoption [1]. Moreover from the business decision point of view, the community as a whole is interesting, not just the usage.

**Surveys.** Surveying community members is a suitable method for gaining information from different interest groups within the community [9,4,11]. However, surveys don't necessarily reach people, whose input would be most valuable. In addition, surveys cannot be used as a mean of continuous analysis as the likelihood of people answering them decreases over time.

**Appearance.** A common method of comparing projects and making decisions on the project to use is not based on any kind of measuring as there often is no time to undergo a vast analysis. The appearance of and the feeling one gets from the community are the driving factors instead of a more formal approach. Results beyond a blatant guess are needed, and therefore the need to evaluate the community further is apparent.

**Onion-Based Measuring.** Open source communities can be modeled with an onion model introduced in [7]. In the model each member of the community has a distinct role. The community is seen as an onion-like structure, where the most influential community members occupy the core layers, while the outer layers hold the less influential ones.

## 3   Constructing an Onion-Based Model: BULB

In this section a measuring model for onion-based measuring of open source communities, BULB, is given. The structure of the community and how the community members fall on layers in the onion is valuable information about the community and its current state in making business decisions. BULB has been developed for this very purpose. The theoretical base of the model has already been introduced in [6]. In it, the traditional onion model for open source communities is substituted with two onions, one for the size of the community and another for the amount of activity on the onion layers. The traditional size onion is produced by simply assessing the number of people on each layer of the onion. Data from several data sources is combined to get a picture of the structure of the community according to the onion model. Data used at this

**Fig. 1.** The data sources on the onion

point range from the version control system information to the number of web hits for relevant sites. Some of the data sources have more effect on the onion built out of the data than others. This is taken into account when constructing the onion model for the community. The data sources used and the onion layers affected by each source are depicted in Figure 1.

The onion is seen as a vector, where each element contains the relevant information about the layer, for example in the size onion the number of people on the corresponding onion layer. Each metric used is measured daily and a vector representing its distribution on the onion layers is created by multiplying it with a coefficient vector that indicates how influential the metric is on each layer. The coefficient values of the layers add up to 1.0. If we denote the set of metrics used with $M$ the distribution vector $d_i$ of each metric is calculated:

$$\forall m_i \in M : d_i = m_i v_i \tag{1}$$

where $v_i$ is the coefficient vector of $m_i$. The example vectors used in the case studyfor distributing the numbers of bug reports and feature requests over the onion are shown in Figure 2. The different data sources can have significant

**Fig. 2.** Example coefficient vectors

differences in their relative values as one can be very large while the other occurs more rarely and is thus smaller. To compensate this the distribution vector of each metric is multiplied with a balancing coefficient $b_i$. We get a partial onion vector:

$$p_i = b_i d_i \tag{2}$$

The significance of the metric on the onion can also be scaled through this coefficient. In the example measurements, the balancing coefficients used were 7.0 for bug reports and 9.0 for feature requests.

After the balancing the complete onion is created by simply adding the values on each onion layer in the partial onion vectors together in order to create the final onion, i.e.

$$o = \sum p_i \tag{3}$$

The traditional onion alone is not able to accurately depict how active the community members on the different layers of the onion are but is simply focused on the size and structure of the community. The activity may vary over time although the size of the community has not changed. Thus the variation in activity on the different layers should be taken into account in addition to the development of the size of the community. As some of the metrics used may give information about the current level of activity on a given layer BULB suggests a second onion similar to the size onion to be used for depicting layer activity. The activity onion is built like the size onion only based on the metrics that measure activity. The distribution vectors and the balancing coefficients are naturally adjusted suitably. In the example case, the balancing coefficients change to 100.0 for bug reports and 140.0 for feature requests as they are clear indications on activity.

## 4    Experimenting BULB with the Vaadin Community

The BULB model is in fact a generic method of depicting the evolution of an open source community. The data values in the onion vectors can be changed to a new community characteristic and the model is still applicable. To study the applicability of the framework, we have experimented it with the developer community of Vaadin [8]. The measurements were carried out from May 1 2009 to Nov 30 2009. The onion model was instantiated to the Vaadin community as shown in Figure 1.

Vaadin is a server-side AJAX web application development framework developed by Oy IT Mill Ltd [5]. The framework is used for developing rich Internet applications with the Java programming language. Vaadin framework was released as open source in December 2007. The business model of the company is based on consulting services and the development of Vaadin. As Vaadin is open source, IT Mill needs up to date information about the Vaadin community. So far the main source of information has been the number of downloads and the number of messages post to the discussion forum. Figure 3 illustrates the measured activity in the Vaadin community over the measurement window. It visualizes the effect of events that have impact in the community. The size data however needs to be filtered to lessen the weekly variation in the raw measures. The size onion of the Vaadin community after filtering the data with a Gaussian filter is shown in Figure 4. The window size of the filter was 31, $\mu = 0$, $\sigma = 31/4 = 7.75$.

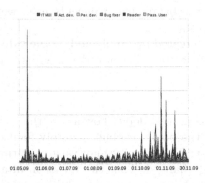

**Fig. 3.** The activity onion of the Vaadin community

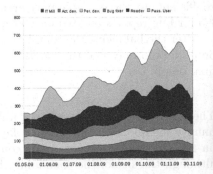

**Fig. 4.** Size onion of the Vaadin community

# 5   Conclusions

We have developed a new onion-based model, BULB, for measuring open source communities. We applied the model in an industrial case to measure the layer sizes in the Vaadin community onion and the activity on the layers. The model was developed in cooperation with industry to make an easy-to-use and fast way for digging valuable information on an open source community out of the available data.

The board and other stakeholders of an open-source company or of companies thinking of adopting an open source product often base their decision to a limited amount of information. With BULB these decisions can be based on more fresh and divergent information than before. We have shown that the described model works and produces sufficiently precise information fast enough to be useful and support decision making.

# References

1. Wiggins, A., Howison, J., Crowston, K.: Heartbeat: Measuring Active User Base and Potential User Interes in FLOSS Projects. In: Open Source Ecosystems: Diverse Communities Interacting. IFIP Advances in Information and Communication Technology, vol. 299, pp. 94–104. Springer, Heidelberg (2009)
2. BRR, http://www.openbrr.org/ (Last visited December 2009)
3. Del Rosso, C.: Comprehend and analyze knowledge networks to improve softaware evolution. Journal of Software Maintenance and Evolution: Research and Practice 21(3), 189–215 (2009)
4. Capra, E., Fancalanci, C., Merlo, F., Rossi Lamastra, C.: A Survey on Firms' Participation in Open Source Community Projects. In: Open Source Ecosystems: Diverse Communities Interacting. IFIP Advances in Information and Communication Technology, vol. 299, pp. 225–236. Springer, Heidelberg (2009)
5. Oy IT Mill Ltd., http://www.itmill.com/ (Last visited December 2009)
6. Heinimäki, T.J., Aaltonen, T.: An onion is not enough: Living in the multi-onion world. In: Proceedings of the Open Source Workshop - OSW 2009 In conjunction with the 4th IEEE Systems and Software Week (SASW 2009), Skövde (October 2009)
7. Nakakoji, K., Yamamoto, Y., Nishinaka, Y., Kishida, K., Ye, Y.: Evolution Pattern of Open-Source Software Systems and Communities. In: IWPSE 2002: Proceedings of the International Workshop on Principles of Software Evolution (2002), pp. 76–85. ACM Press, New York (2002)
8. Grönroos, M.: Book of Vaadin: Vaadin 6. Oy IT Mill Ltd. (2009)
9. Ghosh, R.A., Glott, R., Krieger, B., Robles, G.: Free/Libre and Open Source Software: Survey and Study. In: International Institute of Infonomics (2002)
10. Wasserman, S., Faust, K.: Social Network Analysis: Methods and Applications. Cambridge University Press, Cambridge (1994)
11. Mikkonen, T., Vainio, N., Vadén, T.: Survey on four oss communities: description, analysis and typology. Empirical insights on open source software business (2006)

# A Network of FLOSS Competence Centres

Jean-Pierre Laisné[1], Nelson Lago[2], Fabio Kon[3], and Pedro Coca[4]

[1] Open Source Strategy Director for Bull, President of OW2 Consortium
Coordinator of Qualipso Competence Centres, Coordinator of 2020 Floss Roadmap
http://www.ow2.org/
http://www.qualipso.org
http://2020Flossroadmap.org

[2] Techical director at the Centro de Competência em Software Livre Qualipso at
University of São Paulo
http://ccsl.ime.usp.br

[3] Professor at the Department of Computer Science at, University of São Paulo
Head of Centro de Competência em Software Livre Qualipso at
University of São Paulo
http://www.ime.usp.br
http://ccsl.ime.usp.br

[4] Univeristy Rey Juan Carlos at Madrid, Morfeo Qualipso Competence Centre
http://www.urjc.es/
http://cc.libresoft.es/

**Abstract.** The goal of a Network of Competence Centers is to provide to FLOSS users, developers, and consumers, high-quality resources and expertise on the various topics related to FLOSS. This may be achieved via education, training, consulting, hosting, and certification not only in terms of tools and platforms but also methodologies, studies, and best practices. Based on the experience of QualiPSo Competence Centres, we observe how such a Network is working as a mechanism for sharing success stories, failures, questions, recommendations, best practices, and any kind of information that could help the establishment of a solid international collaborative environment for supporting quality in FLOSS. New Competence Centres are invited to the QualiPSo Network after their proposals are evaluated by the QualiPSo Competence Centres Board to ensure that the prospective Competence Centre is compliant with the QualiPSo Network Agreement, sharing a common vision and ethics. Each Competence Centre acts in its geographical region to increase the awareness of FLOSS and to better prepare the IT workforce for developing and using FLOSS based solutions. As of 2009, the process for Competence Centre creation is sustainable and reusable; guidelines for establishing proposals and opening new Competence Centres have been created, and promotion of Qualipso Competence Centres is done world wide from India to USA thanks to key initiatives such as the Open World Forum and the FLOSS Competence Centre Summit. This lecture will expose how these Competence Centres relate to each other, which governance model is used and, based on existing experiences, will describe how they currently operate in Europe and Brazil and what is planned in Italy, Belgium, Japan, and China for 2010.

P. Ågerfalk et al. (Eds.): OSS 2010, IFIP AICT 319, pp. 348–353, 2010.

# 1  FLOSS Adoption: It's All about TRUST

There are significant signs of broad dissemination of FLOSS concepts both within industries and within governments. Nevertheless, there is still reluctance to massive adoption of FLOSS development, mainly due to lack of confidence. Several "grey areas" around FLOSS cause concerns: legal uncertainties, such as IP protection and indemnification; quality guarantees, such as development life cycle, documentation, support, reliability, and performance; and, finally, business issues such as business models capable of maintaining sustainability.

All these concerns can be summarised in one word: *trust.* And trust is not a quality that can be claimed without being proved. It also relies on perception, on non technical questions such as "Who is behind FLOSS?", "Why be confident in FLOSS?", or even "How to be confident in FLOSS?".

## 1.1  A Distributed Network of Trustworthy and Highly Skilled Resources

It is commonly known that "people talk to people". We want to make use of this fact to establish confidence in FLOSS by offering independent and qualified support and services by means of Competence Centres disseminated around Europe, Brazil, and China. Interconnected among them, these Competence Centres represent a network that openly offers access to skilled resources and promotes trust in FLOSS: It enables end users, ISVs (Independent Software Vendors), developers, etc. to find answers to their questions and use FLOSS in their operations in a reliable and trusted manner.

These services are delivered thanks to collaborative platforms, tools, and process developed mainly by the QualiPSo project[1]. Each Qualipso Competence Centre, having a basic set of functionalities, represents an aggregation point of technologies, skills, and policies. While these competence centres may differ one from the other by their level of expertise on specific domains, all together the Qualipso competence centres form a network of expertise sharing the same ethics and values.

## 1.2  In FLOSS We Trust

Each competence centre is a set of physical resources (bricks & mortar) and virtual sites dedicated to maximize the reusability of tools, processes, and shared knowledge by offering services and training. These services will make documents, tools and platforms openly available to organizations needing to assess the legality, sustainability, robustness, and interoperability of their critical applications running on top of FLOSS.

Qualipso Competence Centres will also favor the reusability of the results of the R&D effort of the Qualipso Project not only in terms of tools and platforms

---

[1] A research project that aims to work on several issues related to FLOSS in order to foster its adoption by the industry at large.

but also of studies, best practices, and other miscellaneous information. They will make these results available to both Community and Industry and will promote them at large through communication programs (public and press relations, white papers, seminars, courses, workshops, conferences, etc.).

The Qualipso Competence Centres are designed to be replicable but replication can only work if all components of the competence centre are reusable from legal to technical aspects. Therefore, all information concerning competence centres is documented and this documentation is freely available to anyone. Transparency of the process must insure that there is neither hidden agendas nor Trojan horses in the competence centre model. The replication aims to provide in each country the same level of information, the same homogeneous tools and processes, and also some dedicated local resources which act as national FLOSS experts specialized on identified specific topics.

QualiPSo competence centres are to be distributed worldwide, and may be instantiated by geographical location, e.g., Paris, Berlin, Madrid, São Paulo, Tokyo, Guangzhou; or by organizations, such as OW2, Morfeo; or even by large companies, such as Telefonica or eGovernment-oriented companies such as SERPRO in Brazil.

## 2    General Description

The goal of the Qualipso Network is to federate all Qualipso Competence Centres sharing the same vision, goal, ethics, methods, and tools. So, at the heart of the Network, stands the *Qualipso Network Agreement*. This document describes how these principles are implemented and how all components of the Qualipso Network are governed.

Each Competence Centre has to define its policy concerning aspects such as its mission statement and scope, market and geographical area of interest, technological specialization, legal and funding model, communication plan etc. A more detailed description will be given during the lecture for attendees to evaluate the compliance of these elements with the FLOSS culture. This description is based on the content of Qualipso Network Agreement avilable at http://www.qualipso.org/sites/default/files/Qualipso — D8.2 Network Agreement V1.17.pdf.

### 2.1    Structure of the Network and Its Components

At the most basic level, the Qualipso Network represents the network of Qualipso Competence Centres. It is a distributed network of trustworthy and highly skilled resources that reuse technologies, procedures, and policies produced by the Qualipso Project. The Qualipso Network federates all Competence Centres and its goal is to protect Qualipso commonly produced assets, intellectual property, and brands. At its head, we find the Qualipso Network Board, which is composed of representatives of all Qualipso Competence Centres. This board manages conflicts of interest and decides on matters related to: registration/cancellation of

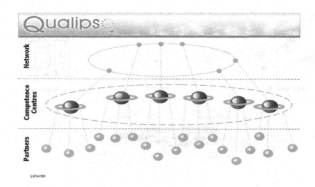

**Fig. 1.** Qualipso competence centres network architecture

Competence Centres, protection of common assets, approval of new Qualipso services, and updates of the Qualipso Network Agreement.

At the next level, Qualipso Competence Centres facilitate reusability of the results of the Qualipso project. Competence Centres provide expertise and services in the form of a set of independent and vendor neutral basic services, tools, and methods on topics addressed by the Qualipso Project from legal advice to benchmark results. Specific services may also be created. Each Competence Centre shall have its own legal model in compliance with defined principles, in order to ensure sustainability of the activities, fairness among partners, openness to new partners, and to define liabilities, responsibilities, and territory. Each Competence Centre shall also comply with the Qualipso Network Agreement and be self-sustainable, i.e., manage its own revenues, define its own funding model, and ensure its efficiency. In short, Competence Centres act locally and cooperate globally.

Finally, Partners are organizations that decide to create a Competence Centre according to some common interest (geography, language, technology, etc.). Partners are committed to provide the necessary resources for each Competence Centre to achieve its goals. There are two types of Partners: Active Partners, which contribute to daily activities, and Associate Partners, which act as sponsors.

## 2.2   Benefits of Belonging to the Qualipso Network

As part of the Qualipso Network, Competence Centres are able to share success stories, failures, questions, recommendations, best practices, and any kind of information that could help the establishment of a solid international collaborative environment for supporting quality Open Source Software. Therefore, Competence Centres and their worldwide network will support the continuing development of the Qualipso vision, helping to provide a sustainable quality platform for Open Source Software.

To summarize, the key benefits of the Qualipso Network of Competence Centres are:

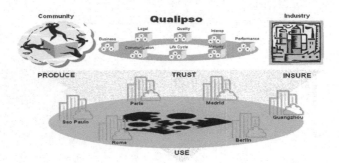

**Fig. 2.** Interaction between FLOSS comunities, the QualiPSo research project, industry, and competence centres

- Global synergies on marketing and communication.
- Multinational shared expertise and knowledge base.
- Local presence through a global network and access to different markets.

## 3   Competence Centres at Work

The Qualipso Competence Centre in Spain (http://cc.libresoft.es/) is a joint effort of the GSyC/LibreSoft research group at the Universidad Rey Juan Carlos (URJC), the CETTICO research group at the Universidad Politécnica de Madrid (UPM) and Telefonica R&D. These are founding members of the MORFEO community, with which the Competence Centre shares a close link. Teófilo Romera Otero, from the GSyC/LibreSoft group, states: "The creation of the Competence Centre is an opportunity to uplift our activities to a higher and institutionalized level, allowing a better impact for the creation of knowledge and the services we have been already offering for years".

The Qualipso Competence Centre in Brazil (http://ccsl.ime.usp.br) is located at the University of São Paulo and started its activities in December, 2008. According to Prof. Fabio Kon, the São Paulo Competence Centre director, "the Competence Centre is an excellent means for the university to communicate with the Brazilian society and software industry; it contributes not only sharing the knowledge developed by our research groups but also working as a meeting point for students, researchers, and practitioners from our region and abroad; we also expect to benefit from being part of an international collaborative network focusing on the quality of Free and Open Source Software".

The German Competence Centre is located at the Fraunhofer Institute for Open Communication Systems (FOKUS) in Berlin and started its activities in the first quarter of 2009. Its main emphasis is on providing a factory for FLOSS projects and on Qualipso services in the context of technical, semantic, and organizational interoperability between Open Source systems as well as between Free/Libre/Open Source systems and closed source systems. Governmental organizations as well as industry and research will benefit from these contributions and the Qualipso network.

# 4 Beyond Qualipso Competence Centres: FLOSS CC Manifesto

As a result of the First International FLOSS Competence Centres Summit, organized by Qualipso as part of the Open World Forum, 11 leading FLOSS promotion organizations worldwide (Berlin, Chennai, Guangzhou, Madrid, Maribor, Newry, Paris, Portadown, Raleigh, São Paulo, Tokyo) joined forces to create a solid international network of FLOSS Competence centers to share experiences, define best practices, strengthen synergies, and collaborate on the promotion of FLOSS.

As a first step, this network provides a Manifesto for FLOSS Competence Centres, written and signed by all network members, exposing the view described in this text. The Manifesto is available at http://www.flosscc.org .

Michael Tiemann, President of the Open Source Initiative and Vice President of Open Source Affairs at Red Hat, confirmed that: *"This unique initiative will further enhance FLOSS initiatives in a sustainable manner world wide. All together the Competence Centres will form a network which will enable to share experiences about FLOSS acquired from different perspectives, different cultures and different visions. And all final results will be made freely and openly available to the entire FLOSS community by people talking the same language and sharing the same problems. This represents a notable step towards a true global knowledge economy."*

# 5 Conclusion

By sharing a common ethics and culture of collaboration, Competence Centres promote synergies among educational institutions, industry, government, and communities, helping the dissemination and application of knowledge on open standards and technologies, and promoting the development of Information Technology in a way that benefits the entire human society. Furthermore, Competence Centres aim to stay one step ahead in FLOSS market and trends, providing a point of contact among the industrial, academic, and community parts of the FLOSS movement, encouraging the effective use of FLOSS technologies. In short, Competence Centres have the ambition to be significant players of an Information Society based on knowledge sharing.

# Profiling F/OSS Adoption Modes: An Interpretive Approach

David López[1], Carmen de Pablos[2], and Roberto Santos[3]

[1] University of León, Spain
david.lopez@unileon.es
[2] Rey Juan Carlos University, Spain
carmen.depablos@urjc.es
[3] Telefónica de España, Spain
roberto.santossantos@telefonica.es

**Abstract.** This article presents the findings of a research aimed at characterizing F/OSS migration initiatives, in total 30 experiences have been considered, 19 of which have been conducted by public administrations and the rest by private firms, operating different industries in eight different countries.

Open source migration projects is a recent research topic, more so when considering it from a managerial perspective. To overcome the lack of theoretical models an empirical approach relying on grounded theory has been adopted as this inductive approach allows theory building and hypothesis formulation.

According to the results, migrating from proprietary into open source is dependent on contextual and organizational factors, as for example, the need of the change itself, the political support for the change, the suitability of IT, the organizational climate, the motivation of the human resources, the kind of leadership for the project or the firm complexity. Besides, migration efforts imply strategic and organizational consequences that the organization must evaluate well in advance.

## 1 Introduction

F/OSS as a radical approach to software development started in the early seventies consolidating itself as an alternative business model in the nineties. Since then F/OSS has been thoroughly studied from a technical perspective [1],[2] and as an emerging economic market [4],[5].

F/OSS is arguably a cost-effective solution in public administrations or education in which large investments in hardware are often required [3],[6] moreover the application of F/OSS tools promotes innovation and industry development worldwide [7],[8],[9]. F/OSS migration initiatives are gaining momentum in the international arena as many developing countries are embracing this model for their ICT policies [10],[11].

Notwithstanding the benefits that F/OSS poses for administrations and large companies, there is still some reluctance among organizations. Migrating into

P. Ågerfalk et al. (Eds.): OSS 2010, IFIP AICT 319, pp. 354–360, 2010.

F/OSS is, as Mr. Schiel technical leader at Munich city hall claims: "LiMux is not a technical project", he says. "Initially, the team approached the migration as a classical IT problem, but the real issues turned out to be different. "It's all about managing change for and with people." At this point we are able to confirm this quote as some of our interviewees do agree with Mr Schiel.

The purposes of this paper are twofold. In the first place presenting results regarding international F/OSS migration experiences. In the second place providing insights into open source adoption by organizations.

The latter goal is paramount to define adequate public policies aimed at fostering F/OSS adoption whereas the former goal provides practitioners with guidelines to maximize value in migration initiatives.

## 2   Qualitative Inductive Research Approach

Grounded theory belongs to the set of qualitative research methods aimed at developing theory grounded in existing data gathered from real scenarios [12]. Being grounded theory an inductive, discovery methodology it allows to produce emergent theory in areas in which there is still knowledge gaps.

Grounded theory is specially suited in context-based, process-oriented descriptions of organizational phenomena [13]. Initially applied in psychology [14],[15] it has been successfully applied in the information systems area for some time now [17] including open source research [18].

Some scholars have studied technology adoption processes [19], [20] notwithstanding the usefulness of modeling user's behavior to technology adoption for our purposes a broader perspective is required. Adopting open source and subsequent migration of existing services is not a process conducted by a single person not even by a single business unit.

According to the above mentioned purposes of this study a microanalysis approach, that is organization centered, is mandatory. This approach departs from quantitative, positivist approaches [21],[22] in the sense that no previous hypothesis are formulated with regards to F/OSS adoption, on the contrary theory is being developed incrementally according to revealed data. We believe both micro -or firm level- and macro analysis complement each other reinforcing together into a global perspective of F/OSS.

### 2.1   Data Collection

In total 30 migration projects have been collected. Following Glaser and Strauss' technique of theoretical sampling, migration projects have been selected according to maximum variability in terms of context, size, purpose, ICT intensity among others. Having many several factors intervening at different intensity levels is key to define concepts, categories and relationships i.e emergent theory [15].

Nineteen of the total documented experiences correspond to public administrations from eight different countries (Brazil, France, Germany, UK, Spain,

Finland) and ten for-profit organizations from small medium companies to large multinationals such as Audi or Peugeot. For every organization the following aspects have been initially considered:

1. Migration objectives.
2. Migration timeframe
3. Type of software or service migrated.
4. Migration cost.
5. Migration critical success factors.
6. Migration critical failure factors.
7. Migration outcome.
8. Perceived benefits.

Data collection in grounded theory is conducted iteratively [14] starting with a general exploratory character and later on more focused approaches towards relevant topics and structured interviews.

Whenever possible personal contact were established with project leaders in order to gather as much information as possible. This information was complemented with existing information in journals, magazines and technical reports in specialized repositories such as OSOR.eu or epractice.eu[1].

Alongside this qualitative approach a survey among public administrations in Spain and Spanish software providers was conducted to further inquire into how market perceived open source as an alternative. Triangulating both qualitative and quantitative data provides more robust conclusions[12].

## 2.2 Data Analysis

Inductive analysis aims at finding relevant concepts and relationships among them. Initially in the so called open coding researchers thoroughly analyze existing data and summarize it into categories [15]. In a second stage, axial coding, initial categories are further developed into subcategories representing variations along dimensions (i.e. axial coding), these subcategories often match questions such as: when, how, why, where and what for.

Given the intra-firm analysis perspective adopted in this research F/OSS migration experiences are considered as internal processes of change. We believe this process-centered approach reflects real organizations' structures therefore facilitating its adoption in existing business units.

As figure 1 shows, F/OSS migration projects are consisted of three main stages: -or main categories in Grounded theory parlance- Adoption Process, Migration Process and finally Migration Results. Then, for each category, several subcategories representing intervening factors are identified.

---

[1] Further information is available at:
http://cenatic.interoperabilidad.org:8080/web/guest/grandes-empresas
http://cenatic.interoperabilidad.org:8080/web/guest/
administraciones-publicas

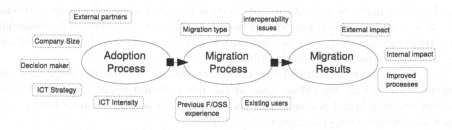

**Fig. 1.** F/OSS migration life cycle

# 3   Research Results

Three main modes of F/OSS adoption emerge from the analysis: Organizations concerned with vendor independence which consider open source software as a strategic asset. Organizations embracing open source as a better option than privative one in terms of performance or new features. Finally, in the third adoption mode, organizations striving for budget constrain consider open source as a cost-effective option in terms of licensing and hardware obsolesce.

**First mode: Strategic movers.** In this adoption mode, organizations have faced vendor issues such as forced migrations, ceasing product support or spiraling ICT costs. Removing vendor lock-ins is considered a priority by top executives who define strategic plans to evolve into the F/OSS paradigm.

These organizations tend to be large both in human resources and ICT budget. They have specific staff dedicated to ICT development and support.

Being intensive users of ICT services, for instance public administrations, they face interoperability issues with existing services as well as users' resistance to change established routines and procedures. They adopt a gradual approach to software migration usually with schedules in the order of years to complete. Sometimes rescheduling is mandatory.

Migration processes in this case often require external partners with technical expertise and large scale deployment experience specially if the organization has not previous experience or faces challenging scenarios of non-interoperable technologies and information.

The outcome of migration initiatives in this case are positive in general with positive results in terms of improved internal processes. Besides they usually create positive externalities in other business units. Due to the large scale nature of this initiatives often they impact society in general either by providing new services or improved software publicly available. Clear examples of this adoption mode are: Munich city hall, French police or USA postal service.

**Second mode: Feature seekers.** This category is consisted mainly of medium to large organizations with intensive use of ICT in their business processes looking for new opportunities to improve current existing IT capabilities, for instance supercomputing or embedded systems. Examples of this mode are Audi or Peugeot.

Migration projects are initiated by internal experts taking advantage of recent developments, either software or hardware. An IT expert working for a financial institution claims that: "We are not fans of F/OSS we stick to the best solution given the context and existing options"

In spite of having technical expertise they usually collaborate with external partners for specific expertise. Due to the non-core nature of affected services interoperability is not a concern neither are end users who remain oblivious of backend reengineering programs.

Contrary to the previous category, migration timeframes remain within months with defined procedures well in advance. Success criteria are clearly determined. The results of this initiative are seldom transformative in the sense of producing impacts just on specific internal services.

**Third mode: Budget optimizers.** Small to medium companies often adopt open source as the best option in terms of price performance ratio. They usually implement new, non-core services, with open source as an inexpensive approach. The typical profile is a local administration or SME operating in a non-intensive ICT market.

## 4    Implications for F/OSS Migration Processes

As a general rule it seems a good practice to look for interoperable technologies in existing or future software deployments in order to ensure optimal vendor independence and flexibility to adapt to business requirements.

Some companies adopt a migration strategy relying on early adopters to test prototypes adapting them according to received feedback. We consider this approach quite adequate for it ensures business continuity while involving users from start, this reinforces the idea that F/OSS migration initiatives entail technical, organizational and business process reengineering [23].

According to our data most of the times large organizations hire external companies providing technical expertise or migration experience. Furthermore there exist some cases, mainly in very specialized areas such as mathematical modeling or embedded hardware, in which several partners collaborate to come up with an adequate solution.

No matter the F/OSS adoption mode -either strategic, functional or optimizing- companies start considering open source as an alternative to specific issues and gradually have it adopted in subsequent initiatives. This behavior based on reinforced trust is consistent with technology acceptance models [19].

## 5    Future Work

F/OSS adoption offers good opportunities for reaching efficiency, flexibility and security in organizational processes, but it also poses challenging questions. Modeling users' response to technology changes is paramount to integrate new software into already existing organizations, at this point recent results enhance former Technology Acceptance Models (TAM) by considering software as a social actor within the organization able to interact with employees at increasing sophistication levels [20]. We believe that further research could provide interesting results complementing this interaction-centric models with constructs valid at a firm and intrafirm level.

In every migration project there is always an internal sponsor or a group of people leading the initiative. Further research into leadership aspects of open source projects may serve practitioners to identify best organizational patterns to induce F/OSS adoption.

An interesting fact that emerged during the present research is that some organizations, mainly large multinationals, are able to generate positive externalities in other units or even contribute with their own developments back to the open source community. Being public administrations large ICT consumers providing further insight into effective means to induce innovation, software reutilization for instance, would be an interesting research topic.

There is international consensus on the importance that ICT has on education and development, 10 percent of broadband penetration increases GDP in developed economies up to 1.2 percent [24]. Being F/OSS capable of provisioning computers at lower costs there may be incentives by governments to engage in national initiatives to promote digital literacy relying on the F/OSS paradigm. Documenting exemplary initiatives in this matter is encouraged.

**Acknowledgment.** The authors gratefully acknowledge the support of CENATIC, the Spanish agency in charge of open source promotion and Telefónica.

# References

1. Raymond: The Cathedral and the bazaar. Knowledge. Technology and Policy 12(3) (1999)
2. Hunter: Open Source Data Base Driven Web Development. Chandos, Oxford
3. Lerner, Tirole: The simple economics of Open Source. Journal of Industrial Economics 50(2), 197–234
4. Lerner, Tirole: The Economics of Technology Sharing: Open Source and Beyond. Journal of Economic Perspectives 19(2), 99–120
5. Riehle: The Economic Motivation of Open Source: Stakeholder Perspectives. IEEE Computer 40(4) (2007)
6. Lakhan, Jhunjhunwala: Open Source in Education. Educause Quarterly 31(2), 32–40
7. Shiff: The Economics of Open Source Software: a survey of the early literature. Review of Network Economics 1(1), 66–74
8. Hippel, Krogh: Open source software and the private-collective innovation model: Issues for organization science. Organization Science 14(2), 209–223
9. Osterloh, Rota: Open source software development, just another case of collective invention. Research Policy 36(2), 157–171
10. Ahmed: Migrating from proprietary to Open Source: Learning Content Management Systems, Doctoral Dissertation, Department of Systems and Computer Engineering, Carleton University, Ottawa, Ontario, Canada

11. UOC Report: The use of open source in Public Administrations in Spain, Universitat Oberta de Calalunya, Report
12. Myers: Qualitative Research in Business and Management. Sage, Thousand Oaks (2009)
13. Myers: Qualitative Research in information systems. MIS Quaterly 21(2), 241–242 (1997)
14. Glaser, Strauss: The discovery of Grounded Theory: Strategies for Qualitative Research. Aldine Publishing Company, New York (1967)
15. Strauss, Corbin: Basics of Qualitative Research: Grounded theory, Procedures and Techniques. Sage, Thousand Oaks (1990)
16. Orlikowski: Information Technology and the Structuring of Organizations. Research Approaches and Assumptions. Information Systems Research 2(1) (1991)
17. Orlikowski: CASE Tools as Organizational Change: Investigating Incremental and Radical Changes in Systems Development. MIS Quaterly (Septermber 1993)
18. Dedrick, West: Why Firms adopt Open Source Platforms: A grounded Theory of Innovation and Standards Adoption. MIS Quaterly. Special Issue on Standard Making (2005)
19. Venkatesh, V., et al.: User acceptance of information technology: Toward a unified view. MIS Quarterly 27(3) (2003)
20. Sameh, Izak: The Adoption and Use of IT Artifacts: A New Interaction-Centric Model for the Study of User-Artifact Relationships. Journal of the Association for Information Systems 10662(9) (2009)
21. Gonzalez-Barahona: About free Software, Rey Juan Carlos University-Dykinson (2004)
22. Wheeler: Why Open Source Software. Look at the Numbers
23. Hammer: Reengineering Work: Do not Automate, Obliterate. Harvard Business Review (2000)
24. Berkman Center: Berkman Center for internet and society. Broadband study for FCC (2009)

# Introducing Automated Unit Testing
# into Open Source Projects

Christopher Oezbek

Freie Universität Berlin
Institut für Informatik
Takustr. 9, 14195 Berlin, Germany
`christopher.oezbek@fu-berlin.de`

**Abstract.** To learn how to introduce automated unit testing into existing medium scale Open Source projects, a long-term field experiment was performed with the Open Source project FreeCol. Results indicate that (1) introducing testing is both beneficial to the project and feasible for an outside innovator, (2) testing can enhance communication between developers, (3) an active stance is important for engaging the project participants to fill a newly vacant position left by a withdrawal of the innovator.

## 1 Introduction

The Open Source development paradigm based on copyleft licenses, global distributed development and volunteer participation has become an alternative development model for software, competing on par with proprietary solutions in many areas. Open Source software especially has established a good track record related to quality measures such as number of post-release defects or time to resolution for bug reports [8,10] based on its open access to source code, openness to participation and use of peer review [13].

The present study originated in the question how to further improve a project's ability to produce high-quality software. From a software engineering perspective the answer proposed in previous work was to introduce innovative processes and tools into Open Source projects [9]. But is such introduction feasible? How must an innovator act to achieve adoption of the introduced innovation? The present study is a first exploration on these questions.

Quality assurance was chosen as the area for improvement and automated unit testing [14] as the innovation, because it represents a well-known and established quality assurance practice from industry, which should easily provide benefit to Open Source projects. Methodologically, an introduction conducted by a researcher is in-between action research [1] and a field experiment [5], because the researcher is interacting in the field but using his own agenda.

The study proceeded in four steps: First, a theoretical model was built of how to introduce automated testing to make the process reproducible by others. This model prescribes activities and goals for lurking [11], joining and acting [2,12],

P. Ågerfalk et al. (Eds.): OSS 2010, IFIP AICT 319, pp. 361–366, 2010.

collaborating and phase-out of the innovator and is shown in Figure 1. Second, the project FreeCol was selected from the project hoster SourceForge.net based on several criteria such as being medium-sized and open for outside participation to ensure interesting interaction and relevant results. FreeCol was started in March 2002, trying to recreate the turn-based strategy game Colonization[1]. FreeCol is a client-server application written in Java and regularly ranked in the top 50 of Open Source projects at the SourceForge.net with on average 16,500 copies downloaded per month. The project has 60 members enlisted on the project page[2] of which 46 are designated as developers and 13 of which are deemed active[3]. The project already had one test case using JUnit at the beginning of this study. Third, testing was introduced into this project, which took place in April and May 2007 following the phase model shown in Figure 1 and resulting in 57 test-cases. In September 2007 the test-suite was broken by a large scale refactoring and the project maintainers asked for a repair, which was performed as a last activity in the project. Fourth and last, the outcome of the introduction was analyzed post-hoc in September 2009 by means of (1) data mining the source code repository [6] of the project for test coverage and test failures [15] and (2) qualitatively analyzing the mailing-list communication on testing[4].

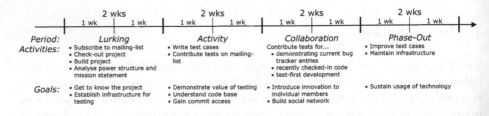

**Fig. 1.** Phases in the introduction process

## 2   Results

Looking back from April 2007 to August 2009 we find the introduction a success based on four quantitative indicators: (1) On average 9.9 test cases were being added per month, raising their number from 73 at the departure of the innovator to 277 in August 2009 (Figure 2) covering 23% of the source code (see Figure 2); a respectable figure for a UI-oriented application such as FreeCol. (2) The percentage of commits affecting test cases is stable between 10.0% to 15.1% per month with 95% confidence. (3) The source code passed the tests

---

[1] http://www.freecol.org/history.html

[2] http://sourceforge.net/project/memberlist.php?group_id=43225

[3] http://www.freecol.org/team-and-credits.html

[4] All scripts used for producing the results in this study as well as intermediate data to reproduce the statistical analysis are available at
http://www.inf.fu berlin.de/inst/ag-se/pubs/test-intro2009data.zip

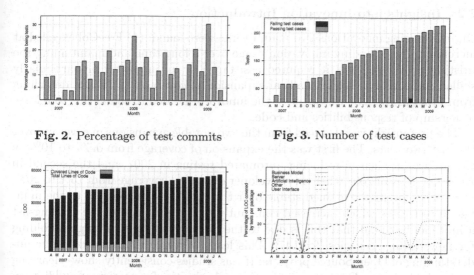

**Fig. 2.** Percentage of test commits    **Fig. 3.** Number of test cases

**Fig. 4.** Lines covered    **Fig. 5.** Coverage per modules

in most months (Figure 2). (4) Of the 32 developers who have ever committed to the FreeCol source code, 16 participated in testing. On the mailing-list several developers voiced their positive attitude towards testing, e.g. [fc:2518][5] and [fc:3351].

## 2.1 Insights into Automated Testing

The most interesting insight regarding the *use* of testing is that test-cases have been used in FreeCol to enhance communication in two ways: (1) If facing a defect without knowledge to repair or understand it, we have seen developers write failing test cases which reproduce the failure and use the test-case as a more concise alternative for communicating the failure (for instance [fc:2606] [fc:2610] [fc:2640] [fc:2696] [fc:3983]). (2) When facing ambiguity about how FreeCol should behave, we have seen developers codify their opinion as test cases [fc:3276] [fc:3056] or existing tests being the starting-point for discussions about how FreeCol should behave [fc:1935]. This is a second major advantage beside the regression detecting abilities of having a test suite (see for instance [fc:3961] or [fc:4431]).

As a second insight we found that testing varied largely by module. While the business logic including the game objects attained more than 50% coverage, other areas such as the server module at 40% and the artificial intelligence module at 22% are less tested and UI testing is completely absent from FreeCol (see Figure 2). How to expand the coverage of underrepresented modules remains an open question.

---

[5] Citations such as [fc:2518] are hyperlinks to e-mails from the Freecol Developer Mailing-list and are numbered in the order they were posted.

## 2.2  Insights into Innovation Introduction

On introducing innovations two main results were found: (1) FreeCol excelled at incrementally expanding innovation usage over a long time and maintaining the existing code base. Yet, it required assistance by an innovator or particularly skilled individual to achieve radical expansion of coverage. (2) When detaching from FreeCol Open Source project, the innovator needed to signal this to release ownership of responsibilities and code.

The first insight was deduced from the two notable expansions in coverage over the last two years. The first was the expansion of coverage from 0.5% to 10% by the innovator when introducing automated testing in 2007, and the second in April and May of 2008 when one developer expanded coverage from 13% to 20%. Otherwise coverage remained stable over the two years, in contrast to the number of test cases which increased constantly (see Figure 2). On the mailing-list a hint can be found that this is due to the difficulty of constructing scaffolding for new testing scenarios [fc:4147]. This leads to a question regarding our understanding of Open Source projects: If—as studies consistently show—learning ranks highly among Open Source developers' priorities for participation [3], then why is it that coverage expansion was conducted by just two project participants? It seems that the innovator and the one developer both brought existing knowledge about testing into the project and that project participants' affinity for testing and their knowledge about it expanded only very slowly. A similar result was reported by Hahsler who studied adoption of design patterns. He found for most projects that only one developer used patterns [4, p.121]. This should strike us as strange, if sharing of best practices and knowledge did occur frequently.

The second insight for innovation introduction resulted from phasing-out the innovator's involvement in May and September 2007. The first attempt in May failed and the test suite was unmaintained during a large-scale refactoring and soon "spectacularly broken", as one maintainer put it. Comparing this with the second more successful departure in September, which resulted in the tests being maintained by one of the maintainers, we find that the primary difference in behavior is one of signaling and ownership. When the innovator first detached, ownership was neither considered nor was the withdrawal communicated to the project. Yet, as Mockus et al. found in their case study of Apache and Mozilla, code ownership is achieved implicitly for code the developer is "known to have created or to have maintained consistently" [8, p.318]. And while such code ownership "doesn't give them [the owners] any special rights over change control", it stipulates a barrier for other developers to engage with the code.[6]

Only when the test suite broke completely after the refactoring, did it become apparent that it was unmaintained. Thus, when phasing-out the innovator's engagement again after fixing the test suite in September 2007, one discussion (see [fc:2182]) was sufficient to create an understanding of shared code ownership in testing. When the innovator disengaged, one of the maintainers picked up the

---

[6] See for instance [7] for a discussion on code ownership as an important part of the mental model of developers.

role of maintaining the test cases successfully, keeping the percentage of test affecting commits at around 10% of the total commits (see Figure 2), until another developer assumed a more active role in testing.

When analyzing the contributions of developers to the testing effort, we find that besides the innovator and the maintainer there were two individuals who contributed extensively to testing. Interestingly, as their contribution increased and waned over time, the maintainer who had already picked up the testing effort initially seemed to adjust his own contribution accordingly. As contributions of the other developers never exceeded five testing commits per month, it seems that the project adopted a flexible code ownership strategy. In this approach, the role of a "test master" exists who contributes heavily to testing and is pivotal to the expansion of test coverage and development of knowledge regarding testing. This role is not formally but rather implicitly assigned and acknowledged explicitly in the project only for instance when a core developer — stumped by a difficulty regarding testing — asked: "Any suggestions, particularly from the resident test expert [name of developer]?" [fc:4446].

## 3  Limitations and Conclusion

This study presents a first exploration into the research area of actively improving an Open Source project and, as a single case using unit testing as the innovation, can not generalize far. Other projects might have different attitudes towards testing, the domain of the software might make testing more difficult, or the researcher as the innovator could have introduced a noticeable bias. For future work, more projects, other innovations and more data source per project should thus be studied, though an active approach like in this study can not be scaled very far due to the effort associated with each case.

To conclude, this study has shown that the introduction of a code-centric process innovation such as automated testing is feasible for an outside innovator using a four-stage model. Regarding automated testing this study has found (1) a number of episodes in which test cases were used for communicating bug reports, and (2) a lack of the state of the practice regarding automated testing. The results for the innovator are that (1) external participants are important for the radical expansion of innovation use, and (2) signaling the departure of the innovator is important even for an innovation which has an explicit signaling mechanisms such as test cases failures. Open questions were raised about the extent to which participants are able to learn about new innovations.

**Acknowledgments.** Dan Delorey provided the author with a list of all java projects on Sourceforge.net that had more than 5 active developers over the course of 2006. Many thanks also to Gesine Milde, Florian Thiel, Lutz Prechelt, the FreeCol maintainers and test masters, and two anonymous reviewers who read early versions of this paper.

# References

1. Avison, D.E., Lau, F., Myers, M.D., Nielsen, P.A.: Action research. Commun. ACM 42(1), 94–97 (1999)
2. Ducheneaut, N.: Socialization in an Open Source Software community: A socio-technical analysis. Computer Supported Cooperative Work (CSCW) 14(4), 323–368 (2005)
3. Ghosh, R.A., Glott, R., Krieger, B., Robles, G.: Free/Libre and Open Source Software: Survey and study – FLOSS – Part 4: Survey of developers. Final Report, International Institute of Infonomics, University of Maastricht, The Netherlands; Berlecon Research GmbH Berlin, Germany (June 2002)
4. Hahsler, M.: A quantitative study of the adoption of design patterns by Open Source software developers. In: Koch, S. (ed.) Free/Open Source Software Development, ch. 5, pp. 103–123. Idea Group Publishing, USA (2005)
5. Harrison, G.W., List, J.A.: Field experiments. Journal of Economic Literature 42(4), 1009–1055 (2004)
6. Kagdi, H., Collard, M.L., Maletic, J.I.: A survey and taxonomy of approaches for mining software repositories in the context of software evolution. Journal of Software Maintenance and Evolution: Research and Practice 19(2), 77–131 (2007)
7. LaToza, T.D., Venolia, G., DeLine, R.: Maintaining mental models: a study of developer work habits. In: ICSE 2006: Proceedings of the 28th international conference on Software engineering, pp. 492–501. ACM, New York (2006)
8. Mockus, A., Fielding, R.T., Herbsleb, J.: Two case studies of Open Source Software development: Apache and Mozilla. ACM Transactions on Software Engineering and Methodology 11(3), 309–346 (2002)
9. Oezbek, C., Prechelt, L.: On understanding how to introduce an innovation to an Open Source project. In: Proceedings of the 29th International Conference on Software Engineering Workshops (ICSEW 2007), Washington, DC, USA. IEEE Computer Society, Los Alamitos (2007); reprinted in UPGRADE, The European Journal for the Informatics Professional 8(6), 40–44 (December 2007)
10. Paulson, J.W., Succi, G., Eberlein, A.: An empirical study of open-source and closed-source software products. IEEE Transactions on Software Engineering 30(4), 246–256 (2004)
11. Preece, J., Nonnecke, B., Andrews, D.: The top five reasons for lurking: Improving community experiences for everyone. Computers in Human Behavior 20(2), 201–223 (2004); The Compass of Human-Computer Interaction
12. Quintela García, L.: Die Kontaktaufnahme mit Open Source Software-Projekten. Eine Fallstudie. Bachelor thesis, Freie Universität Berlin (2006)
13. Raymond, E.S.: The cathedral and the bazaar. First Monday 3(3) (1998)
14. Whittaker, J.A.: What is software testing? And why is it so hard? IEEE Software 17(1), 70–79 (2000)
15. Zhu, H., Hall, P.A.V., May, J.H.R.: Software unit test coverage and adequacy. ACM Comput. Surv. 29(4), 366–427 (1997)

# A Case Study on the Transformation from Proprietary to Open Source Software

Alma Oručević-Alagić and Martin Höst

Department of Computer Science, Lund University, Sweden
Alma.Orucevic-Alagic@cs.lth.se, Martin.Host@cs.lth.se

**Abstract.** This paper presents an extensive analysis of static software quality metrics changes for an open source enterprise database management system (DBMS), as the software was moved from the proprietary into open source software development environment. The software quality metrics of special interest for the research are cyclomatic complexity, effective lines of code, the degree of system modularity, and the amount of comments in the code.

## 1 Introduction

Popularization of OSS has influenced the conventional way in which companies perceive commercial value of software. The companies recognized that commercial value of product can come from other sources rather than conventional sale of software licenses. This research assesses the impact of source code changes made by OSS community to software that was transitioned from proprietary into open source, in terms of static software quality metrics. The case software analyzed is the Ingres database management system (DBMS) [2], which, according to many, has received a new breath of life after its release into the open source community.

## 2 Background

Stemlos [3] conducted code quality analysis in open source development for 100 applications written for Linux. It was determined that some open source products have lower quality of code produced in OSS environment then that which is expected as an industry standard.

The very roots of the case software reach back to the 1970s and UC Barkley, when the initial development of the software was started as open source. The same code base was modified and spawned into Sybase and Microsoft SQL server in 1980s. In 1994, the software was acquired by CA (Computer Associates) from the ASK Group, the company that created a proprietary version of the Ingres code. In order to increase the market share, CA decided to transform the product to open source in 2004, by implementing loss-leader/market positioner business model [4]. In November of 2005, Computer Associates and Garnett and Helfirch capital created a new company, Ingres Corporation. The main role of Ingres

P. Ågerfalk et al. (Eds.): OSS 2010, IFIP AICT 319, pp. 367–372, 2010.

Corporation is to oversee the open source development process. Today, Ingres customer base includes 10,000 enterprise customers, among which 136 belong to the Fortune 500 companies like 3M, Bea Systems, and Lufthansa [6].

The high level architecture of the case software is grouped into four major components:

**Front End:** Functionality covers user interface facilities.
**Back End:** Functionality covers DBMS server functionality.
**Common:** Functionality covers connectivity and communications between the front end the back end.
**Utility:** Functionality covers utility libraries that interact with operating system.

## 3   Research Approach

The following research questions were investigated during the research:

1. What parts of the Ingres DBMS software components went through the most source code changes in terms of source files added, changed, and deleted?
2. How did Ingres DBMS code base change under the OS community process in terms of static source code metrics?

It is important to highlight the objective of this study, i.e. to understand what changes that have been carried out, and not to assess or compare the case software to any other software. The study is conducted as a case study [5].

### 3.1   Data Collection

In order to analyze and compare code metrics of the most recent proprietary version, further referred as 2004v, and open source version, further referred as 2008v, of Ingres, the 2004v was obtained by directly contacting the Ingres Corporation. The 2008v was downloaded from the Ingres Open Source community web site in November of 2008.

A program that parses through the 2004v and 2008v code base was created, or more specifically, the files and subdirectories under the main src directory that contains all of the source files. The files were compared between the two code bases in order classify all files according to the following:

- File type 0 : Source files that can be found only in 2004v
- File type 1 : Source files identical - unchanged between the 2004v and 2008v
- File type 2: Source files that were changed between the 2004v and 2008v
- File type 3: Source files that were added in 2008v

Metrics were measured in both versions are: lines of code ($LOC$), effective lines of code ($ELOC$), comment lines ($C$), total cyclomatic complexity ($TCC$), and file functions count ($FFC$). All metrics are calculated on file level.

For coding purposes developers often use braces or parenthesis to make code more readable, but this practice can inflate $LOC$ metrics [7]. The $ELOC$ metric takes into consideration all lines of code except blank only or comment only lines as well as the lines containing only standalone braces or parenthesis ({, }, (, )) Thus, lines counted by the $ELOC$ metric are a subset of the lines counted by the $LOC$ metric.

$C$ denotes the number of comment lines. The comment lines can appear by themselves on one physical line of code, or can be co-mingled.

The $TCC$ or total cyclomatic complexity metric, also known as McCabe's cyclomatic complexity, is the degree of logical branching per source file.

$FFC$, or total number of file functions, within a source file determines the modularity of the file. The $FFC$ metric combined with $ELOC$ metric produces average number of effective lines of code, $AELOC = ELOC/FFC$. In the same way, the average cyclomatic complexity is calculated as $ACC = TCC/FFC$.

In addition to the above metrics, the amount of comments are of interest. Therefore a metric describing the relative number of comments in each file is calculated as $RC = C/(ELOC + C)$.

Metrics for each file were derived with a metrics tool and stored in a database together with file type information for analysis.

## 3.2 Analysis

Analysis with respect to research question 1 was conducted by determining the percentage changes in terms of file type 0, file type 1, file type 2, and file type 3 per major components of the source code.

When analysing research question 2, the differences between the different versions of the case software were investigated with hypothesis tests. The null hypotheses state that the code changes made to 2004v, resulting in 2008v, had no impact on code metrics.

Let $T = \{0, 1, 2, 3\}$ denote file types according to above and let $M = \{AELOC, ACC, RC\}$ denote the different metrics of interest, so that $\mu_m(v, T_s)$ represents the expected mean of metric $m \in M$ for all files of types $T_s \subseteq T$ in version $v$. Then the following null hypotheses have been defined:

$$H0_{m,changed} : \mu_m(2004v, \{2\}) = \mu_m(2008v, \{2\})$$
$$H0_{m,new} : \mu_m(2004v, \{0, 1, 2\}) = \mu_m(2008v, \{3\})$$
$$H0_{m,all} : \mu_m(2004v, \{0, 1, 2\}) = \mu_m(2008v, \{1, 2, 3\})$$

That is, three null hypotheses have been formulated for each metric in $M$ so that there is one concerning only the changed files ($H0_{m,changed}$), one concerning all files from 2004v and only newly added files to 2008v ($H0_{m,new}$), and one concerning all the files in 2004v and 2008v ($H0_{m,all}$). This means that $|M| \times 3 = 3 \times 3 = 9$ null hypotheses and equally many alternative hypotheses have been defined in total.

Analysis of data for distributions of metrics results for version 2004v and 2008v were performed and it was determined that data for the metrics do not follow normal distribution. Hence in order to compare distribution of the metrics, non parametric tests, Mann-Whitney and Wilcoxon were performed. The Wilcoxon

Signed-Rank Test for matched pairs was used in order to compare paired data sets (i.e., in analysis of $H0_{m,changed}$), and the Mann-Whitney U test was used to compare un-paired data (i.e., in analysis of $H0_{m,all}$ and $H0_{m,new}$).

## 4    Results

Not all of the source code subdirectories will be analyzed in more detail, but only front, back, common, gl, and cl, since these directories contain almost 95% of the code. Hence, the most of the source files are located under /src/front directory or 54.7% of all 2008v. In the second place is src/cl directory housing 15.40% of source files in 2008v, followed by the src/back and src/common, housing 14.06% and 10.44% of all 2008v source files, respectively. Thus, these four directories contain 94.6% of 2008v source files.

Under the src/front directory the components that belong to the front end layer of the software are stored. Over 50% of all changes in the front end layer are due to the addition of the new source file components (type 3). Another 33% of changes are due to changes (file type 2). Thus, around 88% of front end source files have been changed since the case software went open source. Under src/cl library source files for Ingres Compatibility Library are housed. This library grew 69% between 2004v and 2008v, that is it contains 69% of file type 3 files. The src/back end components are deemed very important as the proper functioning of these components significantly affects database performance. The back end components went through the least amount of source code changes and additions, having 67.5% of code unchanged (file type 1) between the 2004v and 2008v. It also contains the least number of file additions (file type 3), thus only having 2.92% of the total number of the source files added (file type 3). Finally the src/common contains components used by both, front and back end. The common components contain 49.05% of file type 1, or almost half of its components are same for 2004v and 2008v. It can be observed that 19.5% of its file were of file type 3, or newly added components.

Table 1 displays code metric statistics summarized for the entire source code base of 2004v and 2008v. Hence, it can be observed that the number of file functions, lines of code and effective lines of code has increased. As one would expect, the higher number of functions and lines of code produced higher values for total cyclomatic complexity of 2008v code compared to 2004v.

The results of hypothesis testing for the stated hypotheses are presented in Table 2 (significance level 0.05).

Concerning $AELOC$, this metric is somewhat increased for changed files, meaning that when files are changed the functions in the files have become somewhat larger. For new files the metric is much lower than for old files, meaning that functions in new files are smaller than in older files. In total, looking at all files, the metric is higher in the new version than in the older version. The differences are statistically different for changed code and new code compared to old code, but not for all code.

**Table 1.** Summary of source code metrics for the whole system

| Code Metric | 2004v | 2008v |
|---|---|---|
| Total $LOC$ | 840,502 | 1,442,225 |
| Total $ELOC$ | 650,055 | 1,110,261 |
| Total $C$ | 484,349 | 630,635 |
| Total $TCC$ | 167,753 | 300,493 |
| Total $FFC$ | 15,588 | 45,216 |

**Table 2.** Mean values and results of hypothesis tests

| $H0$ | mean 2004 | mean 2008 | $p$ | reject $H0$ |
|---|---|---|---|---|
| $H0_{AELOC,changed}$ | 41.35 | 41.69 | $< 0.001$ | yes |
| $H0_{ACC,changed}$ | 10.47 | 10.80 | $< 0.001$ | yes |
| $H0_{RC,changed}$ | 0.53 | 0.54 | 1 | no |
| $H0_{AELOC,new}$ | 23.68 | 11.85 | 0.0042 | yes |
| $H0_{ACC,new}$ | 6.12 | 2.80 | 0.01 | yes |
| $H0_{RC,new}$ | 0.56 | 0.42 | $< 0.001$ | yes |
| $H0_{AELOC,all}$ | 23.68 | 19.02 | 0.1383 | no |
| $H0_{ACC,all}$ | 6.12 | 4.85 | 0.1841 | no |
| $H0_{RC,all}$ | 0.56 | 0.50 | $< 0.001$ | yes |

Concerning $ACC$ the same type of observation as for $AELOC$ can be made. For changed files the complexity is slightly higher and for new files the complexity is much lower. For $RC$ there is no significant difference for changed code, but for new code there are significantly less comments. In total there is relatively less comments in the new version compared to the old version.

## 5 Conclusions

The conducted analysis have shown that over half of the changes made to the case source code were made in the front end group of source code components, while the least of the changes were seen in the back end components. There can be many reasons for this, e.g. simply that more changes were needed in these components, but another reason may be that these are nearer to the interest of the new community that was formed during the open source transition process.

The results of comparison of code quality metrics between all files in 2004v and new files in 2008v show significant and large decrease in $ACC$ and $AELOC$, that is, significant and large increase in quality metrics for code developed by the OSS community. The code quality decrease in metrics smaller than the increase of the changed files, and as a result the code quality metrics for 2008v are higher than those of the 2004v, but this increase in code quality is not significant. This means that the overall code quality metrics, in terms of average cyclomatic complexity and the average effective lines of code per function has increased somewhat for changed code, and decreased rather much for new code. This can be interpreted as an improvement for added code. The number of comment lines per effective lines of code $ACC$ has decrased and there are significantly less comments in newly added code. At the same time the number of comments per effective lines of code ($RC$) has seen significant decrease between the 2004v and 2008v of source code base. Hence, while there was a small improvement in $ACC$ and $AELOC$, the lower number of comments per effective lines of code suggests that code in OSS community was not documented as much as in closed source environment.

The transition of the software was also accompanied by 100% increase in customer base, out of which some 138 customers belong to the Fortune 500 group, and 32% revenue increase reported for the 2008.

For companies planning to go open source, this study can provide an example on how the OSS community can impact static software quality metrics.

## Acknowledgments

The authors would like to express their gratitude to the Ingres Corporation for providing us with a last proprietary version of the software. This work was partly funded by the Industrial Excellence Center EASE – Embedded Applications Software Engineering, (http://ease.cs.lth.se).

## References

1. Fenton, N.E., Pfleeger, S.L.: Software Metrics: A Rigorous and Practical Approach, Revised. PWS Publishing Company, ITP International Thomson Publishing Company (1998)
2. IngresWebSite. Official web site of ingres corporation (2009), http://ingres.com/
3. Oikonomou, A., Stamelos, I., Angelis, L.: Code quality analysis in open source development. Information Systems Journal 12(1), 43–60 (2002)
4. Raymond, E.S.: The Cathedral and the Baazar. O'Reilly Media, Inc., Sebastopol (2001)
5. Runeson, P., Höst, M.: Guidelines for conducting and reporting case study research in software engineering. Empirical Software Engineering 14, 131–164 (2008)
6. Assay, M.: Web server survey (February 2009),
   http://news.cnet.com/8301-13505_3-10156188-16.html
7. RSM Effective lines of code eloc metrics for popular open source software linux kernel 2.6.17, firefox, apache hppd, mysql, php using rsm (2008),
   http://msquaredtechnologies.com/m2rsm/docs/rsm_metrics_narration.htm

# High-Level Debugging Facilities and Interfaces: Design and Developement of a Debug-Oriented I.D.E.

Nick Papoylias

Technical University of Crete, Chania, Greece
npapoylias@isc.tuc.gr
http://www.softnet.tuc.gr/~whoneedselta/misha

**Abstract.** While debugging in general is an essential part of the development cycle, debuggers have not themselves evolved over the years as other development tools have through the advancement of Integrated Development Environments. In this free-software research project we propose a way to overcome this problem by introducing, designing and developing a high-level debugging system.

*High-Level debugging systems* are systems that integrate a source - level debugger with other technologies as to extent both the facilities and the interfaces of the debugging cycle. We designed and developed such a system in a debugging-centric IDE, *Misha*. Misha, introduces among other things: *syntax-aware navigation, data-displaying and editing, reverse execution, debugging scripting* and *inter-language evaluation* through the integration of its source-level debugger (gdb) with a full-fledged source parser, data visualisation tools and other free software technologies.

# 1 Introduction

## 1.1 Problem Statement

Today's advancement in IDEs although constantly offering new programming tools and levels of sophistication, has left debuggers where they were a decade or more ago, mainly giving the programmer the ability to pinpoint source-lines of interest, stepping through subsequent lines of source-code, and monitoring certain expressions as he goes along. Of course the underlying technologies in the debugging backend often offer some additional number of tools - in the same line of thinking - which are nevertheless rarely "embedded" in IDEs and used by the programmer, if - that is - any debugging tools are embedded or used at all.

Given the importance of software monitoring and debugging as it is expressed in scientific publications concerning *effort estimation*[12] and *project management* which on average assert that testing and debugging cover roughly 50 % percent of the development time [5], we propose - both theoretically and technologically - a possible route for the evolution of debuggers that would hopefully meet the current needs of software engineering.

P. Ågerfalk et al. (Eds.): OSS 2010, IFIP AICT 319, pp. 373–379, 2010.
© IFIP International Federation for Information Processing 2010

# 2    Related Work

## 2.1    Published Work

As far as published work is concerned there are *high-level* debugging systems that have been proposed and concern a *domain-specific extension language* that leaves on top of legacy debugging systems. That is the case with *Duel* [11] and *Opium* [8]. In the case of *Duel* we have a high-level debugging system targeting the C language that during normal execution interacts with *gdb* providing new expression evaluations through a domain specific query language. In the case of *Opium* on the other hand the domain-specific query language (based on *Prolog*) analyses traces of program execution for post-mortem analysis. But we believe that there is a catch here given the fact that since their proposals debugging technology didn't catch up with these ideas even though for example *Duel* that was developed in Princeton is now part of *Microsoft Research* bibliography.

In essence a domain-specific language for debugging no matter how powerful and extensible, adds immensely to the complexity of the resulting development environment, and learning such a new language may seem like the last thing a programmer will want to do. In contrast maybe to a widely used and understood general purpose language that provides the same functionality without the burden of learning a debugging-specific one.

In addition there is the thriving field of *reversible and replay debugging*. We regard the ability to debug backwards in time one of the key components of high-level debugging systems, and so does the free software community [10]. In terms of published work some of these approaches can be found in [16], [15] and [1].

Last but certainly not least for reasons that we will discuss below when dealing with our syntax parser, our work is also related to the work of the Harmonia Project in Berkeley (see [4] and [2]) which deals with *high-level interactive software development*, *Language-Aware programming tools* and *programmer-computer interaction* although to our knowledge their work has yet to be expanded in debugging.

## 2.2    Technological Advancements

Besides expanding basic multi-threading support which appears in both major source-level debuggers[1], the gdb development team has lately taken a step further giving a lot of attention in the aforementioned facilities of reverse/replay debugging, and scripting extensibility [18], [19]. Our work relies heavily on some experimental work done for gdb [20] for the first subject but we have taken a very different architectural approach on the second. Nevertheless this convergence on experimental choices strengthens our belief that we are on the right track.

We now turn our attention to advancement in debugging aids through IDEs. Starting with industry standard environments, some related and interesting work appears in *Visual Studio's data visualisers* [6] were data in html, xml or image

---

[1] Gdb and MVSD.

form is according to semantics visualised for the programmer during debugging. Then there is the *high-level debugger* of Mathematica [17] which supports arbitrary computation at breakpoints in it's own language, including some visualisation of intermediate results, mainly mathematical formulae.

For the end an independent - but proprietary - project that we would like to mention is the high-level debugger JBixbe [7] which has some advanced capabilities in terms of *call-graph visualisation* and also a basic support for visualising data, like the popular front-end *DDD* [22] does.

# 3   Our Approach

## 3.1   Rethinking the Debugging Information Flow

All features and facilities of debugging systems depend on the amount and nature of information that is available in the debugger and concern the equivalency of source code with the running process. In most such systems in current use today, this kind of information is usually embedded by the compiler or interpreted in the executable or intermediate byte-code respectively. This is done according to some predefined standard such as the *pdx* stabs format [13] which anticipates specific uses for the kind of information that it embeds.

In our work in order to support current development and future uses of debugging systems other than the ones offered by today's technology we *expanded* the nature and amount of information available to the debugging system by providing it with direct access to a semantically annotated parse tree of the source code. Our choice alters the classical debugging information flow as seen in Fig. 1 (grey area).

Now as seen in Fig. 1 in order to construct this semantically annotated parsing tree and provide additional information to the debugger, we designed and

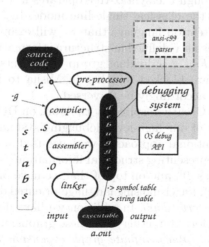

**Fig. 1.** Expanding the information available to the debugging system

developed a seperate parser as part of our debugging system. This parser provided us the means to develop and support features like *syntax-aware navigation* among other things. To our knowledge this is the first time that debugging information is enhanced in such a way rather than simply being embedded in the intermediate or machine code.

In addition such a parser can also be used as a crucial building block for a lot of technologies that are in current use today in IDEs or have been proposed for their development. Some examples include *symbol-browsing, unit-testing, documentation extraction, syntax-completion and refactoring*. To some extend these other uses are the aim of the *Harmonia Project* in Berkeley [3]. Their architectural approach also includes a seperate parser to support these facilities rather than using the first pass of the compiler itself. This desicion seems mandatory for the time being, due to the architectural structure of current compilers which favors syntax-trees in intermediate languages for optimization purposes. In the future we may be able to support these functions directly from the compiler itself, see [21] and [14].

## 3.2    The Five Pillars of High-Level Debugging

**Syntax-Aware Debugging:** This first feature is intented to be the workhorse of the overall effort, and is based directly on the afformentioned extention of available debugging information. The implication here is that by using the parser to analyze source code, *debugging and execution navigation can take place in terms of specific syntactic structures* having different "template" information readily available to the user according to syntactic and semantic information of the target language. The programer is thus able to pinpoint structures of interest as a whole, and not just source-lines, while debugging can take place both as stepping through a "logical-unit" of evolution and as watching the execution flow over time, freezing the program when needed. The navigation through the syntax-tree operates in two modes *breadth and depth first* besides the classical single-line mode. In addition, through the general purpose extention language that we will examine later, conditional debugging as well as user-defined in-structure information can be supported. As we can see in *Fig. 2* individual group statements, if, while and other syntax structures are blocked together in Misha to form logical units of execution that can accordingly be traversed.

**Data Visualisation:** Greatly inspired by the work on DDD, data visualisation is an essential part of our high-level debugging system. Taking things a bit further than conventional approaches we have used and integrated software which is used for representing structural information as diagrams of abstract graphs and networks [9], and on top of that we have provided a comprehensive and generic API for visualising language-oriented datatypes *(containers, strings, integers, interlations)*. In addition we have developed from scratch a graphical widget for interacting with these graphs, which supports *editing and updating graph values, infinite graph expansion via menus and depth settings, layout capabilites* and other features.

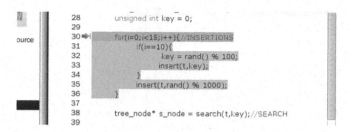

**Fig. 2.** Syntax Aware Navigation

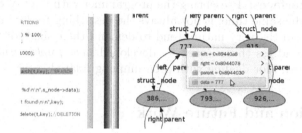

**Fig. 3.** Data Displaying, Positioning and Editing in graphs

**General-Purpose Extention Language:** Our third step was to integrate a general purpose extention language to our debugging system, which will be able to control our parser, the visualisation subsystem, the symbolic debugger as well as the "high-level" debugging facilities. We chose *python* which is a widely used and understood high-level language, distancing ourselves from the domain specific approaches that we saw earlier in related work. Part of what we have achieved here (controlling the symbolic debugger via python) is also a goal for gdb, which aims to use it's extention language as a separate platform for writting usefull tools [19]. In our approach besides being able to control all of the different subsystems (and not just the symbolic debugger) from the debugging console and *in-project* python scripts, thus being able to extend both the debugging system and the IDE, there is the ability to *directly* and *seamlessly* call each project's C functions from within python as seen in Fig 4. This feature besides being usefull for unit testing and code benchmarking purposes, encourages a multi-language approach in software enginnering which is a critical aspect of our future intentions for *Misha*.

**Reverse Debugging:** Stepping backwards in time while debugging is a valuable tool that cannot be absent from our research effort. It is also a community proposal, listed in the high priority project list of the *Free Software Foundation*. In responce to this interest and based upon the still expreri-mental work done for *i386* native reverse execution [18], we integrated and enhanced the execution record facility of *gdb* with our syntax-aware navigation system so that it is able to execute back in terms of complete syntactic structures, just as the programmer using the forward execution will have expected.

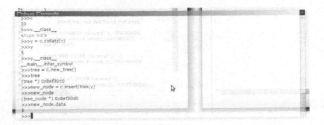

**Fig. 4.** Python to C, seamless inter-language calls

**Innovative Interfaces:** Presenting the programmer with a lot of data and options all at the same time, is not always the best thing to do, but debuggers and IDEs from the very nature tend to demand their share of the desktop. In order to address these issues we developed *new graphic widgets* for the gnome platform, for dealing with programming related issues.

## 4    Conclusion and Future Work

We will like to see our system expanding to the thrieving field of *multi-threaded debugging*. As mentioned earlier the basic operations are already implemented for such an expantion, but there are other posibilities as well. Static code analysis for example that uses our versetile parser can be implement to automatically deduce various *race conditions* between different threads.

In the same line of thinking, our data display system can be expanded to incorporate *call-graph representations* from which a more intuitive interface for setting breakpoints can emerge.

Finally the core implementation of our parser can be enhanced to read source code *incrementally*, giving the possibility among other things to graphically monitor source code changes as they happen.

Apart from the experience and knowledge gained in the course of this work, a lot of new ideas that transent debugging systems have emerged. Especially the multi-language testing and development facilities that we have developed, made us think of the possibility of integrating more than two languages that seamlessly interconnect (without the programmer's intervention through glue-code) in a single and unified environment. Without of course the need of a common intermediate representation[2].

## References

1. Akgul, T., Mooney, V.J.: Instruction-level reverse execution for debugging (2002)
2. Begel, A., Graham, S.L.: An assessment of a speech-based programming environment. In: VLHCC 2006: Proceedings of the Visual Languages and Human-Centric Computing, pp. 116–120. IEEE Computer Society, Los Alamitos (2006)

---

[2] As in .net or jython environements for example where there is a common byte-code backend.

3. Berkeley. Harmonia project (2009),
   http://harmonia.cs.berkeley.edu/harmonia/index.html
4. Boshernitsan, M., Graham, S.L., Hearst, M.A.: Aligning development tools with
   the way programmers think about code changes. In: CHI 2007: Proceedings of the
   SIGCHI conference on Human factors in computing systems, pp. 567–576. ACM,
   New York (2007)
5. Brooks Jr., F.P.: The mythical man-month (anniversary ed.). Addison-Wesley
   Longman Publishing Co., Inc., Boston (1995)
6. Visual Studio Developer Center. Visualisers (2009),
   http://msdn.microsoft.com/en-us/library/zayyhzts.aspx
7. Ds-emedia. Jbixle, high-level debugger (2006), http://www.jbixbe.com/
8. Ducassé, M., Emde, A.-M.: Opium: a debugging environment for prolog development and debugging research. SIGSOFT Softw. Eng. Notes 16(1), 54–59 (1991)
9. Ellson, J.: Graphviz, graph visualization software (2009),
   http://www.graphviz.org/
10. Free Software Foundation. High-priority projects (2009),
    http://www.fsf.org/campaigns/priority.html
11. Golan, M., Hanson, D.R.: Duel - a very high-level debugging language (1993)
12. Gorla, N., Benander, A.C., Benander, B.A.: Debugging effort estimation using
    software metrics. IEEE Trans. Software Eng. 16(2), 223–231 (1990)
13. Menapace, D.M.J., Kingdon, J.: The "stabs" debug format. Cygnus Support (2004)
14. Kitware, B.K.: Gcc xml, extention (2009),
    http://www.gccxml.org/HTML/Index.html
15. Lewis, S.A.: Techniques for efficiently recording state changes of a computer envi-
    ronment to support reversible debugging (2001)
16. Narayanasamy, S., Pokam, G., Calder, B.: Bugnet: Continuously recording program
    execution for deterministic replay debugging. In: ISCA, pp. 284–295 (2005)
17. Wolfram Research. Instant high-level debugging (2009),
    http://www.wolfram.com/technology/guide/InstantHighLevelDebugging/
18. Gdb Development Team. Gdbreversible. Debugging (2009),
    http://sourceware.org/gdb/wiki/Reversible
19. Gdb Development Team. Pythongdb (2009),
    http://sourceware.org/gdb/wiki/PythonGdb
20. Teawater. Gdb record patch (2009),
    http://sourceforge.net/projects/record/
21. Tromey, T.: Interview: Gcc as an incremental compile server (2007),
    http://spindazzle.org/greenblog/index.php?/archives/
    74-Interview-GCC-as-an-incremental-compile-server.html
22. Zeller, A.: Debugging with DDD. Gnu Press (2004)

# To Rule and Be Ruled: Governance and Participation in FOSS Projects

Zegaye Seifu[1] and Prodromos Tsiavos[2]

[1] University of Oslo, Norway
[2] London School of Economics, UK

**Abstract.** Free and Open Source Software (FOSS) Development has evoked images of full participation, emancipation and flat organization. Despite such rhetoric, some recent studies and practices reveal the re-emergence of hierarchical structures in one form or another as an almost inevitable aspect of the software development process. The objective of this paper is to investigate, both theoretically and empirically, the reasons behind this reappearance of hierarchy and its impact on the participation patterns of open source projects.

## 1 Introduction

Since its inception, Free and Open Source Software (FOSS) Development has evoked images of full participation, emancipation and flat organization. However, despite the licensing framework and rhetoric surrounding FOSS projects [7,10], recent analysis of the software development process reveals a more complex picture: the vast majority of contributions come from a small group of core developers who control the architecture and direction of development [5]. In addition, most participants typically contribute to just a single module, though some modules may include patches or modifications contributed by hundreds of contributors [8].

It seems, thus, that while FOSS as a philosophy and licensing practice invites maximum participation in the development process, in practice, the hierarchical structures re-emerge in one form or another as an almost inevitable aspect of the software development process. The objective of this paper is to investigate, both theoretically and empirically, the reasons behind this reappearance of hierarchy and its impact on the participation patterns of open source projects. More generally, it explores the ways in which governance mechanisms in different FOSS projects play a role in the structuring of participation and contribute to the openness of a project. Our effort is to explore the ways in which governance structures contribute to the deepening and widening of participation in the FOSS context.

A multiple case study approach that links to the conceptual framework allows us to offer the foundations for appreciating the role of participation and governance of FOSS projects. The three cases investigated here are complementary and contribute differently to the different theoretical perspectives supporting our

P. Ågerfalk et al. (Eds.): OSS 2010, IFIP AICT 319, pp. 380–388, 2010.

analysis. Data for the study was obtained from two main sources i) mailing list archives ii) formal documents and websites of the projects.

## 2    Analytical Concepts

### 2.1    FOSS Governance

Markus [4] defined FOSS governance as the means of achieving the direction, control, and coordination of wholly or partially autonomous individuals and organizations on behalf of a FOSS development project to which they jointly contribute. She also indicated that FOSS governance can be informal (i.e., enacted through shared norms and social control), formally documented (in written policies and constitutions) or encoded in technology (like mailing lists and version control software).

Related studies so far have been interested in the question of participation in FOSS projects from the perspective of motivational factors: why do talented developers voluntarily contribute? [3]. However, as indicated above, only a minority of developers is responsible for the majority of the contributions. This could be partly attributed to the governance mechanisms of the related projects. Governance structures directly affect motivation and participation of developers, as well as the type and quality of their contributions [9]. The mere availability of source code in open source projects practically doesn't guarantee presence of excess participation and participants. Making progress in analyzing and understanding the nature of collaboration in software development would lead us to develop better understanding of on line and digital communities as participative systems [6].

Markus suggested six elements of formal and informal structures and rules as dimensions and analytical parameters of FOSS governance: ownership of assets (includes intellectual property licenses and formal legal organizational structures), chartering the project (like statements of vision, what the software product should look like, etc.), community management (involves rules about participation and participant), software development processes (including structures that address the important operational tasks of development), conflict resolution and rule changing (including procedures for resolving conflict and for creating new rules) and use of information and tools (which relates to rules about how information will be communicated and managed and how tools and repositories will be used). These categories provide an analytical base for comparing the governance mechanisms displayed in the three projects here.

### 2.2    CBPP and Participation

Commons based peer production happens over the digitally networked environment as large and medium scale collaboration among individuals that are organized without markets or managerial hierarchies. This phenomenon is emerging everywhere in a decentralized way in the information and cultural production systems. The model was first brought in to light by Yochai Benkler [1,2]. The fact that CBPP takes place in the ubiquitous digital networks and the wide distribution of low cost physical capital, results in a conducive situation for creativity and mass participation.

Benkler lays out three characteristics of successful peer production and participation: the project/artifact must be modular, the modules should be predominately finegrained or small size and there should be low cost integration mechanisms. These properties determine the number of participants, the scope of varied investments (heterogeneity), the minimal investment required to participate and the simplicity level of integration. He stressed that a project that is at its best with these features can potentially attract many users.

Another key element of Benklers CBPP model as complemented by the recent work of Tsiavos [11] is that of excess capacity, both in the level of the artifact and that of the peer. Excess capacity in the level of the artifact relates to the non rivalerous nature of the artifact that is to be produced: the use of the produced artifact by one user should not hinder the enjoyment of another. This feature of the produced artifact allows maximum dissemination and parallel production and hence contributes to the increasing of the user participation. However, it needs to be complemented by what Tsiavos [11] describes as peer excess capacity, i.e. whoever participates in the development needs to have the skills and the time to do so and these skills and time need to persist over time. Besides the motivational factors indicated in various FOSS related studies, Benkler emphasizes that the more excess time and skills a peer has the more she is likely to be able to participate to a FOSS project.

The legal openness that the licenses ensure and the unrestricted access to a non-rivalerous good, as the software is, are not enough for maintaining a flat organizational structure. The nature of FOSS development is result oriented and meritocratic: what is important is that running code is written and that the best solution is preferred. As a result, the individuals that because of their experience and knowledge are able to contribute the best code are more likely to acquire greater power than individuals that lack such characteristics.

### 2.3    FOSS and Participation

According to general Intellectual Property Rights theory: in the classic copyright exploitation model, the creator is granted by the law a monopoly right as an incentive to produce a work. In the FOSS model, the focus is not on the work as a whole but rather on the level of the individual contribution. Here, the contributor is not attracted by a monopoly right but rather by other rewards she is herself to define: the reasons why a developer contributes to a FOSS project may be hedonistic, related to self-esteem, peer recognition or may relate to other indirect rewards. In any case, since the rewards are to be identified by the individual itself, the function of the legal regulatory instrument, i.e. the FOSS license, is to reduce friction for such micro-rewards to smoothly function as motivational factor.

It is also important to note that since from the perspective of producing a final product it is not the individual contributions but rather the accumulative result of all participants effort that is of interest, this sustained effort to reduce frictions and maintain excess capacity does not necessarily focus on the individual. In that sense, the operation of the regulatory means supporting FOSS development can focus on the following three objectives: (a) allowing a single individual to participate repetitively and to make contributions of increasingly greater quantity and quality (deepening participation), (b) on increasing the range and diversity of participants, allowing thus the project to scale, (widening participation) (c) or, ideally, on both.

The principal pure FOSS model assumes that the contributions will be small and resulting from a wide range of participants. Practice and more recent theoretical work however, indicate that such a model is the exception rather than the rule. Small size projects or ecologies of such projects are more likely to be the case. In addition, the emergence of hybrid proprietary and FOSS models are a good indication that the result oriented nature of FOSS is its key characteristic and the licensing model only a symptomatic feature of how such objectives may be best achieved. Since the objective of producing high quality running code can be achieved by deepening or widening participation, organizers of software production are likely to encourage both types of participation in order to achieve the best possible results depending on the specific circumstances surrounding the development of a project.

## 3   Case Description

**Skolelinux** is a community-managed Custom Debian Distribution (CDD) aimed at schools. It is by now a Debian sub project to make an overall computer solution and the best distribution for educational purposes. It was initiated in 2001 by a group of four programmers in Norway. For the project in Norway, there is a board consisting of elected members from users and developers every two years. The funding comes from private companies but the government had given most of the project finance at the beginning.

GPL is the preferred license and new contributions are encouraged to stick to it. However any solution that is being considered to be part of the system must live up to the Debian Free Software Guideline (DFSG), which contains the same ideals as in Open source Definition (OSI) and Free Software Foundation. Skolelinux is used in more than 450 schools worldwide - mostly Europe but also in Africa. Regarding the participation of users it is written on their WebPages that the project is community driven, having an ongoing exchange between users and developers.

**Varnish** is hybrid open source software released under the revised BSD (Berkley Software Distribution) license. The project is handled by a company called Linpro in Norway. Technically, the Varnish software is an HTTP accelerator on web servers. Most web sites or content management systems (CMS) present dynamic web pages consisting of a number of different elements. Combining these elements is both time-consuming and CPU intensive, and the process is repeated for every individual user, even though the content is often identical. Thus, Varnish temporarily stores the most frequently requested pages in cache memory and effectively presents these pages from cache. As a consequence, users are offered improved services, and server requirements are reduced by a huge percentage.

It was started in early 2006 and all the accompanying tools in the project are open source Subversion, TRAC, Mailman and GNU author tools. Linpro run its business by giving services to customers and through sponsorship. From the mailing list it is possible to trace and tell that the community is comprised of individuals all over the world. Currently, there are up to 200 people registered in the mailing list of varnish.

**HISP** is a globally distributed open source software development which was initiated in South Africa around 1994 and is based on collaboration between academic institutions, health authorities, and private organizations. The goal of the project is enabling south-south and also south-south-north collaboration. The funding comes from various governmental and non governmental donors, though mainly the Norwegian Agency for Development (NORAD) and the European Commission (EC).

The project develops District Health Information System (DHIS), a for collecting, processing, and analyzing health information for health administration purposes. DHIS 2.0 (and upward versions) is a web-based software package released under the BSD license. It is developed using Java frameworks and supported by other open source tools. It has been implemented in many developing countries in Africa and Asia. The HISP project in general involves academicians from universities doing action research through development and implementation. Thus the DHIS software gets the benefits in terms of resources from the academic environments.

## 4    Findings and Discussions

### 4.1    Overall Picture of Participation in the Three Projects

The development stage and history of the three projects obviously is different. This also determines the amount of data available in their mailing lists and even impacts the number of participants that each could have. We are aware of the fact that the type and purpose of the software also determines the pool of participants. Instead it is actually the trend that we wanted to show here and that helps most to get a comparative view. Accordingly, we considered the data of all the projects since their respective starting period. Table 1 summarizes the period, number of messages, number of threads, the number of participants and the size of data considered for each project.

### 4.2    Governance Mechanisms of the Three Projects

Based on the analytical framework discussed in section 2, the governance mechanisms of the three projects were analyzed. Table 2 presents their similarities

**Table 1.** Overall picture of participation in the three projects

| Project | Year | #of Messages | #of Threads | Size(KB) | #of Participants |
|---|---|---|---|---|---|
| Varnish | Feb 2006-Dec 2009 | 1224 | 340 | 908 | 82 |
| Skolelinux | Jan 2002-Dec 2009 | 15548 | 6344 | 66000 | 170 |
| HISP | Nov 2008-Dec 2009 | 3984 | 2018 | 9000 | 74 |

**Table 2.** Dimensions of OSS governance mechanisms

| Rules | Vanish | Skolelinux | HISP |
|---|---|---|---|
| Ownership of assets | company-manageed; it is BSD licensed | Community-managed; run by a board of elected volunteers; any OSS license | Developed by academic institution supported through action research; BSD licensed |
| Chartering the project | Governed by rules of the owning company and share its vision | The vision is promoting freedom and sharing, starting from schools; governed by the Debian charter | Intends to promote south-south cooperation and finding ways of supporting the developing world |
| Community management and development process | Open for anyone interested; commit patches under the supervision of the core developers | Open for anyone with outright privilege to make changes | Open for anyone with outright privilege to make changes |
| Conflict resolution | No written rules; on discussion in the mailing list with lead of the core developers | Negotiation and voting | Discussions through lists and in person |
| Use of information and tools | Information is openly accessible; use open source tools | Information is openly accessible; use open source tools | Information is openly accessible; use open source tools |

and differences. In all the three cases we notice a relevant consistency between the different modalities of governance and regulation used in order to achieve the desired modes of production. For instance, as in Benklers original model the individual identification of incentives appears to be complemented with a licensing scheme reducing barriers to participation. Thus the dispersed and minimal excess capacity of multiple individuals may be collected and organized. The investigation of the social norms reveals some interesting patterns. In the Varnish case, there are no clear procedural or decision taking rules. However, the decisions are made and the guidance is provided de facto by the core developers who are sponsored by the companies driving the development. In like manner, in the HISP project a similar situation emerges: the core developers are Oslo University staff and research students and they are the ones driving the process and taking the decisions in the absence of any clearly stated rules. In the Skolelinux case, there is a flatter structure with more well defined procedures. Where a kind of hierarchy emerges this is the result of the meritocratic process through which contributions are made. It is not that in the case of HISP and Varnish a meritocratic approach is not followed, but it is the sponsoring organizations behind the development that influence the development direction and effectively control the whole process.

This is a very important finding as it indicates that the control structures are not to be found in a first but rather in a second layer. In the first layer all three project resemble in terms of legal and technological structure and in that sense they all appear as meritocratic. However, in a second layer, there are substantial differences related to the time and expertise sponsoring provided by extra-FOSS organizations. This in turn has a direct effect in the formation of the social norms that will then operate as the main day-to-day governance mechanism for the FOSS projects.

## 4.3    Participation Patterns and Governance

In the cases of Varnish and HISP, we observe hybrid models and it is extremely helpful to see where this imbroglio-like nature lies. In our opinion the hybridity of the model is in the way excess capacity is handled: in these two cases, the capacity of the core developers is sponsored in a traditional fashion, though without using a copyright contract, since there is no reason to restrict access to the source code. Linpro pays for the time of the core developers that make the maximum contributions, while the rest of the participants play a complementary role making contributions that require only minimal excess capacity and for which the classic CBPP model is applicable. Similar is the situation in the case of HISP. The core developers excess capacity is sponsored by Oslo University, while the rest of the developers make marginal contributions requiring again little excess capacity.

On the contrary, in the case of Skolelinux, there is no clearly seen equivalent centralized and focused sponsoring and hence the contributions are more evenly spread and the decision making process is much more democratic. This last observation brings us to the issue of social norms formation. In the former two cases the prevailing social norms assign particular relevance to the core developers and there are no strong and detailed community norms. In the Skolelinux case, on the contrary, the decision making process is much more democratic and inclusive.

The interactions between different governance layers are partially only reflected in the participation patterns. In the Varnish and Skolelinux the participation falls dramatically after an initial phase of wide participation, while in the HISP case we see a different pattern. The changes in participation cannot be attributed merely to the governance structures. The maturity of a project, for instance, plays a crucial role: as the project matures, the original easy to be identified and solved problems give their position to more difficult problems, thus requiring greater levels of capacity and hence limiting participation only to the more knowledgeable and dedicated participants.

The three cases, here, though superficially substantially differ between each other, a second reading present some commonalities as to how the problem of excess capacity is addressed. In the Varnish case, as the project matures, the interest becomes not one of widening but rather of deepening the participation. Since the core developers are being sponsored by Linpro the problem of excess capacity is solved by ensuring that they will keep developing software seeing the community participation as added value to work that is already being conducted by the company. Here, the incentive of Linpro to widen participation is not big enough to lead to any specific measures so that non-paid professionals are able to increase or even maintain the quantity and quality of their contributions.

In the Skolelinux case, while there is an incentive to sustain quality and quantity of contribution, there is no direct ability to sponsor the time and expertise of participants and hence the project inevitably shrinks in size. However, the remaining participants increase the depth of their participation by being able to make more substantial contributions and sustaining a steady number of participants.

Finally, the HISP case is possibly the most interesting one, as it takes a series of conscious steps in increasing the excess capacity of the participating individuals in order to be able to contribute to the project and this is reflected in the number of participants as the project matures. The main mechanism employed by HISP is the use of an educational network that operates as an additional overlay on the FOSS network that supports both financially and educationally the participants. Also, by employing more a more modular and thus efficient software architecture and bug reporting methods, it allows the capturing of even finer granules of excess capacity, thus increasing the participation both in width and depth.

## 5   Conclusion

Traditionally, FOSS has been associated with openness and participation. Organizational studies have indicated that there are hierarchical elements present in the structuring of the FOSS development process, but have not provided a comprehensive account of why this is the case. In this paper we have presented a first account of why this is the case, indicating, however, that even where participation seems to be reducing its scope, it may be increasing its depth. Wherever hybrid models appear, these are the result of the need to sustain capacity that could not be attracted using only the FOSS model and to achieve value added corrections that could not have been achieved through a classic proprietary copyright management model. These models may lead to a reduction in participation, where there is no provision for increasing the excess capacity of the peers. For these reasons, hybrid network-CBPP models seem more likely to support both widening and deepening of participation in FOSS project.

## References

1. Benkler, Y.: Coase's Penguin, or Linux and the Nature of the Firm. Yale Law Journal 112, 369 (2002)
2. Benkler, Y.: The Wealth of Networks: How Social Production Transforms Markets and Freedom. Yale University Press, New Haven (2006)
3. Lerner, J., Tirole, J.: Some simple economics of open source. Journal of Industrial Economics 52 (2002)
4. Markus, M.: The governance of free/open source software projects: monolithic, multidimensional, or configurational? Journal of Management Governance 11, 151–163 (2007)
5. Mockus, A., Fielding, R., Herbsleb, J.: Two Case Studies of Open Source Software Development: Apache and Mozilla. ACM Trans. Soft. Eng. Methods 11, 309–346 (2002)

6. Nakakoji, K., Yamada, K., Giaccardi, E.: Understanding the Nature of Collaboration in Open-Source Software Development. In: Proceedings of Asia-Pacific Software Engineering Conference. IEEE Computer Society, Los Alamitos (2005)
7. Raymond, E.: The cathedral and the bazaar: musings on linux and open source by an accidental revolutionary (Rev. edn.). O'Reilly, Cambridge (2001)
8. Scacchi, W.: Free/Open Source Software Development: Recent Research Results and Emerging Opportunities. In: Proc. European Software Engineering Conference and ACM SIGSOFT Symposium on the Foundations of Software Engineering (2007)
9. Shah, S.: Motivation, Governance, and the Viability of Hybrid Forms in Open Source Software Development. Management Science 52, 1000–1014 (2006)
10. Stallman, R.: Why Software Should Not Have Owners. In: Richard, M., Stallman, J. (eds.) Free Software, Free Society: Selected Essays. GNU Press (2002)
11. Tsiavos, P., Korn, N.: Case Studies Mapping the Flows of Content, Value and Rights across the UK Public Sector. In: Joint Information Systems Committee, London (2009)

# A Comparison Framework for Open Source Software Evaluation Methods

Klaas-Jan Stol[1] and Muhammad Ali Babar[2]

[1] Lero—University of Limerick, Limerick, Ireland
[2] IT University of Copenhagen, Denmark
klaas-jan.stol@lero.ie, malibaba@itu.dk
http://www.lero.ie, http://www.itu.dk

**Abstract.** The use of Open Source Software (OSS) components has become a viable alternative to Commercial Off-The-Shelf (COTS) components in product development. Since the quality of OSS products varies widely, both industry and the research community have reported several OSS evaluation methods that are tailored to the specific characteristics of OSS. We have performed a systematic identification of these methods, and present a comparison framework to compare these methods.

**Keywords:** open source software, evaluation method, comparison framework.

## 1 Introduction

Open Source Software (OSS) is increasingly being integrated into commercial products [1]. Much cited reasons for using OSS are cost savings, fast time-to-market and high-quality software [2]. OSS products can be used as components as an alternative to Commercial Off-The-Shelf (COTS) components. Like COTS evaluation and selection, one of the main challenges of using OSS is evaluation and selection [3]. For that reason, both the research community and industry have proposed evaluation and selection approaches to help practitioners to select appropriate OSS products. However, research has shown that practitioners rarely use formal selection procedures [4]. Instead, OSS products are often selected based on familiarity or recommendations by colleagues [5]. For practitioners it is difficult to choose a suitable evaluation method. We assert that the lack of adoption of these evaluation approaches by practitioners may be a result of a lack of clarity of the OSS evaluation methods landscape. There has been no systematic comparison of the existing OSS evaluation methods. David A. Wheeler lists a number of evaluation methods in [6], but does not provide a thorough comparison of existing evaluation methods. We are aware of only one paper by Deprez and Alexandre [7] that provides an in-depth comparison of two methods, namely QSOS and OpenBRR. However, it is not feasible to extend their approach to compare a large number of methods. In order to improve the state of practice, we decided to systematically identify proposed OSS evaluation methods. Furthermore, we present a comparison framework that can be used to do a systematic comparison of these OSS evaluation methods.

P. Ågerfalk et al. (Eds.): OSS 2010, IFIP AICT 319, pp. 389–394, 2010.
© IFIP International Federation for Information Processing 2010

## 2 Identification of Evaluation Methods

For the identification of the various OSS evaluation methods, we relied on four different sources. Firstly, we selected a large number of publications following a systematic and rigorous search methodology as part of our ongoing extension of a systematic literature review reported in [8]. The search phase of this extension resulted in a repository of approximately 550 papers related to OSS. We screened these papers to identify any OSS evaluation method. We included all papers reporting a method, framework or any other proposed way of evaluating an OSS product. Papers presenting an approach for selecting COTS (as opposed to OSS components only) were also excluded. Secondly, we inspected the "related work" sections of the selected papers. We also noticed that a number of OSS evaluation methods were not reported in research publications, rather only appeared in books or white papers. Since those methods were often referenced in the "related work" sections of many papers, we decided to include those methods in this research. Thirdly, we manually selected publications reported in the proceedings of the five International Conferences on Open Source Systems (2005 to 2009). Lastly, we used the authors' knowledge of the field in order

**Table 1.** Identified OSS evaluation methods, frameworks and approaches

| No. | Name | Year | Source | Orig. | Method |
|-----|------|------|--------|-------|--------|
| 1 | Capgemini Open Source Maturity Model | 2003 | [9] | I | Yes |
| 2 | Evaluation Framework for Open Source Software | 2004 | [10] | R | No |
| 3 | A Model for Comparative Assessment of Open Source Products | 2004 | [11, 12] | R | Yes |
| 4 | Navica Open Source Maturity Model | 2004 | [13] | I | Yes |
| 5 | Woods and Guliani's OSMM | 2005 | [14] | I | No |
| 6 | Open Business Readiness Rating (OpenBRR) | 2005 | [15, 16] | R/I | Yes |
| 7 | Atos Origin Method for Qualification and Selection of Open Source Software (QSOS) | 2006 | [17] | I | Yes |
| 8 | Evaluation Criteria for Free/Open Source Software Products | 2006 | [18] | R | No |
| 9 | A Quality Model for OSS Selection | 2007 | [19] | R | No |
| 10 | Selection Process of Open Source Software | 2007 | [20] | R | Yes |
| 11 | Observatory for Innovation and Technological transfer on Open Source software (OITOS) | 2007 | [21], [22] | R | Yes |
| 12 | Framework for OS Critical Systems Evaluation (FOCSE) | 2007 | [23] | R | No |
| 13 | Balanced Scorecards for OSS | 2007 | [24] | R | No |
| 14 | Open Business Quality Rating (OpenBQR) | 2007 | [25] | R | Yes |
| 15 | Evaluating OSS through Prototyping | 2007 | [26] | R | Yes |
| 16 | A Comprehensive Approach for Assessing Open Source Projects | 2008 | [27] | R | No |
| 17 | Software Quality Observatory for Open Source Software (SQO-OSS) | 2008 | [28] | R | Yes |
| 18 | An operational approach for selecting open source components in a software development project | 2008 | [29] | R | No |
| 19 | QualiPSo trustworthiness model | 2008 | [30, 31] | R | No |
| 20 | OpenSource Maturity Model (OMM) | 2009 | [32] | R | No |

to identify some approaches. We note that we deliberately did not consider any websites (such as web logs) presenting pragmatic "tips for selecting OSS".

Following the abovementioned search process, we identified 20 approaches for OSS evaluation. Table 1 lists the identified OSS evaluation approaches in chronological order of publication. The column "Source" lists references to papers and reports that reported the method, and can be used by interested readers for further investigation. The column "Orig." indicates whether the initiative came from (I)ndustry or from a (R)esearch setting. We considered it to be an industry initiative if it was associated with a company name; otherwise we considered it to be a researchers' initiative. The column "Method" indicates whether it is a well-defined method outlining the required activities, tasks, inputs, and outputs, as opposed to a mere set of evaluation criteria. As can be seen from the table, only half of the approaches that we identified are methods.

# 3   A Comparison Framework

In order to perform a systematic comparison of the selected OSS evaluation methods, we designed a comparison framework called Framework fOr Comparing Open Source software Evaluation Methods (FOCOSEM), which is presented in Table 2.

**Table 2.** FOCOSEM: a comparison framework for OSS evaluation approaches

| Component | Element | Brief description |
|---|---|---|
| Method Context | Specific goal | What is the particular goal of the method? |
| | Functionality evaluation | Is functionality compliance part of the evaluation method? |
| | Results publicly available | Are evaluations of OSS products stored in a publicly accessible repository? |
| | Relation to other methods | How does the method relate to other methods? I.e. what methods was this method based on? |
| Method User | Required skills | What skills does the user need to use the method? |
| | Intended users | Who are the intended users of the method? |
| Method Process | Method's activities | What are the evaluation method's activities and steps? |
| | Number of criteria | How many criteria are used in the evaluation? |
| | Evaluation categories | What are the method's categories of criteria based on which the OSS product is evaluated? |
| | Output | What are the outputs of the evaluation method? |
| | Tool support | Is the evaluation method supported by a tool? |
| Method Evaluation | Validation | Has the evaluation method been validated? |
| | Maturity stage | What is the maturity stage of the evaluation method? |

FOCOSEM is based on four different sources to justify the selection and formation of its components and elements. The first source is the NIMSAD framework, which is a general framework for understanding and evaluating any methodology [33]. NIMSAD defines four components to evaluate a methodology: the problem context, the problem solver (user), the problem-solving process, and the method's evaluation. Previously, NIMSAD has been used for the development of a number of other comparison frameworks in software engineering [34-36]. Hence, we are quite confident about NIMSAD's ability to provide a solid foundation for building an instrument for comparing and evaluating software engineering methods and tools. The second source for FOCOSEM is FOCSAAM, which is a comparison framework for software architecture analysis methods [34]. The third source is a comparison framework for software product line architecture design methods [36]. As a fourth source, we identified differences and commonalities among various OSS evaluation methods. We note that the objective of FOCOSEM is not to make any judgments about different OSS evaluation methods. Instead, we aim to provide insights that may help practitioners to select a suitable OSS evaluation method.

# 4   Conclusion and Future Work

Open Source Software (OSS) products are increasingly being used in software development. In order to select the most suitable OSS product, various evaluation methods have been proposed. Following a systematic and rigorous search of the literature, we identified 20 different initiatives for OSS product evaluation. Furthermore, we have proposed a Framework fOr Comparing Open Source software Evaluation Methods (FOCOSEM). We emphasize that the framework is not intended to make any judgments about the quality of the studied OSS evaluation methods. In future work, we will demonstrate the application of FOCOSEM by comparing the OSS evaluation methods identified in our review. Furthermore, we do not claim our framework is complete; rather, we consider it as a first step towards systematically providing a comparative analysis of OSS evaluation methods. Additional elements can be added to our framework to compare other aspects of the evaluation methods.

## Acknowledgements

This work is partially funded by IRCSET under grant no. RS/2008/134 and by Science Foundation Ireland grant 03/CE2/I303_1 to Lero—The Irish Software Engineering Research Centre (www.lero.ie).

## References

[1] Hauge, Ø., Sørensen, C.-F., Conradi, R.: Adoption of Open Source in the Software Industry. In: Proc. Fourth IFIP WG 2.13 International Conference on Open Source Systems (OSS 2008), Milano, Italy, September 7-10, pp. 211–221 (2008)
[2] Fitzgerald, B.: A Critical Look at Open Source. Computer 37(7), 92–94 (2004)
[3] Maki-Asiala, P., Matinlassi, M.: Quality Assurance of Open Source Components: Integrator Point of View. In: 30th Annual International Computer Software and Applications Conference, 2006. COMPSAC 2006, pp. 189–194 (2006)
[4] Li, J., Conradi, R., Slyngstad, O.P.N., Bunse, C., Torchiano, M., Morisio, M.: Development with Off-the-Shelf Components: 10 Facts. IEEE Software 26(2) (2009)

[5]  Hauge, Ø., Osterlie, T., Sorensen, C.-F., Gerea, M.: An Empirical Study on Selection of Open Source Software - Preliminary Results. In: Proc. ICSE Workshop on Emerging Trends in FLOSS Research (FLOSS 2009), Vancouver, Canada (2009)

[6]  Wheeler, D.A.: How to Evaluate Open Source Software / Free Software (OSS/FS) Programs, http://www.dwheeler.com/oss_fs_eval.html (accessed September 8, 2009)

[7]  Deprez, J.C., Alexandre, S.: Comparing assessment methodologies for free/open source software: OpenBRR and QSOS. In: Jedlitschka, A., Salo, O. (eds.) PROFES 2008. LNCS, vol. 5089, pp. 189–203. Springer, Heidelberg (2008)

[8]  Stol, K., Ali Babar, M.: Reporting Empirical Research in Open Source Software: The State of Practice. In: Proc. 5th IFIP WG 2.13 International Conference on Open Source Systems, Skövde, Sweden, June 3-6, pp. 156–169 (2009)

[9]  Duijnhouwer, F., Widdows, C.: Open Source Maturity Model. Capgemini Expert Letter (2003)

[10]  Koponen, T., Hotti, V.: Evaluation framework for open source software. In: Proc. Software Engineering and Practice (SERP), Las Vegas, Nevada, USA, June 21-24 (2004)

[11]  Polančič, G., Horvat, R.V.: A Model for Comparative Assessment Of Open Source Products. In: Proc. The 8th World Multi-Conference on Systemics, Cybernetics and Informatics, Orlando, USA (2004)

[12]  Polančič, G., Horvat, R.V., Rozman, T.: Comparative assessment of open source software using easy accessible data. In: Proc. 26th International Conference on Information Technology Interfaces, Cavtat, Croatia, June 7-10, pp. 673–678 (2004)

[13]  Golden, B.: Succeeding with Open Source. Addison-Wesley, Reading (2004)

[14]  Woods, D., Guliani, G.: Open Source for the Enterprise: Managing Risks Reaping Rewards. O'Reilly Media, Inc., Sebastopol (2005)

[15]  Business Readiness Rating for Open Source, RFC 1 (2005), http://www.openbrr.org

[16]  Wasserman, A.I., Pal, M., Chan, C.: The Business Readiness Rating: a Framework for Evaluating Open Source, Technical Report (2006)

[17]  Atos Origin: Method for Qualification and Selection of Open Source software (QSOS) version 1.6, Technical Report (2006)

[18]  Cruz, D., Wieland, T., Ziegler, A.: Evaluation criteria for free/open source software products based on project analysis. Software Process: Improvement and Practice 11(2) (2006)

[19]  Sung, W.J., Kim, J.H., Rhew, S.Y.: A Quality Model for Open Source Software Selection. In: Proc. Sixth International Conference on Advanced Language Processing and Web Information Technology, Luoyang, Henan, China, pp. 515–519 (2007)

[20]  Lee, Y.M., Kim, J.B., Choi, I.W., Rhew, S.Y.: A Study on Selection Process of Open Source Software. In: Proc. Sixth International Conference on Advanced Language Processing and Web Information Technology (ALPIT), Luoyang, Henan, China (2007)

[21]  Cabano, M., Monti, C., Piancastelli, G.: Context-Dependent Evaluation Methodology for Open Source Software. In: Proc. Third IFIP WG 2.13 International Conference on Open Source Systems (OSS 2007), Limerick, Ireland, pp. 301–306 (2007)

[22]  Assessment of the degree of maturity of Open Source open source software, http://www.oitos.it/opencms/opencms/oitos/ Valutazione_di_prodotti/Modello1.2.pdf

[23]  Ardagna, C.A., Damiani, E., Frati, F.: FOCSE: An OWA-based Evaluation Framework for OS Adoption in Critical Environments. In: Proc. Third IFIP WG 2.13 International Conference on Open Source Systems, Limerick, Ireland, pp. 3–16 (2007)

[24] Lavazza, L.: Beyond Total Cost of Ownership: Applying Balanced Scorecards to Open-Source Software. In: Proc. International Conference on Software Engineering Advances (ICSEA) Cap Esterel, French Riviera, France, p. 74 (2007)

[25] Taibi, D., Lavazza, L., Morasca, S.: OpenBQR: a framework for the assessment of OSS. In: Proc. Third IFIP WG 2.13 International Conference on Open Source Systems (OSS 2007), Limerick, Ireland, pp. 173–186 (2007)

[26] Carbon, R., Ciolkowski, M., Heidrich, J., John, I., Muthig, D.: Evaluating Open Source Software through Prototyping. In: St.Amant, K., Still, B. (eds.) Handbook of Research on Open Source Software: Technological, Economic, and Social Perspectives (Information Science Reference, 2007), pp. 269–281 (2007)

[27] Ciolkowski, M., Soto, M.: Towards a Comprehensive Approach for Assessing Open Source Projects. In: Software Process and Product Measurement. Springer, Heidelberg (2008)

[28] Samoladas, I., Gousios, G., Spinellis, D., Stamelos, I.: The SQO-OSS Quality Model: Measurement Based Open Source Software Evaluation. In: Proc. Fourth IFIP WG 2.13 International Conference on Open Source Systems (OSS 2008), Milano, Italy (2008)

[29] Majchrowski, A., Deprez, J.: An operational approach for selecting open source components in a software development project. In: Proc. 15th European Conference, Software Process Improvement (EuroSPI), Dublin, Ireland, September 3-5 (2008)

[30] del Bianco, V., Lavazza, L., Morasca, S., Taibi, D.: Quality of Open Source Software: The QualiPSo Trustworthiness Model. In: Proc. Fifth IFIP WG 2.13 International Conference on Open Source Systems (OSS 2009), Skövde, Sweden, June 3-6 (2009)

[31] del Bianco, V., Lavazza, L., Morasca, S., Taibi, D.: The observed characteristics and relevant factors used for assessing the trustworthiness of OSS products and artefacts, Technical Report no. A5.D1.5.3 (2008)

[32] Petrinja, E., Nambakam, R., Sillitti, A.: Introducing the OpenSource Maturity Model. In: Proc. ICSE Workshop on Emerging Trends in Free/Libre/Open Source Software Research and Development (FLOSS 2009), Vancouver, Canada, pp. 37–41 (2009)

[33] Jayaratna, N.: Understanding and Evaluating Methodologies: NIMSAD, a Systematic Framework. McGraw-Hill, Inc., New York (1994)

[34] Ali Babar, M., Gorton, I.: Comparison of Scenario-Based Software Architecture Evaluation Methods. In: Proc. 11th Asia-Pacific Software Engineering Conference (APSEC 2004), Busan, Korea, November 30-December 3, pp. 600–607 (2004)

[35] Forsell, M., Halttunen, V., Ahonen, J.: Evaluation of Component-Based Software Development Methodologies. In: Proc. Fenno-Ugric Symposium on Software Technology, Tallin, Estonia, pp. 53–63 (1999)

[36] Matinlassi, M.: Comparison of software product line architecture design methods: COPA, FAST, FORM, KobrA and QADA. In: Proc. 26th International Conference on Software Engineering (ICSE), Edingburgh, Scotland, United Kingdom, May 23-28 (2004)

# An Exploratory Long-Term Open Source Activity Analysis: Implications from Empirical Findings on Activity Statistics

Toshihiko Yamakami

ACCESS
Toshihiko.Yamakami@access-company.com

**Abstract.** Open source software (OSS) activities are diverse and difficult to capture. The author attempts a web service-based observation of OSS activities. Small community factor is discussed from a social viewpoint.

## 1 Introduction

OSS is a multi-faceted process including code, license, community, tools, development process, innovation methodology, philosophy, and best practices. The author attempts an exploratory analysis of project mining of publicly available open source activities. The author performs an analysis of open source activity pattern in a chorological dimension. The author provides a perspective for long-term observation of open source software project activities and implications for social aspect of OSS.

## 2 Purpose and Related Works

The purpose of this research is to identify the patterns of open source software activities and its implicationsfor social aspect of OSS.

Raymond discussed open source from the business model perspective in this famous open source work series [4].

Ducheneaut discussed the social analysis on a particular open source project from a dynamism viewpoint, how to retain and reproduce a community [3].

Bird analyzed the source code repository and mailing list archive for Postgress [1]. Bird also analyzed community structures of known successful open source projects [2] with the autonomous subcommittee formation.

## 3 Patterns and Chasms

The patterns are described in Table 1. It should be noted that the many open source projects do not reach the active state. In many projects, they even fail to launch the project, therefore, no source code is available to public. It is an interesting research topic how these different states of project have been derived.

The different states are identified with different chasms. The chasms patterns are depicted in Table 2.

P. Ågerfalk et al. (Eds.): OSS 2010, IFIP AICT 319, pp. 395–400, 2010.

**Table 1.** Patterns

| Stopped | Never launched | Initiated, but no real open source activity took place. |
|---|---|---|
| | Launched and dead | Initiated, and no team was formed. |
| | Launched, high activity and dead (or complete) | An active team was formed and disbanded. |
| Active | Launched and one-person | Initiated, and one person keeps the activity. |
| | Launched and one-company with multiple persons | Multiple person team was formed and the activity continues to be active. |
| | Launched and multiple-company(or individuals) | Multiple company committed and the activity continues to be active. |

**Table 2.** Patterns

| Type | Description | Contributing Factors |
|---|---|---|
| Launch Chasm | Some projects cannot reach successful launch | Lack of experience, enthusiasm, completeness. Competitive components. |
| One-person Chasm | Some projects may be never maintained by more than one person | Lack of universality, documentation, use cases. |
| Use Chasm | Some components may not obtain any real users | Quality, lack of applicability, application use cases. |
| One-company Chasm | One company committed, but no other organization committed. | Lack of ecosystem. No industry-wide support. |

## 4   Long-Term Observation

In order to further examine the current states of open source projects, the author uses web services for long-term open source activity archives. There is a site "ohloh.net" which provides web services interfaces for long-term monthly OSS activity observation.

Observation methodology is as follows:

- code size
- contributing member size

| | regular | irregular |
|---|---|---|
| Patterns | Synchronized release patterns | No patterns |
| Code size | steady growth | frequently flat |

The author analyzed the 77 open source projects automatically stored in the ohlor (http://www.ohlor.net). They include Action Script to XSL Transformation. The ohlor started tracking open source codes dated back to 1989. The mean results are shown in Table 3. 1996.3 in year means $1996 + \frac{3}{10}$ (year).

Contributors per project are depicted in Fig. 1 (a). 53.3 % of projects show $1 <= x < 2$. 32.5% of projects show in $2 <= x < 3$. With further detailed analysis, $1.4 <= x < 2.4$ includes 74.0 % of projects.

**Table 3.** Statistics of open source projects

| Description | Average |
|---|---|
| Project | 3755.7 |
| Contributors per Project | 2.2 |
| Year | 1996.3 |
| Source lines per Contributor | 8755.8 |
| Commitments per Contributor | 44.6 |
| Source lines per Commitments | 254.1 |

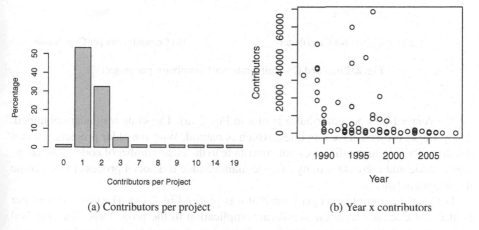

(a) Contributors per project                    (b) Year x contributors

**Fig. 1.** Contributors per project

This follows the following two observations:

- The bazaar-style open source project follows the basic structure that the one contributor per one project, and
- The bazaar-style open source project follows a main contributor with a sub contributor.

This reveals the following social structure of open source projects:

- In the bazaar-style open source project, it is difficult to coordinate the collaborative software development. Therefore, each contributor consists of a separate project. When a project consists of only one contributor, collaboration is simple.
- Even though to minimize the collaboration effort, people cannot perform the software development in an isolated manner. Social respect and attention is needed. This is reflected in the average $1 + alpha$ contributors per project.

The year x contributors is plot in Fig. 1 (b). Many fresh projects have small number of contributors. With the older projects, we can see a large fluctuations among projects. The longer life does not directly contribute to the large number of contributors.

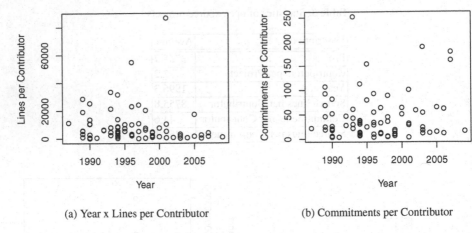

(a) Year x Lines per Contributor　　　　(b) Commitments per Contributor

**Fig. 2.** Lines and Commitments per Contributor per project

The year x Lines per contributor is plot in Fig. 2 (a). The same results are obtained. The younger projects have less lines, which is natural. With the older projects, we can see diversity. The longer life does not contribute to the large number of codes. The adoption of code and increase activity in code maintenance is a social process to overcome the social chasms.

The year x Commitments per contributor is plot in Fig. 2 (b). The commitment per contributor does not have the significant implication to the project age. The survival of long years does not have a significant impact from a statistical viewpoint, which is natural.

## 5  Discussion

### 5.1  Skill to Leverage Open Source Projects

The skills to leverage open source projects are illustrated in Fig. 3. Most projects are considered to remain in the unskilled leadership domain. It needs a systematic management and skill development to foster a productive OSS project. Considering one-man and plus alpha status of many OSS projects, it is considered to be useful to provide case studies of growing projects and skills to harness project growth.

### 5.2  Different Types of Software for Adoption

It should be noted that the type of open source software impacts the adoption. It is influenced by the nature of the source code, whether it is an end user product or a platform product. The success patterns depending on the two types of software are illustrated in Table 4. From the current observation, these two types are not distinguished, but further studies need to examine them.

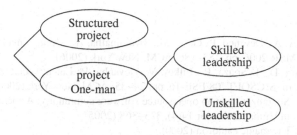

**Fig. 3.** Skills to leverage open source projects

**Table 4.** Success Patterns

| Application Usage | Completeness | Feature completeness. |
|---|---|---|
| | Stability | Stably usable. |
| | Feature Advantage. | Usability of features. |
| Middleware Usage | Combinatorial usage. | Middleware applicability in combination of other components or applications. |
| | Upstream open source application popularity. | Component dependence on popular applications. |

## 5.3  Limitations

This study is based on the external observation of open source project archives. It exhibits an interesting implication for awareness of small-sized fine-grained community in the open source projects, leveraging social respect and self satisfaction. This study is not based on any interviews or detailed motivation analysis, which is a limitation of this study. This study is descriptive, and needs further quantitative analysis on micro-level social adoption.

## 6  Conclusion

The influence of open source software on the software industry continues to increase. In order to identify the open source activity patterns, the author performs a long-term analysis of open source activities in a chorological dimension.

The author obtains the activity data from the publicly available web services to the open source activity archive. The author presents some exploratory results from project mining of publicly available data. Many open source projects consist of a small group with a small number of human relations to maintain a certain level of social respects. The study reveals the minimum level of social ties in many of the open source projects.

It is useful to leverage awareness of this small-society factor in many open source activities. In order to harness many successful open source projects, it is important to raise awareness for learning from successful projects and skills to depart from one-man projects.

# References

1. Bird, C., Gourley, A., Devanbu, P., Gertz, M., Swaminatha, A.: Mining email social networks in postgres. In: MSR 2006, pp. 185–186. ACM, New York (2006)
2. Bird, C., Pattison, D., D'Souza, R., Filkov, V., Devanbu, P.: Latent social structure in open source projects. In: SIGSOFT '08/FSE-16, pp. 24–35. ACM, New York (2008)
3. Ducheneaut, N.: Socialization in an open source software community: A socio-technical analysis. Computers in Human Behavior 14(4), 323–368 (2005)
4. Raymond, E.S.: The magic cauldron (2000),
http://www.catb.org/ esr/writings/cathedral-bazaar/
magic-cauldron/

# Challenges for Mobile Middleware Platform:
# Issues for Embedded Open Source Software Integration

Toshihiko Yamakami

ACCESS
Toshihiko.Yamakami@access-company.com

**Abstract.** Linux is penetrating into mobile software as the basis for the mobile middleware platform. It accelerates the increasing visibility of open source software (OSS) components in the mobile middleware platform. Despite multiple challenges in mobile embedded software engineering, it is crucial to promote open source-aware development of the mobile software platform. The author presents the challenges to open source software integration in embedded software development. The author discusses the open source-aware software development and identifies a path to transition for moving toward it.

## 1 Introduction

Linux has penetrated into a wide range of digital appliances, e.g. mobile handsets, digital TVs, game consoles and HD recorders. It facilitates the reuse of PC-based rich user experience data service software with high speed network capabilities in an embedded software environment. As Linux-based software is widely adopted in digital appliances, the original weak points of Linux in an embedded environment have been addressed e.g. real time processing and battery life capabilities. The author presents these challenges, and suggests a path towards a more open source-aware development model. Then, the author describes how open source-aware development can be deployed from the perspectives of organization and design.

## 2 Purpose and Related Work

The purpose of this research is to identify a path toward open source-aware software development evolution and its implications for software development management.

OSS has been increasing its visibility in embedded softawre [5] [3] [1]. There are several examples of emerging foundation engineering utilizing OSS in the mobile platform software [4] [6] [2]. This has caught industry attention worldwide.

The originality of this paper lies in its discussion of impacts from OSS-based foundation collaboration in the mobile industry.

P. Ågerfalk et al. (Eds.): OSS 2010, IFIP AICT 319, pp. 401–406, 2010.
© IFIP International Federation for Information Processing 2010

# 3  Open Source-Related Landscape

## 3.1  Driving Forces for Embedded Open Source and Issues

There are multiple factors that drive embedded software engineering:

- Market pressure to shorten time to market,
- Market pressure to reduce software development and maintenance cost, and
- Device convergence.

As the many more devices become network-enabled, the size of network and application software grows, which impacts the cost structure of digital appliances. Managing the costs of maintaining this large-scale code base is a critical concern for embedded software development.

Nowadays, advanced mobile handsets consist of (a) Linux kernel, (b) middleware platform, and (c) application code. Each is 5–10 million lines of code.

This is causing a transition from constraint-based specialized software development to OSS-based large-scale heterogeneous software development.

The amount of OSS code is increasing dramatically. Almost 10 million lines of code are used for just the middleware of advanced mobile handsets, excluding the kernel itself and higher level applications. The ratio of OSS code in relation to them is increasing. It is estimated that 80–95 % of the Linux-based mobile middleware platform is originated from OSS. The ratio is expected to increase in the future.

However, there are several constraints in the mobile middleware platform from (a) embedded software, (b) mobile-specific service development, and (c) OSS-based software management.

The combination of these issues provides a significant challenge for today's embedded software development.

## 3.2  Issues from Embedded Software Engineering

Mobile handsets require the inclusion of special hardware management for telephony, which is device-specific.

- Customization for hardware integration: Embedded software depends on some hardware components. These components need hardware-dependent coding.
- Stability for embedded integration: Generally, open source software is based on bazaar-style development. Embedded software has inheritent constraints on software updates after factory shipment. Since the software is written into ROM (read-only memory), it cannot be amended. Embedded software development needs to address this challenge.
- Real-time processing requirements: Mobile phone requires real-time processing of incoming calls. OSS generally does not set performance as the top priority. Device development needs to address device-specific performance requirements.
- Battery life duration requirement: Mobile phone and other battery-driven devices require special consideration to the duration for battery life.

## 3.3 Issues from Mobile-Specific Service Development

Mobile data services are hot issues in mobile business development. This leads to multiple issues that need in mobile middleware platform development.

- Wireless carrier customization: For differentiation, wireless carriers would like to deploy carrier-specific services. Embedded software development needs to address this carrier-specific customization.
- Synchronization for handset launch schedule with OSS roadmaps: Mobile handsets have some seasonal cycle for shipment depending on each country for marketing which is irrelevant to OSS roadmaps.

## 3.4 Issues from OSS Management

There are issues surrounding OSS management in general.

- Design and Customization
  - Coordination among multiple OSS components: Each OSS component has a different release cycle, major version updates and roadmap. In order to facilitate product commercialization, coordination and version selection are crucial.
  - Coordination among multiple dependencies among OSS component sets: When a large number of OSS components are used, different OSS components have different dependencies on some common OSS components.
- Miscellaneous Non-coding Tasks requiring Resources
  - Evaluation to choose components and versions: There are an emerging number of OSS components with different version releases. Commercialization takes care of evaluation and version selection.
  - Synchronization for software development with upstream OSS project roadmaps: When there is visibility of an upcoming a major version up of major OSS component, commercialization needs to synchronize a future development plans and future major version updates.
- Community Coordination and Management
  - Granting back to the OSS community: As a good citizen of the OSS community, general patches and modifications need to be granted back to the original OSS community.
  - Constraint to disclose handset commercialization schedule: During interactions with the OSS community, there are some non-disclosable trade secrets.
- Legal issues
  - GPL contamination: GPL code needs to be carefully managed in order not to disclose any proprietary software in wrong use of GPL code.
  - Export compliance: Some of encryption modules need to be managed in compliance to appropriate import/export compliance procedures.
  - Patent protection: As well as any proprietary software, OSS modules also need to address patent protection.

## 4    An Open Source-Aware Software Development

The management of these issues is leading to a new software design and maintenance paradigm. Open source-oriented embedded software engineering requires the following special expertise and coordination:

- Interface design between OSS, customized, and proprietary components
- Professional service for code and packaging management, and evaluation

A three-stage model of software development transition towards open source-aware development is depicted in Fig. 1. Most of the proprietary software development depends on the assumption of in-house development. This influences the organization, design process, and source control schemes. Considering the fact that the major part of the software development is shifting towards an open source-based one, at least for the middleware level, it is crucial to manage the transition. The transition starts with proprietary development. In the proprietary development, the entire code base is owned by the company. It does not require per-module code and license management.

During OSS penetration, some of the modules may be replaced by OSS codes. However, this is a replacement-oriented process, not one that wholly replaces the development process.

When OSS penetration reaches a certain point, the whole development process needs to be revisited. Code and license management is done on a per-module basis. The OSS professional service needs to be deployed in the development process.

**Fig. 1.** A three-stage model of software development transition towards open source-aware development

The software development team organization with open source awareness is illustrated in Fig. 2.

**Fig. 2.** Open source-aware development organization

The software design process with open source awareness is depicted in Fig. 3. It is difficult to isolate customized OSS components from proprietary components post-process. Therefore, a clear separation of design is necessary during the software design process.

**Fig. 3.** Open source centric design principle

Multipel OSS components and their customization need to be integrated with awareness of code control, version control and license control. When there are conflicting OSS dependencies, it requires minor adjustments, e.g. back-porting et al. Also, the total code base including proprietary and OSS components needs to be integrated with awareness of code control, version control and license control.

## 5 Transition Management

Software vendors are aware of the importance of OSS. However, the challenges posted by less visibility and higher diversity of projects and codes make it difficult for software vendors to make a best-fit blending of multiple OSS components.

Embedded software needs to cope with the time-dimensional issues of software management. There is a lot of hardware-dependent code in the embedded software. This makes the transition from one software framework to another one difficult.

Also, this hardware-dependency puts an unavoidable portion of OSS components in need of case-by-case customization. Transition management needs to address this type of customization in the case of embedded software engineering.

In addition to the requirements to address multiple license terms in different OSS components, each customization needs to be carefully separated from other code to ensure the proper license management and grant back to the upstream OSS versions.

This puts the fundamental heterogeneity into embedded software engineering. Transition management needs to give considerations to organization, process, architecture, and source code management ahead of any transition management.

## 6 Conclusion

Since the major portion of the large-scale software platform consists of OSS components, the software development process needs to revisit this reality in the long run. The mobile industry is one of the areas where this transition is becoming increasingly visible. The author describes the challenges for large-scale software project with a large number of software components. The author proposes an open source-aware software development process. This will bring the procedural and organizational impacts on embedded software development. In the past, embedded software development focused on code-size-aware hardware-specific coding. There is ongoing radical transition of software development towards tens of millions of lines of code in an embedded environment.

Increasing involvement of OSS components leads to open source-aware development. In-depth analysis reveals the new heterogeneity caused by multiple different OSS components in an integrated embedded software context. This heterogeneity needs to be addressed through the advanced design of the entire software development process and its transitions.

# References

1. Barr, M., Massa, A.: Programming Embedded Systems: With C and GNU Development Tools, 2nd edn. O'Reilly Media, Inc., Sebastopol (2006)
2. Google: Android - an open handset alliance project (2007), http://code.google.com/android/
3. Hollabaugh, C.: Embedded Linux: Hardware, Software, and Interfacing. Addison-Wesley Longman Publishing Co., Inc., Boston (2002)
4. LiMo Foundation: LiMo Foundation Home page (2007), http://www.limofoundation.org/
5. Massa, A.J.: Embedded software development with eCos. Prentice-Hall, Englewood Cliffs (2002)
6. Symbian Foundation: Symbian foundation web page (2008), http://www.symbianfoundation.org/

# Open Source Software Developer and Project Networks

Matthew Van Antwerp and Greg Madey

University of Notre Dame
{mvanantw,gmadey}@cse.nd.edu

**Abstract.** This paper outlines complex network concepts and how social networks are built from Open Source Software (OSS) data. We present an initial study of the social networks of three different OSS forges, BerliOS Developer, GNU Savannah, and SourceForge. Much research has been done on snapshot or conflated views of these networks, especially SourceForge, due to the size of the SourceForge community. The degree distribution, connectedness, centrality, and scale-free nature of SourceForge has been presented for the network at particular points in time. However, very little research has been done on how the network grows, how connections were made, especially during its infancy, and how these metrics evolve over time.

## 1  Introduction to Complex Networks

The OSS network is defined as follows. Developers and projects are considered nodes in the graph. If a developer works on a project, there is an edge between the developer and that project. Since developers can only work on projects, the resulting graph will be bipartite, with developers and projects being the two groups having no edges within those groups. This bipartite graph can be easily transformed into a developer network or a project network. From the CVS database [8], users, groups, and timestamps were extracted. The timestamps are the dates (in unix time) of the oldest and most recent commits to that particular project. With this information, even if two users worked on the same project, a tie was only created between them if they worked on the project at the same time, i.e. their time frame windows overlapped.

### 1.1  Previous Work

Xu in her dissertation [13] analyzed many aspects of the developer and project networks in the SourceForge community. Xu examined the SourceForge developer network over time and determined it to be scale-free [12]. Xu also examined the community structure of the SourceForge developer network in [11] using metrics such as modularity [7,6], identifying the largest communities and their populations. Gao examined the diameter, clustering coefficient, centrality, and other metrics of the SourceForge developer network over a timespan of a year and a half [3].

P. Ågerfalk et al. (Eds.): OSS 2010, IFIP AICT 319, pp. 407–412, 2010.

In [4], the authors apply social network analysis to CVS data, graphing network measurements such as degree distribution, clustering coefficient in modules, weighted clustering coefficient, and connection degree of modules for various projects at different time periods in the histories of Apache, Gnome, and KDE. They concluded that both the module network and the developer network exhibit small world behavior.

## 2   SourceForge, GNU Savannah, and BerliOS Developer

SourceForge was launched in November 1999. It is the world's largest OSS hosting site, with over 2.3 million registered users and over 180,000 projects at time of writing. It hosts numerous prominent and popular OSS projects. It is also the most studied hosting platform for the purposes of OSS research. SourceForge data is available at the SourceForge Research Data Archive (http://srda.cse.nd.edu) [9].

Many popular GNU/Linux utilities are or were at some time hosted at GNU Savannah, including gcc, emacs, libc, autoconf, automake, and make. The site has been up since around 1996, although many projects had their CVS logs imported and many of them date back to the early 1980s. They are strict about only hosting free software (SourceForge, for example, allows you to host a project that does not have a free software license). Many prominent OSS figures contribute to projects hosted here, including Richard Stallman, Ulrich Drepper, and Roland McGrath. Despite having far fewer developers than SourceForge, it is a very active community. Despite having not even 4000 developers, they have made nearly 5 million code commits. SourceForge has about 65 million code commits with orders of magnitude more members.

BerliOS Developer is a German website hosting 5,425 projects and 43,708 registered users at time of writing [2]. While difficult to tell exactly how old the site is, the BerliOS project itself was registered in June 2000 and the earliest CVS timestamp dates to 1996, although only a handful of projects have CVS commits dating prior to June 2000 and all of those projects were registered on BerliOS itself after June 2000. These projects likely had previously existing CVS archives imported into the BerliOS hosting platform. It is similar in functionality and services offered to SourceForge, but does not have the worldwide popularity of it. It is about as old as SourceForge as well, so the two share some similarities with BerliOS having a much smaller user base.

## 3   SourceForge Developer Network

73,829 users have made at least one CVS commit. Of those, 47,946 users are connected to at least one other developer (in other words, they are not the sole developer on all of the projects they work on), which is 64.94%. Of these connected users, the largest connected component contains 19,269 users, which is 40.19%, or 26.10% of all users who have made at least one commit. A visualization of a random sample of the developer network is found in figure 1. Just under 2/3 of the user-project ties were sampled to create this network. This resulted in

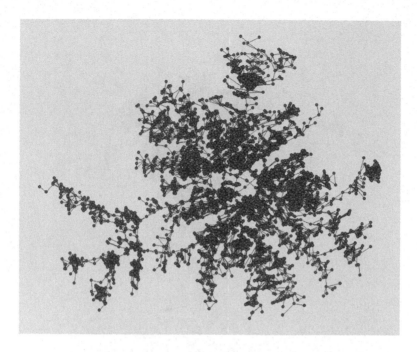

**Fig. 1.** Sample of the SourceForge developer network

37,811 vertices in the network with the largest connected component containing 4687 vertices. This largest connected component is what is displayed in figure 1. There are many clusters of developers, but no central core in this sample. There are many "rings" of developers and towards the outside of the graph, there are linchpin developers where the graph would become disconnected without their presence. Visualizations were developed with Pajek [1].

## 4    Savannah Developer and Project Networks

3889 users have made at least one CVS commit. Of those, 3042 users are connected to at least one other developer (they are not the sole developer on all of the projects they work on), which is 78.22%. Of these connected users, the largest connected component contains 1747 users, which is 57.42%, or 44.92% of all users who have made at least one commit. A visualization is provided in figure 2. In that figure, developers who are the sole developer on all projects they work on are excluded. They would be singletons in the network were they included. The Savannah project network can also be seen in figure 2. The network is well-connected with most projects in one large cluster. This is due to the long life of most Savannah projects, the rarity of new projects hosted at Savannah, and the fact that many developers here work on multiple Savannah projects during their lifetime.

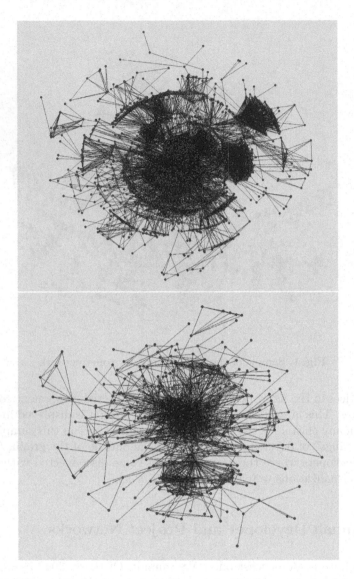

**Fig. 2.** (top) The largest connected component in the Savannah developer network visualized with the Kamada-Kawai algorithm for drawing graphs [5], a force-based algorithm. Distance between two nodes in the figure roughly corresponds with the length of the shortest path between them in the graph. (bottom) The Savannah project network.

## 5   BerliOS Developer and Project Networks

1582 users have made at least one CVS commit. Of those, 1113 users are connected to at least one other developer (they are not the sole developer on all of the projects they work on), which is 70.35%. Of these connected users, the largest connected component contains only 100 users, which is 8.98%, or 6.32% of all users who have made at least one commit. The BerliOS project network is mostly disconnected. There are however a handful of interesting cliques present in this network.

# 6  Repeat Network Connections

The SourceForge developer network, of 396,590 developer-developer ties, only 10,491 are duplicates or the original links that were later duplicated. This comprises only 2.65% of all developer pairs. However, for Savannah, of 46,937 developer pairs, there are 4620 pairs that are duplicates or links that were later duplicated, nearly 10%. For BerliOS, there are 3349 developer ties and 84 of them are repeats or the links that would later be duplicated. This is 2.51% of all pairs, comparable to SourceForge. The phenomena of repeat network connections in developer networks has not been extensively studied. The abundance of presumably fruitful developer ties in Savannah indicates that the projects here were likely successful. This also likely indicates that the typical project on Savannah is more successful than the typical project at SourceForge or BerliOS. This phenomena is examined further in [10].

# 7  Evaluation of the Communities

BerliOS is not a very globally connected developer community. While many developers are connected to someone else, there does not seem to be any sort of small-world effect in this network. SourceForge is a very large community and is better connected than BerliOS. About one quarter of all developers (CVS committers) in SourceForge are in the largest connected component. However, Savannah has nearly half of all developers in the largest connected component, an impressive aspect. A summary of the aforementioned statistics is available in table 1.

**Table 1.** Size of largest connected component in the developer networks

| Hosting Site | Total Size | Number Connected | % Connected | Largest CC | % of total |
|---|---|---|---|---|---|
| SourceForge | 73,829 | 47,946 | 64.94% | 19,269 | 26.10% |
| Savannah | 3889 | 3042 | 78.22% | 1747 | 44.92% |
| BerliOS | 1582 | 1113 | 70.35% | 100 | 6.32% |

# 8  Conclusions

We presented initial statistical analysis of the project and developer networks of three different OSS forges. The evolutionary trends displayed by these networks may offer crucial insight into OSS phenomena. Software versioning logs provide a great resource for building and studying these networks.

## Acknowledgments

Research reported in the paper was supported in part by the National Science Foundation's CISE IIS-Digital Society & Technology program under Grant ISS-0222829 and by the National Science Foundation's CISE Computing Research Infrastructure program under Grant CNS-0751120

## References

1. Batagelj, V., Mrvar, A.: Pajek - program for large network analysis. Connections 21, 47–57 (1998)
2. BerliOS Developer, http://developer.berlios.de
3. Gao, Y.: Computational Discovery in Evolving Complex Networks. PhD thesis, University of Notre Dame (2007)
4. Gregorio, L.L.-F.: Applying social network analysis to the information in cvs repositories. In: Proceedings of the First International Workshop on Mining Software Repositories (MSR 2004), Edinburgh, UK (2004)
5. Kamada, T., Kawai, S.: An algorithm for drawing general undirected graphs. Inf. Process. Lett. 31(1), 7–15 (1989)
6. Newman, M.E.J.: Fast algorithm for detecting community structure in networks. Physical Review E 69, 066133 (2004)
7. Newman, M.E.J., Girvan, M.: Finding and evaluating community structure in networks. Physical Review E 69, 026113 (2004)
8. Van Antwerp, M.: Studying open source versioning metadata. Master's thesis, University of Notre Dame, Notre Dame, IN (April 2009)
9. Van Antwerp, M., Madey, G.: Advances in the sourceforge research data archive. In: Workshop on Public Data about Software Development (WoPDaSD) at The 4th International Conference on Open Source Systems, Milan, Italy (2008)
10. Van Antwerp, M., Madey, G.: The importance of social network structure in the open source software developer community. In: The 43rd Hawaii International Conference on System Sciences (HICSS-43), Hawaii (January 2010)
11. Xu, J., Christley, S., Madey, G.: The open source software community structure. In: NAACSOS 2005, Notre Dame, IN (June 2005)
12. Xu, J., Madey, G.: Exploration of the open source software community. In: NAACSOS 2004, Pittsburgh, PA (June 2004)
13. Xu, J.: Mining and Modeling the Open Source Software Community. PhD thesis, University of Notre Dame (2007)

# Warehousing and Studying Open Source Versioning Metadata

Matthew Van Antwerp and Greg Madey

University of Notre Dame
{mvanantw,gmadey}@cse.nd.edu

**Abstract.** In this paper, we describe the downloading and warehousing of Open Source Software (OSS) versioning metadata from SourceForge, BerliOS Developer, and GNU Savannah. This data enables and supports research in areas such as software engineering, open source phenomena, social network analysis, data mining, and project management. This newly-formed database containing Concurrent Versions System (CVS) and Subversion (SVN) metadata offers new research opportunities for large-scale OSS development analysis. The CVS and SVN data is juxtaposed with the SourceForge.net Research Data Archive [5] for the purpose of performing more powerful and interesting queries. We also present an initial statistical analysis of some of the most active projects.

## 1 Introduction

Versioning programs have been in use by open source software projects for many decades. Publicly available logs offer a development trail ripe for individual and comparative studies. In this paper, we describe the downloading and warehousing of such data. We also present some preliminary data analysis. The process is similar to that done in [2] which described an approach to populating a database with version control and bug tracking system data for individual project study. At Notre Dame, Jin Xu also took an individual project approach to retrieving and studying projects on SourceForge [6]. Xu built a similar retrieval framework however for web pages to gather project statistics.

## 2 SourceForge.net Data

Most of SourceForge's data is stored in a back-end database. The actual source code is stored in a Concurrent Versions System (CVS) or Subversion (SVN) repository. The data stored there includes who is making a change to the code, how the new version of the code differs from the most recent version, the number of removed and new lines of code, a revision number, a comment, and a timestamp. The entire history of a project can be reviewed by walking chronologically through one or more CVS or SVN logs. The logs tell us what changes a project has undergone, when those changes took place, and by whom. We have recently obtained CVS and SVN metadata (everything except the actual code) and built another database that is juxtaposed with the back-end database. This data is available for scholarly research at http://srda.cse.nd.edu .

P. Ågerfalk et al. (Eds.): OSS 2010, IFIP AICT 319, pp. 413–418, 2010.

# 3    Concurrent Versions System

Concurrent Versions System (CVS) is software developed for software version control allowing simultaneous use by multiple users. It is built upon the Revision Control System (RCS) software, which takes care of individual file versioning [4]. CVS provides a layer of abstraction allowing for concurrent access to a particular RCS file with intelligent conflict mediation. CVS also groups files together into a logical entity (a project) and allows tagging of particular file revisions as a logical snapshot (a project release, for example) [1].

We downloaded log data from all projects on SourceForge.net that use CVS for version management and also allow anonymous CVS access. To do so, the following CVS command was used:

```
cvs -d:pserver:anonymous@PROJECT.cvs.sourceforge.net:/cvsroot/PROJECT rlog .
```

Using `rlog` instead of `log` allows the process to run without having to run a time and space-consuming `checkout` command. The dot (.) at the end is used in place of a module name to indicate that log data is being requested for all modules.

Similarly, we downloaded CVS metadata for the projects on the open source hosting platform BerliOS Developer and GNU Savannah.

```
cvs -d:pserver:anonymous@cvs.savannah.gnu.org:/sources/PROJECT rlog .
cvs -d:pserver:anonymous@cvs.berlios.de:/cvsroot/PROJECT rlog .
```

For obtaining SVN data, the following commands were used:

```
svn log --verbose http://PROJECT.svn.sourceforge.net/svnroot/PROJECT
svn log --verbose svn://svn.berlios.de/PROJECT
svn log --verbose svn://svn.savannah.gnu.org/PROJECT
```

# 4    Download Process

The method employed was to make one serial line of requests on multiple machines. Due to the lack of physical machines at our immediate disposal, and the ease with which they can be set up, virtual machines were employed. In addition, if a machine were about to make a request to the same CVS server it just contacted, a stall time was employed. When this occurs, the virtual machine (VM) would sleep for 5 seconds before making the next request.

## 4.1    Job Distribution

We wrote a central server process to handle distributing jobs to each VM. This process spawned a process for each VM which would submit a project name that the VM would download and then return a signal for one of the following: 1) success, 2) initiation failure, or 3) progress failure. Upon return, the handler process would then submit a new project name to the VM it handles and the cycle continues. The information was tracked in a database on the central server.

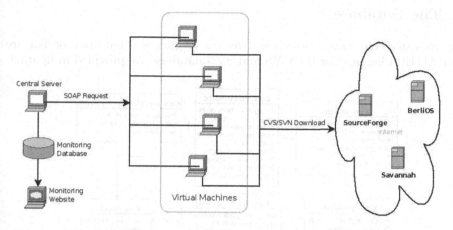

**Fig. 1.** Diagram of the job distribution and download process. This can be easily modified to retrieve other publicly available data or rsync the code instead of just retrieving the metadata.

The database contained the name of the project, the name of the server (VM) the job was deployed to, the timestamp of the submission, the returned signal, the timestamp of the returned signal, and if applicable, the number of lines in the downloaded CVS log. In order to monitor the download progress and be aware of potential problems, a web frontend using AJAX was deployed to monitor the database. A schematic of this process is shown in figure 1.

SourceForge.net CVS data was obtained over the span of about 7 days. Source-Forge SVN data took about 2 days to download. Obtaining CVS and SVN data for BerliOS and Savannah took about 48 hours total. The number of projects successfully downloaded from each site is shown in table 1.

Any changes made to projects since the log files were downloaded are obviously not present in our database. Therefore, continuous updates to the data are necessary, a data warehousing issue brought up in [3]. Two aspects of the database and the log files make this relatively simple to do. CVS contains a filter option to only return log information after a specified time. For each project, we can search in our database to find the most recent timestamp, and then use that as the range specifier and only new updates will be returned. SVN allows a user to specify a range of revisions when running the log command. In this case, we can simply retrieve the number of the latest revision and download all revisions since that one.

**Table 1.** Number of logs downloaded from each hosting site, classified by versioning software

| Hosting Site | CVS | SVN |
|---|---|---|
| SourceForge | 103869 | 24416 |
| BerliOS | 1252 | 1718 |
| Savannah | 1775 | 8 |

## 5    The Database

Due to space constraints, details on the log parsing and database design are omitted. ER diagrams for the CVS and SVN database are provided in figure 2.

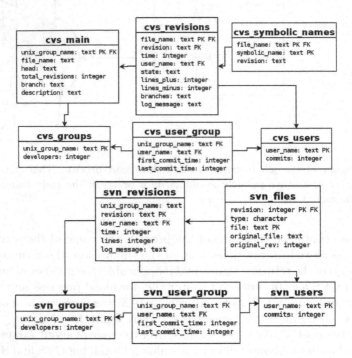

**Fig. 2.** Entity Relation (ER) diagrams for the new CVS and SVN database

## 6    Data Analysis

In this section, we present some quantitative data on gcc and emacs, two extremely mature and long-lived open source projects from GNU Savannah.

The GNU Savannah hosted project gcc is the GNU C compiler. The CVS log begins in 1988 and had 345,723 file commits up until November 2005 when the project was transferred elsewhere. Certain months had nearly 10,000 commits. Nearly 300 people have contributed to the project in its 20 year history. The CVS log file for gcc was the largest of all projects that were downloaded, with a size of about 1.5 GB. The graph is found in figure 3.

Another mature project hosted on Savannah is emacs, the popular editor. 200 users have contributed since its initial CVS checkin in 1985. The most active months had over 2500 file commits, with a total of 122,254 commits over all time. The information is graphed in figure 4.

While these are relatively simple quantitative statistics, right away there are interesting portions that warrant further investigation. Both gcc and emacs have a slight, but noticeable lull in activity before a sustained increase in activity.

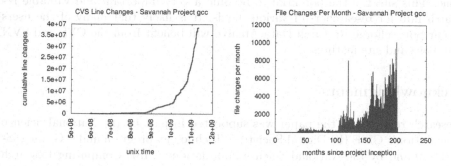

**Fig. 3.** Savannah project gcc

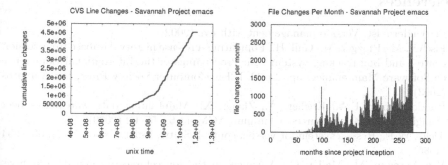

**Fig. 4.** Savannah project emacs

Comparing the cumulative line changes graphs for the projects, the lull occurs for both approximately mid-2001, roughly coinciding the dot-com bubble burst. Another trend noticed in some of the younger and less mature projects was a distinct pattern in the cumulative line changes graphs. These graphs often showed a period of positive acceleration, then an inflection point, then negative acceleration. This would seem to indicate an increasing number of additions to the software, a peak activity period, followed by a level of code maturity where most of the fixes are minor (patches and bug fixes). The patterns in development activity, and comparison of these patterns across different projects will be examined more thoroughly in a future publication.

# 7    Conclusions

This large data set offers a multitude of open source software and social networking research opportunities. We can learn about project development trends and group similar projects together by development similarity. We can examine contribution trends by individual coders. We can see how they migrate from project to project and how the amount and types of contributions differ over time. This site has the potential to become a very important and valuable research hub for researchers of various fields. It is likely that many of the users of our SourceForge Research Data Archive will benefit from the CVS and SVN database and site features.

## Acknowledgments

Research reported in the paper was supported in part by the National Science Foundation's CISE IIS-Digital Society & Technology program under Grant ISS-0222829 and by the National Science Foundation's CISE Computing Research Infrastructure program under Grant CNS-0751120.

## References

1. Per Cederqvist. Version management with cvs (2002)
2. Fischer, M., Pinzger, M., Gall, H.: Populating a release history database from version control and bug tracking systems. In: Proceedings of the International Conference on Software Maintenance, pp. 23–32. IEEE Computer Society Press, Los Alamitos (2003)
3. Rundensteiner, E.A., Koeller, A., Zhang, X.: Maintaining data warehouses over changing information sources. Commun. ACM 43(6), 57–62 (2000)
4. Tichy, W.F.: Rcs—a system for version control. Softw. Pract. Exper. 15(7), 637–654 (1985)
5. Van Antwerp, M., Madey, G.: Advances in the sourceforge research data archive. In: Workshop on Public Data about Software Development (WoPDaSD) at The 4th International Conference on Open Source Systems, Milan, Italy (2008)
6. Xu, J., Huang, Y., Madey, G.: A research support system framework for web datamining research: Workshop on applications, products and services of web-based support systems. In: The Joint International Conference on Web Intelligence (2003 IEEE/WIC) and Intelligent Agent Technology, Halifax, Canada, October 2003, pp. 37–41 (2003)

# Workshop – Open Source Software for Computer Games and Virtual Worlds: Practice and Future

Per Backlund[1], Björn Lundell[1], and Walt Scacchi[2]

[1] University of Skövde, Sweden
[2] University of California, Irvine, USA
{per.backlund,bjorn.lundell}@his.se, wscacchi@ics.uci.edu
http://www.his.se, http://cgvw.ics.uci.edu/

## 1 Introduction

Computer games and virtual worlds are increasingly used throughout our society with people playing on the bus, at home and at work. Computer games thus affect larger and larger number of people and areas in the society of today. There are even scholars who advocate that games or virtual environments create better environments for learning than traditional classrooms. This situation motivates the use of games and game technology for additional purposes, e.g. education, training, health care or marketing.

This new use distinguishes between entertainment games and games for other uses, with the term Serious Games being the most common for the latter category. Although the term itself is well established in both academia and industry, there is no current single definition of the concept. However, a common component of these definitions is that is the addition of pedagogy (activities that educate or instruct, thereby imparting knowledge or skill) that makes games serious. For the purpose of this workshop we define Serious Games as:

> Serious Games are games that engage users in their pursuit, and contribute to the achievement of a defined purpose other than pure entertainment (whether or not the user is consciously aware of it).

A Serious Game can be achieved through a spectrum ranging from the mere utilisation of game technology for non-entertainment purposes to development of specifically designed games for some non-entertainment purpose or the use and/or adaptation of commercial games for non-entertainment purposes. We also propose that any combination of the above would constitute a feasible way to achieve the desired effect.

There are numerous examples of serious games and virtual worlds from various sectors. But what are the possible roles that Open Source Software can take in facilitating the development of a new generation of these technologies or applications? Many new games and virtual worlds are being created through the use of open source game engines and game asset creation tools. Game modding, itself an idea that often relies on retail computer games that are packaged with software development kits to create new game variants, are generally licensed for non-commercial redistribution with game mod source code using a open source software license. But the intersection of Open Source, Games and Virtual Worlds is perhaps just beginning. More ideas are

—. Ågerfalk et al. (Eds.): OSS 2010, IFIP AICT 319, pp. 419–420, 2010.

being pursued, including how to facilitate games that rely on user-created content, or that incorporate social media (e.g., YouTube videos, Flickr photos, audio recording remixes), and social networking services to create new modes of game play. Games, Virtual Worlds, and Open Source Software also help serve the needs of independent game developers and virtual world developers who work with limited resources, outside large commercial enterprises.

## 2  Workshop Aim

The workshop aims to bring people from the Open Source and Serious Games communities together to discuss the current status of the area and to find a common future where the two areas can enrich each other.

Open Source Software for Computer Games and Virtual Worlds: Practice and Futures will feature *position statements* and *presentations* which will be open for discussion.

Suggested topics include, but are not limited to the following:
- Current examples of game or virtual world applications for learning, health, energy, environment, manufacturing or other areas
- Current examples of open source tools and techniques for creating games or virtual world applications
- Practical and theoretical perspectives of open source in games or virtual worlds
- Open source approaches to game modding and user created content
- The intersection of game culture and free/open source software culture
- Innovative combinations of game play mechanics, social media, and social networking, and open source software
- Experience in developing games or virtual worlds using open source software tools, techniques, concepts, or game engine, such as OGRE, OpenSim, Open Croquet, Irrlicht, Doom/Quake, etc.

## Program Committee

- Per Backlund, University of Skövde, Sweden (workshop co-organiser)
- Bjorn Lundell, University of Skövde, Sweden (workshop co-organiser)
- Walt Scacchi, University of California Irvine, USA (workshop co-organiser)
- Rosario de Chiara, University of Salerno, Italy
- Henrik Gustavsson, University of Skövde, Sweden
- Vittorio Scarano, University of Salerno, Italy
- Robert J. Stone University of Birmingham and HFI DTC, UK

# WoPDaSD 2010: 5th Workshop on Public Data about Software Development

Jesús M. González-Barahona[1], Megan Squire[2], and Daniel Izquierdo-Cortazar[1]

[1] GSyC/LibreSoft, Universidad Rey Juan Carlos, Mostoles, Madrid
{jgb,dizquierdo}@libresoft.es
[2] Elon University, North Carolina, The USA
msquire@elon.edu

## 1  Introduction

Projects such as FLOSSmole and FLOSSMetrics are compiling huge quantities of data about libre (free, open source) software development. The availability of these data in formats suitable for analysis by third parties are enabling researchers to focus on the study of the data, and not on data retrieval activities. This is fortunate, since data retrieval from software development repositories is becoming more and more complex, especially when reliable and detailed information from many projects is needed.

The use for research purposes of this kind of data compiled by teams external to the researcher is posing new problems. Annotation of data, exchange formats, traceability and privacy issues, are becoming issues to be addressed. In addition, working with FLOSS projects to easy obtaining their data, and showing them how that can benefit their activities is also of increasing importance.

Despite these open issues, the use of these open datasets is enabling researchers in many ways: reproduction of results is easier; massive analysis (based on data from hundreds or even thousands of projects) is possible; quick obtaining of results is simplified; availability of data for research communities with little experience in retrieving data from software repositories.

Studies and research results based on this kind of dataset have already been presented in workshops, conferences and journals, but rarely the focus is on how to benefit from the datasets, or on the problems derived from their use. In addition, the details of how to use the datasets for different purposes, or specific results from their analysis, are not published elsewhere.

This workshop is once again (for the fifth year in a row) a place to discuss all these topics, and to present research results developed with these ideas in mind: how these large datasets about FLOSS software development are retrieved, how can they be analyzed and mined, how they can be published, exchanged and extended, which lessons are we learning from their use, and which results are being obtained from their analysis.

## 2  Goals

The goal of this workshop is to foster the production and analysis of publicly available data sources about software development and the exchange of data

P. Ågerfalk et al. (Eds.): OSS 2010, IFIP AICT 319, pp. 421–422, 2010.
© IFIP International Federation for Information Processing 2010

between different research groups. The workshop is aimed at the following kinds of studies (although other related studies could also be considered):

- Results based on the analysis of large datasets about software development.This refers mainly to research conducted on FLOSSmole or FLOSS-Metrics data, but also on other similar open source datasets. The analysis should show a methodology to explore the projects, but also it should show explanations to "odd" things that could appear in the data set. For instance, a company-driven project can show different behavior than a community-driven project. The study can be in the field of software engineering, economics, sociology, human resources, and others.
- Retrieval process and exchange formats of publicly available data collections about software development. The data collections presented should be publicly available, based themselves on public data (so that other groups could reproduce the data collection process), and be related to the field of software development. This includes, but is not limited to, data from source code control systems, but tracking systems, mailing lists, websites, source and binary code, quality assurance systems, etc. Although any kind of data collection can be considered, those including information about a large number of projects will be considered especially appropriate.
- Data mining activities and new retrieval tools. Working with a huge quantity of data invites complexity in storage and analysis. Data mining techniques are welcome in this section, provided that papers include some conclusions about a specific set of projects. Again, this analysis should show a methodology to explore the data and explanations about the whole process. Cross-analysis of datasets, and specially of those provided by the organizers (FLOSSMole and FLOSSMetrics databases) is especially welcome. Also, new tools developed to obtain data from several data sources, such as forums, wikis, bug tracking systems and others fit perfectly here.
- Usage of public datasets about software development by new research communities, which until now did little empirical research in this area because they lacked the expertise needed to retrieve information directly from the repositories, but are now empowered by the availability of these datasets. Research results produced by these communities, cases of use, problems found, etc. are possible contributions to the workshop.

# Second International Workshop on Building Sustainable Open Source Communities OSCOMM 2010

Imed Hammouda[1], Timo Aaltonen[1], and Andrea Capiluppi[2]

[1] Tampere University of Technology, Finland
{imed.hammouda,timo.aaltonen}@tut.fi
[2] University of East London, UK
a.capiluppi@uel.ac.uk

**Abstract.** The Second International Workshop on Building Sustainable Open Source Communities (OSCOMM 2010) aims at building a community of researchers and practitioners to share experiences and discuss challenges involved in building and maintaining open source communities.

## 1 Workshop Scope

Open source software is gaining momentum in several forms. In addition to the huge increase in the number of open source projects started and the remarkable rise of FLOSS adoption by companies, new models of participation in the movement are emerging rapidly. For instance, companies are increasingly releasing some of their proprietary software systems as open source on one hand and acquiring open source software on the other hand. For all these forms of involvement, a central question is how to build and maintain a sustainable community of users and developers around the open source projects.

Research findings show that developing and maintaining online communities in general is a complex activity. In the case of open source communities, the situation is worsened as the problem is multi-facet (e.g. legal, social, technical, business) bringing own kinds of challenges. We think that it is the right time for the research community and the industry to discuss the community building problem from its various perspectives by exchanging related experiences, sharing relevant concerns, and proposing guidelines to manage the challenges highlighted earlier. This is vital as more and more companies are moving towards community-driven development models.

## 2 Workshop Theme

Workshop topics include (but are not limited to):

- challenges of building and maintaining open source communities covering concerns related to legal, socio-cultural, business, technical, etc. dimensions;
- organization and interaction schemes in open source communities;

P. Ågerfalk et al. (Eds.): OSS 2010, IFIP AICT 319, pp. 423–424, 2010.
© IFIP International Federation for Information Processing 2010

- models and classification schemes of communities: participation (e.g. volunteer, mixed, company-based), origin (e.g. individual, company), host (e.g. academy, company), scope (e.g. public, corporate);
- practical approaches, best practices, frameworks, methodologies, technologies, tools, and environments to support community building and management;
- industrial involvement in building, managing and interfacing with communities: opening up software platforms and acquiring open source software, motives and cost-benefit models;
- building open source communities: the role of companies, academy, governments, NGOs, and individuals;
- open source communities versus other kinds of communities such as firm-hosted communities, corporate communities, social networks, global software teams;
- experience reports and lessons on building and maintaining open source communities.

## 3  Workshop Goals

The goal of the workshop is to bring together interested academics, practitioners, and enthusiasts to discuss topics related to open source communities. The workshop will offer an opportunity for the participants to share experiences and discuss challenges involved in building and maintaining open source communities. The workshop will also identify key research issues and challenges that lie ahead.

## 4  Further Information

Further information regarding organization and program of the workshop is available at http://tutopen.cs.tut.fi/oscomm10/.

# Open Source Policy and Promotion of IT Industries in East Asia

Tetsuo Noda[1], Sangmook Yi[2], and Dongbin Wang[3]

[1] Shimane University, Faculty of Law and Literature
Matsue, Japan
nodat@soc.shimane-u.ac.jp
[2] Duksung Women's University, Department of Business Administration
Seoul, Korea
endien@gmail.com
[3] Tsinghua University, Canter of China Study
Beijing, China
wdb05@mails.tsinghua.edu.cn

**Abstract.** The development style of open source has a possibility to create new business markets for Regional IT industries. Some local governments are trying to promote their regional IT industries by adopting an open source in their electronic government systems. In this paper, we analyze the data of open source application policy of the Japanese government and case studies of promotion policy of local industries by local governments; for example, Nagasaki Prefecture and Matsue City. And it aims to extract the issues in the open sources application policy of local governments and the promotion policy of regional industries in Japan.

## 1 Introduction

The term "Wikinomics" describes a style of business where companies accumulate huge amounts of information to generate revenue, and is typified by multinationals like Google or Amazon. These businesses maximize their high productivity and earnings by leveraging Open Source Software (OSS), which is built on a cooperative development model.

OSS originates from the West coast of the USA, and is still primarily developed and enhanced by American multinationals. It could even be said that the current technical evolution of OSS is focused mainly on companies originating from the United States. However, the inherent benefits of OSS extend beyond the boundaries of enterprises, organizations and even nations, and it has the potential to foster new business markets in regions other than North America.

East Asia nations have made some progress with this technology, and started to introduce OSS for e-government systems during the early part of this century. Many countries granted it a central role in their policies. The reasons for this include adoption of software based on standard specification, liberation from vender lock-in, or opposition to the market control of proprietary software. However, the primary reason is to reduce adoption costs for e-government systems.

P. Ågerfalk et al. (Eds.): OSS 2010, IFIP AICT 319, pp. 425–426, 2010.

While this policy work is useful, there is a great deal more that needs to be done. The OSS adoption policy in each nation of East Asia must be accompanied by technological progress in domestic IT service industries or US multinationals will expand at the cost of local businesses. If this continues unchecked it will create a new form of lock-in for East Asian nations.

Some Asian nations are trying to promote their domestic IT service industries, putting their OSS adoption policy to practical use, and this workshop will provide case studies of that work. It will also provide a forum for discussing current challenges and opportunities around both policy and practical implementation issues across Asia.

- the history and the current stage of open source introduction policy
- the policy of the human resource development in the field of open source
- the IT solution market using open source and the ratio of public sectors
- the current state of open source technology of IT enterprises
- the existence and activity of open source communities
- the open source policy of each Local Government Unit and the appearance of the introduction of open source
- the results that led to promotion of industry of home country

## 2  Workshop Goals

We intend to extract the issues of open source introduction policy not accompanied by the technological progress of domestic IT service industry. And this will give an indication to the roles of governments not only in East Asia but also other developing countries.

## 3  Program Committee

- Tetsuo Noda, Shimane University, Japan (workshop co- organizer)
- Sangmook Yi, Duksung Women's University, Korea (workshop co- organizer)
- Dongbin Wang, Tsinghua University, China (workshop co- organizer)
- Shane Coughlan, Regional Director Asia, Open Invention Network, Japan
- Jonathan Lewis, Hitotsubashi University, Japan
- Tomoko Yoshida, Kyoto Notre Dame University, Japan
- Terutaka Tansho, Shimane University, Japan

# OSS 2010 Doctoral Consortium (OSS2010DC)

Walt Scacchi[1], Kris Ven[2], and Jan Verelst[2]

[1] University of California, Irvine, USA
[2] University of Antwerp, Belgium
http://www.ua.ac.be/oss2010dc

## Goal

The goal of the Doctoral Consortium is to provide PhD students with an environment in which they can share and discuss their goals, methods and results before completing their research. Participants will be selected based on the quality of the proposed research, its potential significance and contribution to the OSS domain, and the potential benefit of the Doctoral Consortium to the PhD student's research.

The Doctoral Consortium will take place on May 30, allowing participants to attend the OSS 2010 conference after the Doctoral Consortium. This allows PhD students to further discuss their research with other researchers in the following days.

As well, because of the diversity of the communities involved, the Doctoral Consortium will allow PhD students to make connections beyond their own disciplines. As a result, we expect that participation will allow PhD students to develop a better understanding of the different research communities, which we believe will facilitate their participation in future inter-disciplinary research.

We will also invite other faculty members to attend the Doctoral Consortium to stimulate discussion.

## Scope

The scope of research topics of the Doctoral Consortium is the same as for the main conference. We therefore invite submissions related to all aspects of open source software including, but not limited to software engineering perspectives, emerging perspectives, social science, and studies of OSS deployment.

We invite submissions from PhD students in the early stages of their research (e.g., those who are at the end of their first year or in their second year), as well as in the late stages of their research (e.g., those who are close to graduating).

PhD students who apply for the Doctoral Consortium should at least have decided on a research topic or topic area, and have a proposal for an appropriate research method. Preferably, PhD students should still have the time to incorporate the feedback obtained during the Doctoral Consortium in their dissertation.

## Full Papers

All full papers submitted to the Doctoral Consortium will be peer reviewed by at least two independent reviewers. PhD students that are accepted to the Doctoral Consortium,

P. Ågerfalk et al. (Eds.): OSS 2010, IFIP AICT 319, pp. 427–428, 2010.

will give a presentation of their work. We aim to provide sufficient time for discussion (at least 20 minutes) to ensure that PhD students obtain quality feedback from the Doctoral Consortium co-chairs, the members of the program committee, as well as other PhD students. This feedback will allow them to enhance their own research proposal. Subsequently, doctoral students whose advisory committee lacks sufficient expertise with current OSS research may benefit in a number of ways from participating in the Doctoral Consortium with attending faculty.

## Lightning Talks

Similar to last year, we will be hosting a special session of "lightning talks" during the OSS 2010 Doctoral Consortium. During this lightning talks session, multiple PhD students will be able to briefly present their research proposal. Each presenter will be provided with a 3-minute time slot and will have one slide available. The lightning talks session allows PhD students to give a brief presentation of their research, to actively participate in the Doctoral Consortium, and to generate awareness of their topic.

The lightning talks session is primarily targeted towards PhD students who are in the early phases of their research. Attending the discussion on the research proposals of other PhD students may also be beneficial for them, as it provides ideas on what future reactions to their own research may be. In addition, by giving a lightning talk, they are able to generate an interest in their research topic, which allows them to connect to other researchers in related areas and to gain preliminary feedback on their proposal.

## Doctoral Consortium Chairs

| | | |
|---|---|---|
| Walt Scacchi | University of California, Irvine | USA |
| Kris Ven | University of Antwerp | Belgium |
| Jan Verelst | University of Antwerp | Belgium |

## Program Committee

| | | |
|---|---|---|
| Kevin Crowston | Syracuse University | USA |
| Joseph Feller | University College Cork | Ireland |
| Daniel M. German | University of Victoria | Canada |
| Jesus Gonzalez-Barahona | Universidad Rey Juan Carlos | Spain |
| Björn Lundell | University of Skövde | Sweden |
| Maha Shaikh | London School of Economics | UK |

# Student Participation in OSS Projects

Gregory W. Hislop[1], Heidi J.C. Ellis[2], Greg DeKoenigsberg[3], and Darius Jazayeri[4]

[1] Drexel University
hislop@drexel.edu
[2] Western New England College
hellis@wnec.edu
[3] Red Hat Inc.
gdk@redhat.com
[4] OpenMRS
djazayeri@pih.org

**Abstract.** Open Source Software (OSS) is undergoing extraordinary growth. This rapid growth requires an increasing number of software developers working in a variety of areas. Computing education needs to provide students with professional experience, preferably within the context of a large, distributed software project. Educating students within OSS projects provides a solution to both the need for both developers to work on OSS projects as well as the need to provide computing students with professional experience. This panel will discuss the issues involved with educating students using OSS.

## 1 Description

Student participation in Open Source Software has the potential to offer significant benefit both to OSS communities and to educators. For OSS, students can make meaningful contributions, and education that includes OSS experience will help to insure that the number of developers with OSS skills continues to grow. From the education perspective, OSS participation can provide students with exposure to large, real-world projects and help students to understand how to handle the complexity of large, long-lived projects, how to behave in a professional environment, how to communicate effectively in a distributed development environment, and much more.

However, there are a variety of roadblocks to involving students in OSS projects. Instructors perceive high learning curves for projects and development environments, difficulties in obtaining entree into OSS projects, and problems in fitting development within an academic schedule. Students may be concerned with the complexity of the project and being viewed negatively by the OSS community for their lack of experience. OSS participants may be concerned about student inexperience, and the limits of academic term schedules.

This panel will discuss the impact of the growth in OSS on computing education. Questions to be addressed include:

- What is the value of student participation in OSS projects?
- What does it take to involve students in OSS projects including on-ramp issues, student-friendly projects and more?

P. Ågerfalk et al. (Eds.): OSS 2010, IFIP AICT 319, pp. 429–430, 2010.

- What are the variety of ways that students can participate in an OSS project? Student involvement includes both software and services and can range from coding to documentation to providing support for OSS applications.
- What are some examples of successful efforts within courses?
- What are some existing OSS efforts to involve students?

This panel hopes to encourage active discussion of the issues related to educating students with respect to and in the environment of OSS.

## 2  Panelists and Perspectives

Greg Hislop (panel moderator), is an Associate Professor in the College of Information Science and Technology, Drexel University. He is co-PI on the NSF project "SoftHum: Student Participation in the Community of Open Source Software for Humanity," which is investigating the development of course materials to support student open source participation within the classroom (xcitegroup.org/softhum). He is PI on the HumIT project which is developing ways to have students provide infrastructure support for humanitarian OSS projects (xcitegroup.org/humit). Greg will speak from the perspective of curriculum development for incorporating students in OSS.

Heidi Ellis is Associate Professor and Chair of Department of Computer Science and Information Technology, Western New England College. She is PI on the NSF SoftHum project. She has been involved with the Humanitarian Free and Open Source Software (HFOSS) project (hfoss.org). Heidi will speak from the perspective of instructional delivery and course materials for incorporating students in OSS.

Greg DeKoenigsberg is a Senior Community Architect for Red Hat. He is a former chairman of the Fedora Project, an open source software development community with over 10,000 volunteer contributors. He serves on the advisory boards of several open source advocacy organizations, writes about open source issues, and speaks at open source events worldwide. He has been with Red Hat Since 2001. Greg will discuss his experiences with Red Hat in building collaborative  communities that support student involvement in OSS.

Darius Jazayeri is the Lead Software Designer for OpenMRS, an OSS electronic medical record system built by a collaborative that includes the Regenstrief Institute, Inc. and Partners In Health. He has over eight years experience developing open source medical records in developing countries. Darius recently won the 2009 Pizzigati Prize for Public Interest Computing from the Tides Foundation. Darius will discuss his experiences with project management while mentoring students involved in OpenMRS projects.

# Open Source Software/Systems in Humanitarian Applications (H-FOSS)

Greg Madey

Computer Science & Engineering
College of Engineering
University of Notre Dame
gmadey@nd.edu

In the past few years we've seen many catastrophic natural disasters, most recently the Haitian and the Chilean Earthquakes. Others include the 2004 Indian Ocean Earthquake and Tsunami, the 2005 Kashmir earthquake, the 2008 Sichuan earthquake, and Cyclone Nargis that hit Myanmar in 2008. Because these events are rare and often impact poor countries, the development of information systems that support humanitarian and crises response may not be profitable, and thus rarely developed. Systems needed to track medical services to populations of poor nations are often not developed nor deployed because there is no profitable business model for such products. Commercially systems typically require expensive training and hardware not practical in poor underserved places on the planet.

Humanitarian Free and Open Source Software is FOSS developed to support humanitarian, crises response and health care applications. Example H-FOSS projects include the Sahana Disaster Management System, Open MRS Medical Record System, and Crises Commons with its CrisesCamps. This panel will examine this emerging category of FOSS, its trends, challenges, and opportunities. Panelists will come from these and other H-FOSS projects.

P. Ågerfalk et al. (Eds.): OSS 2010, IFIP AICT 319, p. 431, 2010.
© IFIP International Federation for Information Processing 2010

# The FOSS 2010 Community Report

Walt Scacchi[1], Kevin Crowston[2], Greg Madey[3], and Megan Squire[4]

[1] University of California, Irvine, Institute for Software Research,
Irvine, CA, USA
wscacchi@ics.uci.edu
http://www.ics.uci.edu/~wscacchi
[2] Syracuse University, School of Information studies, Syracuse, NY, USA
crowston@syr.edu,
http://crowston.syr.edu/
[3] University of Notre Dame, South Bend, IN, USA
gmadey@nd.edu
http://www.nd.edu/~gmadey/
[4] Elon University, Department of Computing Sciences,
Elon, NC, USA
msquire@elon.edu
http://facstaff.elon.edu/msquire

**Abstract.** The purpose of this panel is to disseminate the findings from the related FOSS workshop, a CCC-sponsored exploratory workshop held at University of California, Irvine in February 2010. At the OSS conference we will give first a report of what was learned at the FOSS workshop, and then we will glean important feedback from community members who were unable to be at the FOSS workshop. The four conveners of the FOSS workshop will be the panelists at the OSS conference.

## 1 The Purpose of the Workshop

The purpose of the FOSS workshop at UC-Irvine was to generate ideas and perspectives from within the free and open source research community about the future of research in the field. To start the workshop, we solicited position papers from our fifty North American attendees about the following subject areas:

- How does FOSS as a diverse socio-technical movement accomplish global software development, without a traditional central authority or source of funding/resources?
- How do distributed groups make decisions? What sort of conflicts are common, and how are conflicts settled?
- What are the differences and similarities between FOSS projects and proprietary (non-FOSS) projects? Is there a taxonomy of characteristics of these two types of projects? Are there hybrid projects, and how are these described?
- How do we measure "success" of a FOSS project? What are the various attributes of a project that might help us measure success? Do we have all the data we need, or are there additional measures that we need to collect?

P. Ågerfalk et al. (Eds.): OSS 2010, IFIP AICT 319, pp. 432–433, 2010.
© IFIP International Federation for Information Processing 2010

- What are the different ways that software developers are incentivized within the various types of FOSS projects? How does this incentive structure compare to proprietary projects? What do the developers themselves report are the best and worst incentives?
- How can the benefits of FOSS be translated into a language technology decision-makers can understand? Are there "best practices" for FOSS technology adoption or for rollovers from proprietary to FOSS models within businesses or governments?
- What are the various techniques and technologies that help self-organized groups to work effectively? How can these self-organizing techniques and technologies be applied to other domains?
- What are the different roles in a FOSS project (e.g., core developer, active user)? What levels of contribution is needed from members in various roles are needed to sustain a project (e.g., how important are active users)?
- How long can such a movement be sustained?
- Are there conditions or events that constitute an inflection point that will mark the decline of FOSS as a socio-technical movement?

## 2  Findings from the Workshop

Based on discussions, debate, and reflection at the FOSS workshop, we were able to synthesize and change the list of questions and focus areas. At the end of the FOSS workshop, our focus areas for the future of FOSS research included:

- Collaboration - how can studying FOSS help us understand how humans collaborate?
- Software Engineering Practice - how can FOSS help us understand the current and future state-of-the-art in software development?
- Transfer, Ecosystem, and Society - what can we learn about other domains from studying FOSS, and from what other domains is FOSS being influenced?
- Learning and Education - can FOSS serve as an educational tool and what are the implications of using FOSS in the classroom?
- Evolution - how do FOSS outcomes, activities, technologies, infrastructures, etc. develop and change over time? Do these changes follow specific patterns or principles, and what evolutionary trajectories are typical and or similar within FOSS when compared to other forms of developing and evolving software?
- Motivational Transformations - how does studying FOSS help us understand the global IT infrastructure and the process of innovation?
- Research Infrastructures - what are the best ways to support the data and analysis needs of the research community?

These interest areas represent what the community believes will be the best way to focus and extend the FOSS research agenda in the coming years.

# The Present and Future of
# FLOSS Data Archives

Megan Squire[1], Jesús M. González-Barahona[2], and Greg Madey[3]

[1] Elon University, Department of Computing Sciences,
Elon, NC, USA
msquire@elon.edu
http://flossmole.org
[2] Libre Software Engineering Lab (GSyC), Universidad Rey Juan Carlos, Madrid, Spain
jgb@gsyc.es
http://flossmetrics.org
[3] University of Notre Dame, South Bend, IN, USA
gmadey@nd.edu
http://www.nd.edu/~gmadey/

**Abstract.** The purpose of this panel will be to discuss the features available in current archives of data about open source projects. The panel will also discuss possible future activities and features to be implemented into these data archives. Community feedback, requests, and questions will also be integrated into this panel discussion.

## 1 Purpose

This panel is made up of some of the leaders of various open source data archiving projects: Megan Squire, Elon University and the FLOSSmole project, Jesus Gonzalez-Barahona, Universidad Rey Juan Carlos and the FLOSSMetrics project, and Greg Madey, University of Notre Dame, and the Sourceforge Research Data Archive.

Panelists will discuss the current and future needs of the research community, and specifically how these needs can be met by existing data archives.

The topics for discussion on the panel will include:

- What are the salient features of each data archive project? What is the mission of each project? How does it differ from the other projects?
- What are the biggest challenges facing each data archive project?
- How has the project helped to address significant research questions or otherwise helped the research community?
- What are some of the common requests from community members for the projects, and how is the project addressing these requests?
- What are the future initiatives of each project?
- What are some ways that open source community members can get involved with the project? Are there particular initiatives that the community can help with?

P. Ågerfalk et al. (Eds.): OSS 2010, IFIP AICT 319, pp. 434–435, 2010.

All of the data archiving projects represented on the panel have different answers to these questions. We expect a lively and fruitful two-way conversation between the panelists and the community members about the features and futures of data archiving projects.

# Author Index